The past forty years have been a time of spectacular development in the study of general relativity and cosmology. A special role in this has been played by the influential research groups led by Dennis Sciama in Cambridge, Oxford and Trieste. In April 1992 many of his ex-students and collaborators came to Trieste (where he is currently Professor) for a review meeting to celebrate his 65th birthday. This book consists of written versions of the talks presented which, taken together, comprise an authoritative overview of developments which have taken place during his career to date. The topics covered include fundamental questions in general relativity and cosmology, black holes, active galactic nuclei, galactic structure, dark matter and large scale structure. The authors are: M. A. Abramowicz, A. M. Anile, J. D. Barrow, J. J. Binney, B. J. Carr, B. Carter, C. J. S. Clarke, N. Dallaporta, G. F. R. Ellis, F. de Felice, G. W. Gibbons, B. J. T. Jones, S. W. Hawking, M. A. H. MacCallum, J. C. Miller, R. Penrose, D. J. Raine, M. J. Rees, V. Romano, W. C. Saslaw, D. W. Sciama and K. P. Tod.

The book will be of interest to graduate students and researchers in cosmology, relativity, astronomy, astrophysics, theoretical physics and applied mathematics.

The Renaissance of General Relativity and Cosmology

DENNIS SCIAMA

The Renaissance of General Relativity and Cosmology

A Survey to Celebrate the 65th Birthday of Dennis Sciama

GEORGE ELLIS
SISSA, Trieste, and University of Cape Town

ANTONIO LANZA
SISSA, Trieste

JOHN MILLER
Osservatorio Astronomico di Trieste

CAMBRIDGE
UNIVERSITY PRESS

CAMBRIDGE UNIVERSITY PRESS
Cambridge, New York, Melbourne, Madrid, Cape Town, Singapore, São Paulo

Cambridge University Press
The Edinburgh Building, Cambridge CB2 2RU, UK

Published in the United States of America by Cambridge University Press, New York

www.cambridge.org
Information on this title: www.cambridge.org/9780521433778

First published 1993
This digitally printed first paperback version 2005

A catalogue record for this publication is available from the British Library

ISBN-13 978-0-521-43377-8 hardback
ISBN-10 0-521-43377-0 hardback

ISBN-13 978-0-521-02108-1 paperback
ISBN-10 0-521-02108-1 paperback

Contents

Author Addresses

Marek A. Abramowicz: Nordisk Institut for Teoretisk Fysik, Blegdamsvej 17, 2100 København, Denmark and International Centre for Theoretical Physics, Strada Costiera 11, 34014 Trieste, Italy.

A. Marcello Anile and Vittorio Romano: Dipartimento di Matematica, Università di Catania, Viale Andrea Doria 6, 95125 Catania, Italy.

John D. Barrow: Astronomy Centre, University of Sussex, Brighton BN1 9QH, U.K.

James J. Binney: Department of Physics, University of Oxford, Keble Road, Oxford OX1 3NP, U.K.

Bernard J. Carr: Astronomy Unit, Queen Mary and Westfield College, University of London, Mile End Road, London E1 4NS, U.K.

Brandon Carter: Département d'Astrophysique Relativiste et de Cosmologie, C.N.R.S., Observatoire de Paris, 92 Meudon, France.

Christopher J.S. Clarke: Faculty of Mathematical Studies, University of Southampton, Southampton SO9 5NH, U.K.

Nicolò Dallaporta: Dipartimento di Astronomia, Università di Padova, Vicolo dell'Osservatorio, 35122 Padova, Italy.

Fernando de Felice: Istituto di Fisica Matematica *J.-L. Lagrange*, Università di Torino, Via C. Alberto 10, 10123 Torino, Italy.

George F.R. Ellis: Department of Applied Mathematics, University of Cape Town, Rondebosch 7700, Cape Town, South Africa and S.I.S.S.A., Via Beirut 2-4, 34013 Trieste, Italy.

Gary W. Gibbons: Department of Applied Mathematics and Theoretical Physics, University of Cambridge, U.K.

Bernard J.T. Jones: Niels Bohr Institute, Blegdamsvej 17, 2100 København, Denmark.

Stephen W. Hawking: Department of Applied Mathematics and Theoretical Physics, University of Cambridge, U.K.

Malcom A.H. MacCallum: School of Mathematical Sciences Queen Mary and Westfield College, University of London Mile End Road, London E1 4NS, U.K.

John C. Miller: Osservatorio Astronomico di Trieste, Via Tiepolo 11, 34131 Trieste, Italy.

Roger Penrose: Mathematical Institute, University of Oxford, St. Giles, Oxford, U.K.

Derek J. Raine: Department of Physics and Astronomy, University of Leicester, Leicester LE1 7RH, U.K.

Martin J. Rees: Institute of Astronomy, University of Cambridge, Madingley Road, Cambridge, U.K.

William C. Saslaw: Astronomy Department, University of Virginia; National Radio Astronomy Observatory, Charlottesville, Virginia and Institute of Astronomy, Cambridge, U.K.

Dennis W. Sciama: S.I.S.S.A., Via Beirut 2-4, 34013 Trieste, Italy.

K. Paul Tod: Mathematical Institute and St John's College, Oxford, U.K.

Introduction

The past 30 years have seen a great revival of General Relativity and Cosmology, and major developments in astrophysics. On the theoretical side this has been centred on the rise of the Hot Big Bang model of cosmology and on our developing understanding of the properties of black holes. On the observational side it has been based on astonishing improvement of detectors and measuring instruments in astronomy and experimental relativity, in particular enabling measurement of the microwave background radiation and extension of astronomical observations to the whole electromagnetic spectrum.

Dennis Sciama has played an important role in these developments, particularly through the research schools he has run at Cambridge, Oxford, and Trieste, supervising and inspiring many research students who have worked on these topics, and challenging his colleagues with penetrating questions about the physics and mathematics involved. The extent of his influence will become apparent on studying the Family Tree of students, and the list of books that have been the product of those who have taken part in these research groups (see below).

Dennis' 65th Birthday was on November 18, 1991. To mark this event, a meeting was held at SISSA, Trieste (Italy) from 13th to 15th April, 1992, under the title *The Renaissance of General Relativity and Cosmology: A survey meeting to celebrate the 65th birthday of Dennis Sciama*. Ex-students and close colleagues of Dennis were invited to give talks at the meeting, aimed at broadly reviewing and discussing developments in relativity, cosmology and astrophysics during this period. A great many of those invited were able to attend, although a few could not, and the resulting discussions gave an excellent survey of these exciting developments. This book contains the texts of those talks, as well as some contributions from invitees who could not attend.

Dennis was born on November 18, 1926, in Manchester. He married Lidia in 1959. She is a social anthropologist at Oxford. Their daughter Susan (born 1962) is a

1

painter in London, and Sonia (born 1964) is studying experimental psychology.

During the war, Dennis worked at T.R.E. on solid state physics, in particular studying the properties of lead sulphide. He obtained a BA at Cambridge University in 1947 and an MA in 1949. He commenced research on statistical mechanics under Dr Temperley, but then changed to work on Mach's Principle with P A M Dirac as supervisor (who therefore qualifies as the Grandfather of the Family Tree), obtaining his Cambridge PhD in 1953. During this time many ideas about relativity and cosmology were aired and shared in late night discussions between Dennis and Herman Bondi, Tommy Gold, Fred Hoyle, and Felix Pirani. Dennis also profoundly influenced the future development of relativity theory by interesting Roger Penrose in the subject.

He was a Junior Research Fellow at Trinity College, Cambridge from 1952 to 1954 and 1956 to 1958, a member of the Institute of Advanced Studies, Princeton, during 1954-1955, an Agassiz Fellow at Harvard University in 1955-1956, Research Associate at King's College, London from 1958 to 1960, and Visiting Professor at Cornell University in 1960-1961. He was appointed Lecturer in Mathematics at Cambridge University in 1961, where he built up a strong Relativity school in the Department of Applied Mathematics and Theoretical Physics (first sited in the Phoenix Wing of the Cavendish Laboratory, Free School Lane, and then at the old Cambridge University Press site at Silver Street). He was Fellow of Peterhouse from 1963 to 1970. He then moved to Oxford, holding the position of Senior Research Fellow at All Soul's College from 1970 to 1985, while building up a research group in the Astrophysics Department. He spent 1977-1978 as the Luce Professor of Cosmology, Mount Holyoke College, Massachusetts, and was a part-time Professor of Physics at the University of Texas, Austin, from 1978 to 1982. In 1983 he became Professor of Astrophysics at the International School of Advanced Studies (SISSA) in Trieste, where he is at present Director of the Astrophysics Sector, while still maintaining strong links with Oxford and Cambridge (during this period he has been a Research Fellow at Churchill College, Cambridge, and in 1990 was elected Emeritus Fellow at All Souls College, Oxford).

He has been the Thomas Gold Lecturer at Cornell University (1990) and the Milne Lecturer at Oxford University (1991). His status has been recognized by election as Foreign Member of the American Philosophical Society (1982), Fellow of the Royal Society of London (1983), Foreign Member of the American Academy of Arts and Sciences (1983) and Foreign Member of the Academia Lincei (1984). In 1991 he was awarded the Guthrie Prize of the British Institute of Physics. He was President of the International Society of General Relativity and Gravitation from 1980 to 1984.

He has been invited to give many popular talks and technical review lectures.

While Dennis is interested in all physics, his research has focused on the major areas discussed at the celebratory meeting. His initial work on Mach's Principle was followed by investigation of the role of torsion in relativity (leading to his name being linked with those of Einstein, Cartan, and Kibble in the ECSK theory of gravity). After a period as a strong supporter of the Steady State Theory, analysis of quasi-stellar object source count observations (carried out with Martin Rees, then a research student) lead him to change viewpoint; he then focused on the physics of hot big bang cosmology and in particular the interactions of matter and radiation in the expanding universe (for example the inter-relation between X-ray and radio emission by intergalactic matter). He was one of the early theorists to consider observational tests for the presence of dark matter, and (with Martin Rees) the effect of clumping of matter on the cosmic microwave background radiation; his influential paper "The Recent Renaissance of Observational Cosmology" (1971) usefully summarised knowledge at the time, and is reflected in the title of the present meeting. He has maintained a strong interest in aspects of quantum theory and the nature of the vacuum. Recently he has been particularly interested in the evolution of dark matter particles in the expanding universe, his present passion being the possibility that decay of massive neutrinos accounts in a unified way for a whole variety of astronomical observations (as described in his article in this volume).

His wide interests have found expression in authorship of books of various kinds. The popular books *The Unity of the Universe* (1959) and *The Physical Foundations of General Relativity* (1969) have been instrumental in interesting many people in these topics. *Modern Cosmology* (1971) remains one of the best descriptions of the standard big bang theory of cosmology, and has been re-issued as the first Cambridge Science Classic (1982). The technical discussions reported in *Quantum Gravity I* (1975) and *Quantum Gravity II* (1981), edited with Chris Isham and Roger Penrose, reflected the current position in the ongoing struggle to understand the nature of a quantum theory of gravity. At present two books are in the process of completion: with Derek Raine on *The Thermodynamics of Black Holes*, and *Modern Cosmology and the Dark Matter Problem*, on the present understanding of dark matter in the universe, with an emphasis on the possible role of massive neutrinos. In addition to all this, Dennis initiated the influential Cambridge University Press series *Cambridge Monographs on Mathematical Physics* (originally conceived of as being focused on relativity theory, but later broadened to include other branches of theoretical physics).

While all this is impressive enough, Dennis' *forte* has been in his care for students, based on his realisation of their importance for the development of the subject.

Through them his influence is widely pervasive in general relativity theory, astrophysics, and cosmology, and extends to quantum theory and string theory. This does not of course mean that he agrees with them or they with him! What he has communicated to them is a passion for physics and for understanding the Universe, and a feeling for what is important and what not: a vision of what can be done, based as far as possible on a solid mathematical foundation, to increase physical understanding of what is happening. The quality of his care for students is illustrated by an early experience related by Douglas Gough (now Professor at the Institute of Astronomy, Cambridge). When he was a young research student, he became embroiled in a serious argument with a senior professor about a topic in fluid dynamics. To follow up a point in the debate, he needed a particular copy of *Monthly Notices of the Royal Astronomical Society*, which happened to be out of the library. On the recommendation of his research supervisor, he went to borrow the relevant journal issue from Dennis, who willingly lent it to him. As he left the office, Dennis said, "I hear you are involved in a major disagreement with X. I don't know what the issues are, but if you feel sure of your case, stick to your guns; and if you need support, I'll give it to you". This support by someone who was not his supervisor was tremendously encouraging to Doug, and made a great impression on him. This quality of caring is the basis of the affection felt for Dennis by his students and associates, which was so clearly expressed by them at the Trieste meeting.

The result of this attitude has been a great influence on the development of theoretical astrophysics and relativity world-wide. This is demonstrated by the list of the students that have completed their PhD's in the various Sciama research groups, and then in turn had their own research students, as shown in the Family Tree below [drawn up initially by Bill Saslaw, and completed by Antonio Lanza]. It is probably not complete, as we have not been able to contact all the people listed there to determine who were in turn their students; however it is sufficient to show the strength and depth of the Sciama influence.

Additionally it is interesting to point out some of the major technical books that have been produced by members of the Sciama family tree, or by close associates. Apart from those written or edited by Dennis (mentioned above), these include *Non-equilibrium Relativistic Kinetic Theory*, J M Stewart (Springer-Verlag, 1971), *The Large Scale Structure of Space-Time*, S W Hawking and G F R Ellis (Cambridge University Press, 1973), *Special Relativity*, W G Dixon (Cambridge University Press, 1978), *Cosmic Dust*, P G Martin (Oxford University Press, 1978), *General Relativity: an Einstein Centenary Survey*, Ed. S W Hawking and W Israel (Cambridge University Press, 1979), *Exact Solutions of Einstein's Field Equations*, D Kramer, H Stephani, M MacCallum and E Herlt, (Cambridge University Press 1981), *The

Isotropic Universe, D J Raine (Adam Hilger, 1981), *Galactic Astronomy*, D Mihalas and J Binney (Freeman, 1981), *The Origin and Evolution of Galaxies*, Ed. B J T Jones and J E Jones (Reidel, 1982), *The Galactic Centre*, Ed. G R Riegler and R D Blandford , (Am. Inst. Phys, 1982), *The Very Early Universe*, Ed. G W Gibbons, S W Hawking and S T C Siklos (Cambridge University Press, 1983), *Spinors and Space-Time: Volume 1*, R Penrose and W Rindler (Cambridge University Press, 1984), *An Introduction to Twistor Theory*, S A Huggett and K P Tod (Cambridge University Press, 1985), *Interacting Binaries*, Ed. P P Eggleton and J E Pringle (Kluwer, 1985), *Interacting Binary Stars*, J E Pringle and A A Wade (Cambridge University Press, 1985), *Spinors and Space-Time: Volume 2*, R Penrose and W Rindler (Cambridge University Press, 1986), *Accretion Power in Astrophysics*, J Frank, A R King, and D J Raine (Cambridge University Press, 1986), *Dynamical Space-time and Numerical Relativity*, Ed. J Centrella (Cambridge University Press, 1986), *The Anthropic Cosmological Principle*, J D Barrow and F J Tipler (Oxford University Press, 1986), *Gravitation in Astrophysics: Cargese 1986*, Ed. B Carter and J Hartle (Plenum 1987), *Galactic Dynamics*, J J Binney and S Tremaine (Princeton University Press, 1987), *Gravitational Physics of Stellar and Galactic Systems*, W C Saslaw (Cambridge University Press, 1987), *General Relativity and Gravitation*, Ed. M A H MacCallum (Cambridge University Press, 1987), *Three Hundred Years of Gravitation*, Ed. S W Hawking and W Israel (Cambridge University Press, 1988), *The Post-recombination Universe*, Ed. N Kaiser and A Lasenby (Kluwer, 1988), *Relativistic Fluids and Magneto-fluids*, A M Anile (Cambridge University Press, 1989), *Relativity on Curved Manifolds*, F de Felice and C J S Clarke (Cambridge University Press, 1990), *Further Advances in Twistor Theory*, L J Mason and K P Tod (Wiley, 1990), *Introduction to General Relativity*, L Hughston and K P Tod (Cambridge University Press, 1991), *Active Galactic Nuclei*, R D Blandford, H Metzger and L Woltjer (Springer, 1990), *Advanced General Relativity*, J M Stewart (Cambridge University Press, 1992).

Popularisations are also of importance, and some significant volumes of this type too have been derived from the Family Tree members. Such books include *The Left Hand of Creation*, J D Barrow and J Silk (Basic Books, 1986), *A Brief History of Time*, S W Hawking (Bantam, 1988), *The World Within the World*, J D Barrow (Oxford University Press, 1988), *Flat and Curved Space Times*, G F R Ellis and R M Williams (Oxford University Press, 1989), *The Emperor's New Mind*, R Penrose (Oxford University Press, 1989), *Theories of Everything*, J D Barrow (Oxford University Press, 1990), *Cosmic Coincidences*, J Gribbin and M J Rees (Black Swan, 1991), *Pi in the Sky: Counting, Thinking, and Being*, J D Barrow (Oxford University Press, 1992).

The celebratory meeting was sponsored and supported financially by SISSA (Director: Daniele Amati), the Trieste Observatory (Director: Giorgio Sedmak), the International Centre for Theoretical Physics (Director: Abdus Salam), and CIRAC, the Consortium of Astronomy and Cosmology Research departments in the Region (Director: Margherita Hack). We thank them for their support. It was organised by a committee consisting of John Miller (Secretary), Antonio Lanza (Treasurer), Aldo Treves, Marek Abramowicz, Martin Rees, and George Ellis (Chairman), with secretarial assistance by Lidia Bogo. It was a pleasure to see the way the committee and the participants cooperated pleasantly to make the meeting a success; thank you all. Finally, as Chairman of the organising Committee, I particularly wish to thank John Miller and Antonio Lanza for the work they did, for they carried the brunt of the organisational load, efficiently arranging all the details of the meeting with a minimum of fuss, despite various problems that placed considerable stress on them.

G.F.R. Ellis

DENNIS W. SCIAMA — Cambridge Period (A)

Year	G.F.R.Ellis	W.G.Dixon	S.W.Hawking	S.A.Orzag	B.J.Burn	G.Lemmer	R.G.McLenaghan	B.Carter	K.R.Johnson	M.J.Rees	M.Stone	W.C.Saslaw	R.Hunt	J.M.Stewart
1961														
1962														
1963														
1964														
1965													R.Hunt	J.M.Stewart
1966											M.Stone			
1967										M.J.Rees		W.C.Saslaw		
1968														
1969	J.M.Stewart		G.W.Gibbons											
1970			C.R.Prior											
1971	M.A.H.MacCallum		B.J.Car				B.Smith			B.J.T.Jones				
1972		J.M.Bird	A.S.Lapedes							P.G.Martin		S.Ward		
1973	B.Treciokas		S.T.C.Siklos							R.D.Blandford				
1974	A.R.King	E.R.Lord	N.J.Toop				E.Bridson	H.Quintana		N.Sanit				
1975	T.J.Gordon		M.J.Perry				D.Kerrighan	D.N.C.Lin		N.C.Smart				
1976	P.D'Eath		P.D'Eath							J.E.Pringle		M.Valtonen		L.R.D.Davis
1977			C.N.Pope											L.O'C.Dury
1978			A.L.Yuille					J.P.Luminet		G.R.Gisler				
1979			M.S.Faweett				G.Fee			R.L.Zaniek		M.Fall		S.M.Stumbles
1980			I.G.Moss							W.R.Stoeger		D.N.C.Lin		J.Porrill
1981	R.Maartens		N.P.Warner							P.S.Wesson				R.W.Corkill
1982	S.D.Nell									J.Frank				
1983	A.Sievers		B.Allen				S.Czapor			M.C.Begelman				
1984							N.Kamran	J.A.Marck		N.A.Sharp				
1985	R.Bruno		E.P.Shellard							S.D.M.White		M.Whitle		
1986	W.Roque		B.Whitt							C.J.Hogan		A.J.S.Hamilton		
1987	G.Tivon		J.Halliwell							D.B.Wilson				
1988	G.Schreiber		R.Laflamme							N.Kaiser				R.A.W.Gregory
1989	M.Jaklitsch		A.Lyons							A.Kashlinsky; E.S.Phinney		H.Lehto		
1990	A.Maurellis		H.F.Dowker					D.Prior		S.L.Stepney				
1991	V.Faraoni		G.Esposito					P.Peter		R.S.Webster		D.Valls-Gabard		
1992	M.Bruni; S.Borgani		G.Lyons							H.M.P.Couchman; M.G.Baring; M.Bithell; D. Scott; D. Syer		R.Sheth		

DENNIS W. SCIAMA — Cambridge Period (B)

Year	P.C.Waylen	G.W.Gibbons	B.C.Bramson / C.J.S.Clarke	F.de Felice	M.A.H.MacCallum / S.H.Stobbs	D.J.Raine	M.J.Khan	B.J.T.Jones	P.G.Martin
1961									
1962									
1963									
1964									
1965									
1966									
1967									
1968					S.H.Stobbs				
1969					M.A.H.MacCallum		M.J.Khan		
1970			C.J.S.Clarke						
1971						D.J.Raine			
1972								B.J.T.Jones	P.G.Martin
1973									
1974					A.R. King				
1975						C.P.Winlove			
1976		J.Richer							
1977		D.Lohiya							B.Everson
1978						P.Candelas			J.Maza
1979	G.J.Weir								
1980		C.M.Hull	J.A.Vickers					K.L.Chan	C.Rogers
								J.M.Centrella	
1981					A.A.Coley			N.A.Sharp	
								G.P.Efstathiou	
								C.S.Frenk	
1982								R.Wyse	G.Clayton
1983		S.B.Davis		Yu Yunqiang	R.S.Harness			P.L.Palmer	
1984		P.Ruback						D.Heath	
1985		D.Wiltshire						V.de Lapparent	
								Liu Xiang Dong	
1986					A.J.Galloway			J.Hessleberg	
1987		D.Giulini			D.W.Kitchingham			J.Louko	
1988			J.Mannistre	L.Sigalotti	G.C.Joly			P.Lilje	
1989			S.A.Hayward		M.E.de Araujo	J.Mardeljevic		V.Martinez	
1990		M.Ortiz	G.Hollier		R.A.Sussman			E.Martinez-Gonzales	
					J.Castejon-Amenedo				
1991					M.B.Ribeiro			L.Appel	M.E.Mandy
								S.Rugh	
1992		C.Gundlach			P.K.S.Dunsby			R.van de Weygaert	

DENNIS W. SCIAMA — Oxford Period

1971	
1972	
1973	
1974	A.M.Anile; J.C.Miller; I.A.Holmes
1975	K.P.Tod
1976	J.J.Binney; L.P.Hughston
1977	P.Candelas; A.J.Wickett
1978	T.N.Palmer; J.D.Barrow
1979	P.A.Connors; A.N.Hall
1980	R.F.Stark; D.Deutsch; R.F.Stark; R.Holder
1981	E.P.T.Jones; O.Strimpel
1982	P.J.Mann; A.Melott (Austin)
1983	J.D.Porter; D.Sonoda; A.A.Naimi
1984	A.L.H.Smith
1985	R.Kelly; K.Howard; J.Stein-Schabes; A.L.H.Smith; C.L.N.Ruggles; A.C.Ottewill
1986	A.Newton; B.Jensen; A.Lanza; E.H.Ling
1987	L.Pallister; A.Lutken; O.Pantano; G.Russo
1988	C.McGill; C.Ordonez; A.Burd; D.Willmes
1989	P.Binks; P.Saha; M.Pfionis
1990	P.Coles; J.McLaughlin
1991	P.Tribble; M.Madsen; P.J.Cuddeford; M.Sammartino
1992	A.Dougan; S.Hawthorne; S.Cotsakis; P.Saich; R.Splinter; J.Pauls; V.Romano

DENNIS W. SCIAMA   Trieste Period

1983							
1984	R.Valdarnini						
1985							
1986							
1987	S.Manorama	P.Salucci	M.R.Dubal	P.Molaro			
1988		R.Scaramella	L.Denizman	L.Sigalotti			
1989							
1990		S.Mollerach	A.Pisani	A.Romeo	S.Sonego		
1991			Y.Anini	P.J.Cuddeford	C.M.Raiteri	R.K.Gulati	
1992			V.Antonuccio	M.Arnaboldi	S.Borgani	A.Carlini	M.Girardi	Y.P.Jing	A.Valentini

Publication List of Dennis Sciama

PAPERS

[1] "On the origin of inertia" (1953). *Mon. Not. R. astr. Soc.*, **113**, 34.

[2] "A note on the reported color-index effect of distant galaxies" (1954). *Ap. J.*, **120**, 597 (with H. Bondi & T. Gold).

[3] "On the formation of galaxies in a steady state universe" (1955). *Mon. Not. R. astr. Soc.*, **115**, 3.

[4] "Evolutionary processes in cosmology" (1955). *Adv. in Sci.*, **12**, 38.

[5] "A model for the formation of galaxies" (1955). In *IAU Symposium No. 2: Gas Dynamics of Cosmic Clouds*, ed. J.M. Burgers & H.C. van der Hulst, p. 175.

[6] "Charges and parities of elementary particles" (1957). *Phys. Rev.*, **107**, 632.

[7] "Inertia" (1957). *Scientific American*, **196** (February), 99.

[8] "On a geometrical theory of the electromagnetic field" (1958). *Nuovo Cim.*, **8**, 417.

[9] "On a nonsymmetric theory of the pure gravitational field" (1958). *Proc. Camb. Phil. Soc.*, **54**, 72.

[10] "Le principe de Mach" (1961). *Ann. Inst. H. Poincaré*, **17**, 1.

[11] "Les trois lois de la cosmologie" (1961). *Ann. Inst. H. Poincaré*, **17**, 13.

[12] "L'observation et la cosmologie" (1961). *Ann. Inst. H. Poincaré*, **17**, 25.

[13] "On the nonsymmetric theory of the gravitational field" (1961). *J. Math. Phys*, **2**, 472.

[14] "Recurrent radiation in general relativity" (1961). *Proc. Camb. Phil. Soc.*, **57**, 436.

[15] "Faraday rotation effects associated with the radio source Centaurus A" (1962). *Nature*, **196**, 760.

[16] "Consequences of an open model of the galactic magnetic field" (1962). *Mon. Not. R. astr. Soc.*, **123**, 317.

[17] "On the analogy between charge and spin in general relativity" (1962). In *Recent Developments in General Relativity* (Pergamon), p. 415.

[18] "Retarded potentials and the expansion of the universe" (1963). *Proc. Roy. Soc.*

A, **273**, 484.

[19] "On the interpretation of radio source counts" (1963). *Mon. Not. R. astr. Soc.*, **126**, 195.

[20] "Cosmic X- and infra red rays as tools for exploring the large scale structure of the universe" (1964). *Ap. J.*, **140**, 1634 (with R.J. Gould).

[21] "The physical structure of general relativity" (1964). *Rev. Mod. Phys.*, **36**, 463.

[22] "On the interpretation of Faraday rotation effects associated with radio sources" (1964). In *Physics of nonthermal radio sources* ed. S.P. Maran & A.G.W. Cameron (NASA Goddard Institute for Space Studies), p. 139.

[23] "On a possible class of galactic radio sources" (1964). *Mon. Not. R. astr. Soc.*, **128**, 49.

[24] "On intergalactic gas as a possible absorber of extragalactic radio noise" (1964). *Nature*, **204**, 767.

[25] "On the formation of galaxies and their magnetic fields in a steady state universe" (1964). *Quart. Journ. R. astr. Soc.*, **5**, 196.

[26] "Proposal for the detection of dispersion in radio-wave propagation through intergalactic space" (1965). *Phys. Rev. Lett.*, **14**, 1007 (with F.T. Haddock).

[27] "The structure of the universe" (1965). *Scientia*, **58**, 1.

[28] "The red shift" (1965). *Sci. J.*, **9**, 52.

[29] "Radio astronomy and cosmology" (1965). *Sci. Progr.*, **53**, 1.

[30] "A model of the quasi-stellar radio variable CTA 102" (1965). *Nature*, **207**, 738 (with M.J. Rees).

[31] "Structure of the quasi-stellar radio source 3C 273B" (1965). *Nature*, **208**, 371 (with M.J. Rees).

[32] "Recent results in cosmology" (1966). *Adv. Sci.*, **22**, 593.

[33] "Distribution of quasi-stellar radio sources in steady state cosmology" (1966). *Nature*, **210**, 348 (with W.C. Saslaw).

[34] "On the origin of the microwave background radiation" (1966). *Nature*, **211**, 277.

[35] "Inverse Compton effect in quasars" (1966). *Nature*, **211**, 805 (with M.J. Rees).

[36] "Cosmological significance of the relation between redshift and flux density for quasars" (1966). *Nature*, **211**, 1283 (with M.J. Rees).

[37] "Cosmological aspects of high-energy astrophysics" (1966). In *High energy astrophysics* – Enrico Fermi summer school No. 35 ed. L. Gratton, p. 418.

[38] "Absorption spectrum of 3C 9" (1966). *Nature*, **212**, 1001 (with M.J. Rees).

[39] "The kinetic temperature and ionization level of intergalactic hydrogen in a steady state universe" (1966). *Ap. J.*, **145**, 6 (with M.J. Rees).

[40] "On a class of universes satisfying the perfect cosmological principle" (1966). In *Perspectives in General Relativity (Festschrift for V. Hlavaty)*, ed. B. Hoffmann (Indiana University Press, Bloomington), p. 150 (with G.F.R. Ellis).

[41] "Peculiar velocity of the sun and the cosmic microwave background" (1967). *Phys. Rev. Lett.*, **18**, 1065.

[42] "Possible large scale clustering of quasars" (1967). *Nature*, **213**, 374 (with M.J. Rees).

[43] "Possible circular polarization of compact quasars" (1967). *Nature*, **216**, 147 (with M.J. Rees).

[44] "Peculiar velocity of the sun and its relation to the cosmic microwave background" (1967). *Nature*, **216**, 748 (with J.M. Stewart).

[45] "The detection of heavy elements in intergalactic space" (1967). *Ap. J.*, **147**, 353 (with M.J. Rees).

[46] "Extragalactic soft X-ray astronomy" (1968). *Nature*, **217**, 326 (with M.J. Rees & G. Setti).

[47] "Large scale density inhomogeneities in the universe" (1968). *Nature*, **217**, 511 (with M.J. Rees).

[48] "Is interstellar hydrogen capable of maser action at 21 cm?" (1968). *Nature*, **217**, 1237 (with S.H. Storer).

[49] "Metastable helium in interstellar and intergalactic space" (1968). *Astrophys. Lett.*, **2**, 243 (with M.J. Rees & S.H. Stobbs).

[50] "Singularities in collapsing stars and expanding universes" (1969). *Comm. Astrophys. and Space Physics*, **1**, 1 (with S.W. Hawking).

[51] "The free electron concentration in HI regions of the Galaxy" (1969). *Comm. Astrophys. and Space Physics*, **1**, 35 (with M.J. Rees).

[52] "The evolution of density fluctuations in the universe, I Dissipative processes in the early stages" (1969). *Comm. Astrophys. and Space Physics*, **1**, 140 (with M.J. Rees).

[53] "The evolution of density fluctuations in the universe, II The formation of galaxies" (1969). *Comm. Astrophys. and Space Physics*, **1**, 153 (with M.J. Rees).

[54] "The astronomical significance of mass loss by gravitational radiation" (1969). *Comm. Astrophys. and Space Physics*, **1**, 187 (with G.B. Field & M.J. Rees).

[55] "Quasi-stellar objects" (1969). *Phil. Trans. Roy. Soc. A*, **264**, 263.

[56] "Interpretation of the cosmic X-ray background" (1969). *Proc. Roy. Soc. A*, **313**, 349.

[57] "Upper limit to the radiation of mass energy derived from expansion of galaxy" (1969). *Phys. Rev. Lett.*, **23**, 1514 (with G.B. Field & M.J. Rees).

[58] "Is the Galaxy losing mass on a timescale of a billion years?" (1969). *Nature*, **224**, 1263.

[59] "Generally covariant integral formulation of Einstein's field equations" (1969). *Phys. Rev.*, **187**, 1762 (with P.C. Waylen & R.C. Gilman).

[60] "A proposal for an X-ray analysis of interstellar grains" (1970). *Astrophys. Lett.*, **5**, 193 (with P.G. Martin).

[61] "Thermodynamics and cosmology" (1970). *Comm. Astrophys. and Space Physics*, **2**, 206 (with J.M. Stewart & M.A.H. MacCallum).

[62] "Massenverlust durch gravitationsstrahlung?" (1971). *Umschau, 71 Jahrgang*, p. 944.

[63] "Gravitational wave astronomy: an interim survey" (1971). *Radio and Electronic Engineer*, **42**, 391.

[64] "Astrophysical cosmology" (1971). *Course 47, Enrico Fermi Summer Schools*, p. 183.

[65] "Space research and observational cosmology" (1971). *E.S.R.O. S.P. No. 52*, p. 5.

[66] "The expansion of the Galaxy" (1971). *E.S.R.O. S.P. No. 52*, p. 29 (with G.B. Field & M.J. Rees).

[67] "The recent renaissance of observational cosmology" (1971). In *Relativity and Gravitation*, ed C.G. Kuiper & A. Peres (Haifa), p. 265.

[68] "Static gravitational multipoles – the connection between field and source in general relativity" (1971). *Gen. Rel. and Grav.*, **2**, 331 (with C.J.S. Clarke).

[69] "Eppur si muove" (1972). *Comm. Astrophys. and Space Physics*, **4**, 35.

[70] "Soft X-ray emission from intergalactic gas in the neighbourhood of the Galaxy" (1972). *Mon. Not. R. astr. Soc.*, **157**, 335 (with R. Hunt).

[71] "Global and non-global problems in cosmology" (1972). In *General Relativity, Papers in honour of J.L. Synge*, ed. L. O'Raifeartaigh, p. 35 (with G.F.R. Ellis).

[72] "Recent developments in the theory of gravitational radiation" (1972). *Gen. Rel. and Grav.*, **3**, 149.

[73] "X-ray emission from intergalactic gas in the neighbourhood of galaxies" (1972). *Nature*, **238**, 320 (with R. Hunt).

[74] "On a possible interstellar galactic chromosphere" (1972). *Nature*, **240**, 456.

[75] "The universe as a whole" (1973). In *The Physicist's Conception of Nature*, ed. J. Mehra (Reidel, Dordrecht), p. 17.

[76] "Early stages of the universe" (1974). *Highlights of astronomy*, **3**, 21.

[77] "Gravitational waves and Mach's Principle" (1974). In *Ondes et Radiations Gravitationelles* (Colloque International CNRS No. 220), p. 267.

[78] "The physics and astronomy of black holes" (1975). *Quart. Journ. R. astr. Soc.*, **16**, 1.

[79] "Black holes and their thermodynamics" (1976). *Vistas in Astron.*, **19**, 385.

[80] "Irreversible thermodynamics of black holes" (1977). *Phys. Rev. Lett.*, **38**, 1372 (with P. Candelas).

[81] "The thermodynamics of black holes" (1977). *Ann. New York Acad. Sci.*, **302**, 161.

[82] "Cosmological tests of general relativity" (1977). *Atti Convegni Lincei*, **34**, 431.

[83] "The irreversible thermodynamics of black holes" (1978). *Gen. Rel. and Grav.*, **9**, 183 (with P. Candelas).

[84] "The angular broadening of compact radio sources observed through ionised gas in a rich cluster of galaxies" (1979). *Ap. J.*, **228**, L15 (with A.N. Hall).

[85] "Cosmological implications of the 3 K background" (1979). *Proc. Roy. Soc. A*, **368**, 17.

[86] "Black holes and fluctuations of quantum particles: an Einstein synthesis" (1979). In *Relativity, quanta and cosmology*, ed. M. Pantaleo & F. de Finis (Johnson Reprint Corporation), p. 681.

[87] "Detailed ultraviolet observations of the quasar 3C 273 with IUE" (1980). *Mon. Not. R. astr. Soc.*, **192**, 561 (with M.H. Ulrich et al.).

[88] "Quantum processes at large redshifts" (1980). *Phys. Scr.*, **21**, 769.

[89] "The origin of the universe" (1980). In *The state of the universe*, ed. G.T. Bath (Oxford University Press), p. 3.

[90] "Gravitational collapse to the black hole state" (1980). In *General Relativity and Gravitation*, ed. A. Held (Plenum), p. 359 (with J.C. Miller).

[91] "The anisotropy of the cosmic black body radiation and its meaning" (1980). In *Astrophysics and elementary particles, common problems*, ed N. Cabibo et al. (Contributi del Centro Linceo Interdisciplinare No. 53 – Accademia Nazionale dei Lincei, Roma), p. 141.

[92] "The distribution of interstellar CIV in the Galaxy" (1980). *Second European IUE Conference* (ESA SP – 157), p. 345 (with G.E. Bromage & A.H. Gabriel).

[93] "The ultraviolet spectrum of the anomalous EUV source halo star HD 192273" (1980). *Second European IUE Conference* (ESA SP – 157), p. 353 (with G.E. Bromage & A.H. Gabriel).

[94] "Issues in cosmology" (1980). In *Some strangeness in the proportion (Einstein centennial)*, ed. H. Woolf (Addison-Wesley), p. 387.

[95] "Quantum field theory, horizons and thermodynamics" (1981). *Adv. Phys.*, **30**, 327 (with P. Candelas & D. Deutsch).

[96] "Neutrino lifetime constraints from neutral hydrogen in the galactic halo" (1981). Phys. Rev. Lett., **46**, 1369 (with A.L. Melott).

[97] "Massive neutrino decay and the photoionization of the intergalactic medium" (1982). *Mon. Not. R. astr. Soc.*, **198**, 1P.

[98] "On the ionization balance of helium in the intergalactic medium" (1982). *Mon. Not. R. astr. Soc.*, **200**, 13P.

[99] "Massive neutrinos in cosmology and galactic astronomy" (1982). In *Astrophysical cosmology*, ed. H.A. Bruck, G.V. Coyne & M.S. Longair (Pontif. Acad. Sci.), p. 529.

[100] "Decaying neutrinos as a photoionization source in galactic halos" (1982). *Phys. Rev. D*, **25**, 2214 (with A.L. Melott).

[101] "Massive photinos and ultraviolet astronomy" (1982). *Phys. Lett.*, **112B**, 211.

[102] "Massive neutrinos and ultraviolet astronomy" (1982). In *Progress in cosmology*, ed. A.W. Wolfendale (Reidel), p.75.

[103] "Evolutionary aspects of the cosmic black body radiation" (1982). In *Evolution in the universe* (Inauguration of the ESO Building, Garching), p. 105.

[104] "On an Einstein – de Sitter universe of radiating massive photinos" (1982). *Phys. Lett.*, **114B**, 19.

[105] "Cosmological constraints on broken supersymmetry" (1982). *Phys. Lett.*, **118B**, 327.

[106] "Negative energy radiation and the second law of thermodynamics" (1982). *Phys. Lett.*, **119B**, 72 (with D. Deutsch & A.C. Ottewill).

[107] "Black hole explosions" (1982). In *Cosmology and Astrophysics* (Essays in honor of Thomas Gold), ed. Y. Terzian & E.M. Bilson (Cornell University Press), p. 83.

[108] "On the production of NV and OVI in galactic halos by photons emitted by massive photinos" (1983). *Phys. Lett.*, **121B**, 119.

[109] "Massive neutrinos and ultraviolet astronomy" (1983). In *IAU Symposium No. 104: Early evolution of the universe and its present structure*, ed. G.O. Abell & G. Chincarini (Reidel), p. 313.

[110] "The role of particle physics in cosmology and galactic astronomy" (1983). In *IAU Symposium No. 104: Early evolution of the universe and its present structure*, ed. G.O. Abell & G. Chincarini (Reidel), p. 493.

[111] "Is there a quantum equivalence principle?" (1983). *Phys. Rev. D*, **27**, 1715 (with P. Candelas).

[112] "Massive neutrinos and photinos in cosmology and galactic astronomy" (1983). In *The very early universe* (Nuffield Workshop), ed. G.W. Gibbons, S.W. Hawking & S.T.C. Siklos (Cambridge University Press), p. 399.

[113] "Cosmological consequences of a low mass for the selectron" (1984). *Phys. Lett.*, **137B**, 169.

[114] "Is there a quantum equivalence principle?" (1984). In *Quantum theory of Gravity* (Essays in honor of B. de Witt), ed. S. Christensen (Adam Hilger), p. 78 (with P. Candelas).

[115] "Cosmology, galactic astronomy and elementary particle physics" (1984). *Proc. Roy. Soc. A*, **394**, 1.

[116] Introductory survey, First ESO-CERN Symposium on Large scale structure of the universe, Cosmology and Fundamental physics, CERN (1984). p. 3.

[117] "Massive neutrinos and photinos in cosmology and galactic astronomy" (1984). In *The Big Bang and Georges Lemaitre*, ed. A. Berger (Reidel), p. 31.

[118] "Cosmology, galactic astronomy and elementary particle physics" (1984). *ESA SP-207*, p. 171.

[119] "Paul Adrien Maurice Dirac" (1986). *Quart. J. R. astr. Soc.*, **27**, 1.

[120] "Particle physics and cosmology" (1986). In *Superstrings, Supergravity and Unified Theories*, ed. G. Furlan (World Scientific), p. 428.

[121] "Cosmology, astronomy and fundamental physics" (1986). In *Proceedings of the ESO/CERN Conference*, ed. G. Setti & L. van Hove (ESO), p. 321.

[122] "Photino decay and the ionisation of Lyman alpha clouds" (1988). *Mon. Not. R. astr. Soc.*, **230**, 13P.

[123] "Low-mass photinos and Supernova 1987A" (1988). *Phys. Lett.*, **215B**, 404 (with J. Ellis, K.A. Olive & S. Sarkar).

[124] "N_ν from NS and SN" (1989). In *Theoretical and Phenomenological Aspects of Underground Physics*, p. 249.

[125] "Cosmological implications of Supernova 1987A" (1989). *Atti Acad. Naz. Lincei*, **LXXII**, 395.

[126] "Problems for light Higgsinos" (1989). *Phys. Lett.*, **220B**, 586 (with M. Drees, J. Ellis & P. Jetzer).

[127] "Dark matter decay and the ionisation of HI regions of the Galaxy" (1990). *Ap. J.*, **364**, 549.

[128] "On the role of a strongly flattened galactic halo in the decaying dark matter hypothesis" (1990). *Mon. Not. R. astr. Soc.*, **244**, 1P.

[129] "Decaying dark matter and the mass model of the Galaxy" (1990). *Mon. Not. R. astr. Soc.*, **244**, 9P (with P. Salucci).

[130] "On the particle mass in the decaying dark matter hypothesis" (1990). *Mon. Not. R. astr. Soc.*, **246**, 191.

[131] "Dark matter decay and the spiral galaxy NGC 891" (1990). *Mon. Not. R. astr. Soc.*, **247**, 506 (with P. Salucci).

[132] "Free electron density in HI regions as an indicator of ionising dark matter" (1990). *Nature*, **346**, 40.

[133] "Precision estimate of cosmological and particle parameters in the decaying dark matter hypothesis" (1990). *Phys. Rev. Lett.*, **65**, 2839.

[134] "Consistent neutrino masses from cosmology and solar physics" (1990). *Nature*, **348**, 617.

[135] "The impact of the CMB discovery on theoretical cosmology" (1990). In *The Cosmic Microwave Background: 25 years later*, ed. N. Mandolesi & N. Vittorio (Kluwer), p. 1.

[136] "Dark matter decay and the ionisation of hydrogen throughout the universe" (1990). *Comments in Astrophysics*, **15**, 71.

[137] "Dark matter decay and the ionisation of hydrogen" (1990). In *LEP and the universe*, ed. J. Ellis (CERN), p. 65.

[138] "The role of photinos in Friedmann universes" (1990). In *A.A. Friedmann: Centenary Volume*, p. 169.

[139] "ν_τ decay and the ionisation of hydrogen in the universe" (1991). In *"Neutrino 1990" – Nucl. Phys. B Suppl.*, **19**, 138.

[140] "Dark matter decay and the temperature of Lyman alpha clouds" (1991). *Ap. J.*, **367**, L39.

[141] "Decaying dark matter and the value of the Hubble constant" (1991). *Astron. Astrophys.*, **243**, 341 (with P. Salucci).

[142] "Dark matter decay and the ionisation of the local interstellar medium" (1991). *Astron. Astrophys.*, **245**, 243.

[143] "Radiative neutrino decay in SUSY and the ionisation of hydrogen throughout the universe" (1991). *Phys. Lett.*, **259B**, 323 (with F. Gabbiani & A. Masiero).

[144] "Dark matter decay and the heating and ionisation of HI regions" (1991). In *IAU Symp. No. 144, The interstellar disk-halo connection in galaxies*, ed. H. Bloemen (Kluwer), p. 77.

[145] "IUE observations of the quasar 3C 263 constrain the ionising photon luminosity of decaying dark matter" (1991). *Mon. Not. R. astr. Soc.*, **249**, 21P (with A.C. Fabian & T. Naylor).

[146] "Neutrino decay and the temperature of Lyman α clouds" (1991). In *ESO Mini-Workshop on Quasar Absorption Lines*, ed. P.A. Shaver, E.J. Wampler & A.M. Wolfe, p. 21.

[147] "Dark matter decay, reionization and microwave background anisotropies" (1991). *Astron. Astrophys.*, **250**, 295 (with D. Scott & M.J. Rees).

[148] "Does a uniformly accelerated quantum oscillator radiate?" (1991). *Proc. Roy. Soc. A*, **435**, 205 (with D.J. Raine & P.G. Grove).

[149] "The physical significance of the vacuum state of a quantum field" (1991). In *The Philosophy of the Vacuum*, ed. S.W. Saunders & H.R. Brown (Oxford University Press), p. 137.

[150] "The cosmological u-v background from decaying neutrinos" (1991). In *The Early Observable Universe from Diffuse Backgrounds*, ed B. Rocca-Volmerange, J.M. Deharveng & J. Tran Thanh Van (Editions Frontières), p. 127.

[151] "Neutrino decay and the ionisation of nitrogen" (1992). In *Elements and the Cosmos: in honour of B.E.J. Pagel*, ed. M.G. Edmunds & R. Terlevich (Cambridge University Press), p. 314.

[152] "Decaying neutrino theory" (1992). *Nature*, **348**, 718 (with P. Salucci & M. Persic).

[153] "Dark matter decay and the ionisation of nitrogen in the interstellar medium" (1992). *Int. Journ. Mod. Phys. D*, **1**, 161.

[154] "Time paradoxes in relativity" (1992). In *The Nature of Time*, ed. R. Flood & M. Lockwood (Blackwell, Oxford), p. 6.

[155] "Decaying neutrinos and the nature of the dark matter in galaxy clusters" (1992). *P.A.S.P.*, in press – Dec. 1992 (with M. Persic & P. Salucci).

[156] "The C^0/CO ratio problem - a new precision test of the decaying neutrino theory" (1992). (To be published).

[157] "Is the universe unique?" (1993). In *Die Kosmologie der Gegenwart*, ed. G. Börner & J. Ehlers (Serie Piper) (in press).

BOOKS AUTHORED

The Unity of the Universe, 1959 (Faber & Faber).
The Physical Foundations of General Relativity, 1969 (Doubleday).
Modern Cosmology, 1971 (Cambridge University Press).
Modern Cosmology and the Dark Matter Problem, 1993 (Cambridge University Press).

BOOKS EDITED

(Each of the following was edited in collaboration with C.J. Isham and R. Penrose):

Quantum Gravity: an Oxford symposium, 1975 (Oxford University Press).
Quantum Gravity II: a second Oxford symposium, 1981 (Oxford University Press).

Exact and Inexact Solutions of the Einstein Field Equations

GEORGE F. R. ELLIS

1 INTRODUCTION

There has been a tremendous growth in the understanding of General Relativity and of its relation to experiment in the past 30 years, resulting in its transformation from a subject in the doldrums on the periphery of theoretical physics, to a subject with a considerable experimental wing and and many recognised major theoretical achievements to its credit. The main areas of development have been,
* solar system tests of gravitational theories,
* gravitational radiation theory and detectors,
* black holes and gravitational collapse,
* cosmology and the dynamics of the early universe.
On the theoretical side, this development is based on understanding exact and inexact solutions of the Field Equations (the latter has three different meanings I will discuss later). In this brief review of theoretical developments, there is not space to give full references to all the original papers. Detailed references can be found in previous surveys, in particular 'HE' is Hawking and Ellis (1973), 'TCE' is Tipler Clarke and Ellis (1980), 'HI' is Hawking and Israel (1987), and 'GR13' is the proceedings of the 13th International meeting on General Relativity and Gravitation held in Cordoba, Argentina in 1992. Many of the issues raised here are considered at greater length elsewhere in this book, e.g. in the articles by MacCallum and Tod.

2 EXACT SOLUTIONS AND GENERIC PROPERTIES

The use of an exact approach whenever possible is of great utility in understanding the implications of the Einstein field equations ('EFE'), because of major complexities arising in their solution. This is due to three features (HE, Wald 1984):
* The non-linearity of the equations themselves, considered as differential equations for the metric. When written out in full for a general metric, the field equations involve at least 119,640 terms (in terms of the metric, the components of the Ricci tensor are $6 \times 13280 + 4 \times 9990$ terms, see Pavelle and Wang (1985)). The prime problem is not so much the products of the Christoffels in the Riemann tensor, as the inversion of the metric tensor needed to go from the metric to the Christoffels;

* The non-linear interaction between matter, serving as a source of curvature for the space time through the EFE, and the metric, determining the motion of matter in the space-time through the energy-momentum conservation equations. Thus the matter determines the metric which determines the movement of the matter. Because the conservation equations are identically satisfied as a result of the EFE, despite this major feedback loop the equations are consistent and there is consistent development of a solution from given initial data;
* There is no fixed background space-time. This is essential, because of the principle of equivalence and the resulting representation of gravity by space-time curvature. There is even no fixed *a priori* space-time topology (the large scale connectivity of space-time cannot be taken for granted).
It is because of this interacting set of non-linear features that it pays to use an exact approach wherever possible. I should like to pay tribute here to the early pioneers in this regard: particularly, H P Robertson in the USA, J L Synge in Dublin, the Hamburg school (P Jordan, E Schücking, J Ehlers, W Kundt, R Sachs, and M Trümper), the London group (H Bondi, F Pirani, I Robinson, W Bonnor), A Lichnerowicz in France, and A Z Petrov in Russia. They laid the foundations for what followed. Furthermore, two teachers played a particularly important role by emphasizing to their students the power of exact solutions and exact results that provide reliable models of physical reality: namely J A Wheeler and D W Sciama. It is a pleasure to dedicate this paper to Dennis Sciama in recognition of his overall contribution and his excellent supervision (I was his first research student, and benefited greatly from his enthusiasm for and his knowledge of physics, astronomy, and cosmology, as well as his awareness of the potential to use modern mathematical techniques in the investigation of these areas).

2.1 Exact Solutions
Exact solutions are characterised firstly by their matter content: vacuum, cosmological constant, electromagnetic field, perfect fluid, scalar field, or other (perhaps a combination of sources). It is important to remember here the remark by Synge: without specified matter properties, we do not have a physical solution (given an arbitrary metric we can always run the field equations backwards and determine the stress energy tensor required to satisfy the Einstein equations; but it will almost always not have a satisfactory physical meaning). Furthermore an important issue in a realistic solution is what happens when there is a change of properties of matter (for example, at the surface of a star). This leads to the study of jump conditions at such surfaces of change, and in particular to criteria for existence of surface layers there, and characterisation of their properties (Israel 1966).

Given their matter content, exact solutions almost inevitably (whether or not this is the starting point of the study) have some exact continuous symmetries, characterised by a Lie Algebra of Killing vectors. Particular symmetries that have been extensively investigated are static, stationary, spherically symmetric, plane symmetric, cylindrically symmetric, axially symmetric, stationary axisymmetric and boost-rotation symmetric, and spatially homogeneous space-times (Kramer et al. 1980). Discrete symmetries have been much less explored. Other approaches to exact solutions (apart from simply assuming a particular metric form and calculating away) include specifying the algebraic (Petrov) type of the Weyl tensor; assuming separability properties of geodesic integrals; and specifying imbedding properties of the space-time.

One of the main problems is to avoid simply rediscovering solutions already known, possibly in quite different coordinates (A. Krasinski is currently compiling a comprehensive survey of exact cosmological solutions that reveals many rediscoveries of the same relatively few solutions). The rest of this section summarises properties of important exact solutions discovered so far.

A: Minkowski Spacetime This spacetime (HE:118-124) is the basis of special relativity, determining the metric behaviour underlying modern theoretical physics, and thus implying the fundamental properties of time dilation, length contraction, relativity of simultaneity, and transformation properties of physical quantities. However it cannot represent adequately a local non-uniform gravitational field (because it is locally flat); it is only consistent with the field equations in the vacuum case. Nevertheless it has interesting properties:

* In Rindler coordinates (based on uniformly accelerating observers), it has many properties similar to those of black holes; for example (Kay and Wald 1991) such observers experience event-horizons, leading to prediction of Unruh radiation measured by accelerating observers.

* By cutting and pasting, one can generate locally Minkowski space-times with spacelike, timelike, or null conical singularities (Ellis and Schmidt 1975); then, although the space-time is locally flat, there are non-local curvature and gravitational effects. This is the basis of the Regge calculus approximation to General Relativity (Regge 1961, Williams and Tuckey 1992), and of the pure-gravitational version of cosmic strings (Vilenkin in HI); when there are several such strings moving relative to one another, interesting interactions can occur and causal violations become possible (Allen and Simon 1992).

Minkowski space-time has a non-trivial conformal boundary, providing the basis for studies of asymptotically flat space-times (Penrose 1968, HE:221-225).

B: The Schwarzschild exterior solution This vacuum spherically symmetric spacetime (HE: 149-156, TCE:126-127) is asymptotically flat in its outer region ($r > 2m$), and is necessarily static there. It represents the exterior field of spherical stars (whether

static or not) because of Birkhoff's theorem (HE:369-372). It provides our basic understanding of planetary motion and the bending of light (and so for example implies gravitational lensing). It is also the basis of study of relativistic spherical accretion and of accretion disks around spherically symmetric objects and black holes (Rees in HI), providing the foundations of understanding of AGNs and QSO's.
Its maximal (vacuum) extension is a Black Hole (de Witt and de Witt 1973), with an unusual topology: two asymptotically flat spaces back to back joined through a Schwarzschild 'throat' (Penrose 1968, HE:149-156). There is a change of symmetry at $r = 2m$ which is a Killing horizon, where the timelike Killing vector of the exterior region changes to spacelike, so the spacetime changes from being static to being spatially homogeneous there. This is also the event horizon of the Black Hole, separating those regions from which one can reach or signal to infinity from those where this is not possible. This horizon surrounds a curvature singularity at the centre. This singularity is not based on a timelike world-line, as one might think at first; causally it is a spacelike surface. Boyer-Ehlers bifurcation (Boyer 1969) of the event horizon at the Schwarzschild throat is the basis of the Hawking radiation emitted by the Black Hole (Kay and Wald 1991).

C: Static spherically symmetric fluid body Static spherically symmetric interior star solutions can be joined onto the Schwarzschild exterior solution at constant r values, provided $r > 2m$ and the interior and exterior masses are matched, giving matter sources for the exterior gravitational field. The simplest such solution is the Schwarzschild interior solution (with constant mass density throughout the interior). Examination of such solutions gives exact limits on the solution parameters, which imply limits on star sizes in relation to their mass (Buchdahl 1959, Wald 1981:125-135), and consequently put limits on gravitational redshift that can be observed from a spherical object of given size (this is important for interpretation of QSO redshifts).

D: Dynamic Spherically symmetric fluid bodies Study of dynamic spherically symmetric stellar evolution and gravitational collapse is based on spherically symmetric fluid solutions (again joined on to a Schwarzschild exterior solution), which in general can only be integrated numerically. However the pressure-free case corresponds to the Lemaître-Tolman-Bondi spherically symmetric exact solutions, giving the basic picture of spherical gravitational collapse to form a black hole (HE:299-310, Israel in HI).

E: The Kerr solution This axially symmetric stationary vacuum solution (de Witt and de Witt 1973, HE:161-168) extends the Schwarzschild exterior solution to the case of a rotating system. This is believed to give the basic picture of the exterior field of a rotating black hole, however no fully satisfactory interior solution has been obtained that matches onto it. The maximal vacuum extension gives the unique family of rotating black hole solutions (de Witt and de Witt 1973, Mazur 1987), again with an event horizon limiting the region of space-time from where one can send signals to infinity.

It has a non-trivial global topology, including ring singularities with a two-sheeted structure, and many asymptotically flat regions. Furthermore causal horizons separate an interior region where causal violations can occur from the exterior asymptotically flat region. Because the Killing horizon and event horizon no longer coincide, it has an ergosphere region lying between the two; in principle one can extract rotational energy from the black hole by processes carried out in this region.

F: FLRW Universe models   The fluid-filled Friedmann-Lemaître-Robertson-Walker (FLRW) evolving cosmological models (HE:134-142, Ellis 1987) are locally isotropic everywhere, and consequently are spatially homogeneous (with spatial sections of constant curvature). These are the standard models at the basis of modern cosmology (Weinberg 1972)), underlying the concept of an expanding universe with a hot big bang origin, experimentally vindicated through the successful prediction of primeval element abundances (based on calculations of nucleosynthesis in the early universe), together with the observation of relic radiation from the hot early phase in the form of Cosmic Microwave Blackbody Radiation (CMBR).

These universes start at an initial singularity where not only all the matter in the universe, but also space and time themselves originate. Most of the matter in the universe is hidden from us by particle horizons, which also limit the possibilities of causal physical interactions in the early universe, thereby leading to the prediction of domain walls, cosmic strings, and other textures resulting from symmetry-breaking as the universe expands and previous unconnected regions come into causal contact with each other (Blau and Guth in HI).

FLRW dynamics is usefully summarised by various forms of phase diagrams. The static form is *the Einstein Static universe* (with cosmological constant), which was the first to demonstrate the possibility of a universe with closed (finite) spatial sections, the 'no-boundary' property strongly advocated by Einstein and Wheeler for Machian reasons. Particular simple expanding cases are the spatially flat *Einstein-de Sitter* models, the linearly expanding (empty) *Milne* universes, and the de Sitter universe.

G: de Sitter Universe   Although this stationary empty space-time with cosmological constant (HE:124-131) is just a spacetime of constant curvature, it is of interest in demonstrating the different views of a single space-time given by different coordinate coverings. The full hyperboloid is geodesically complete but evolving; the Steady State Universe proposed as a cosmological model by Bondi, Gold, and Hoyle, is based on half of this hyperboloid, represented in FLRW coordinates. It has recently been revived by Guth as the basis of inflationary universe models with a scalar field matter source (Gibbons et al. 1983, Blau and Guth in HI, Linde in HI).

H: Swiss cheese (Einstein-Strauss) models These are made by combining matched sections of the FLRW expanding universes with spherically symmetric (Schwarzschild) vacuum or (Lemaître-Bondi-Tolman) dust solutions, to form an inhomogeneous universe model representing in an exact way growth of inhomogeneities and galaxy formation in a cosmology with a spatially homogeneous and isotropic background, as pointed out by Tolman (see Harwitt (1992) for a recent perspective). These exact models can also be used to examine issues such as gravitational lensing and CMBR anisotropies arising from inhomogeneities in the expanding universe.

I: Spatially homogeneous universes The spatially homogeneous fluid-filled cosmological models comprise the Kantowski-Sachs rotationally symmetric universes and the Bianchi models (HE:142-149, MacCallum in Hawking and Israel 1980). They allow extension of cosmological investigations to distorting and rotating universes, giving estimation of effects of anisotropy on primordial element production and on the measured CMBR spectrum and anisotropy.
The simplest such models are the Kasner (spatially flat) universes, permitting pancake and cigar singularities at the origin of the universe. Other interesting cases are the mixmaster (Bianchi IX) models, which show chaotic behaviour at early times, and the tilted (Class B) universe models that allow a non-scalar initial singularity. The dynamics of special classes of these universes is illuminated by phase diagrams showing their qualitative behaviour.

J: Plane waves Exact vacuum plane wave solutions are a large class including homogeneous and sandwich waves (HE: 178-179, Kramer et al. 1980). They demonstrate the existence of exact gravitational wave solutions of the field equations (it must be remembered that even the existence of gravitational waves has been a controversial issue in the past), and clarify properties such as how such waves affect the motion of test particles, in particular confirming the exact tensor-transverse (quadrupole) nature of gravitational radiation; there is no monopole, dipole, or longitudinal component (Smarr 1979, Thorne in HI).
In general they have curious global properties including the existence of Cauchy (prediction) horizons, and focusing of timelike and null curves (Penrose 1965); these in turn underlie some of the properties of colliding plane wave exact solutions, which give rise to singularities with non-trivial global character (Griffiths 1991).

K: Further solutions A variety of further solutions are of interest in understanding the implications of general relativity theory and the nature of curved space-times (although they have no immediate physical application). Amongst them are,

a: The *Anti-de Sitter universe* (HE:131-134, TCE:171-173) is a space-time of constant curvature with interesting geodesic connectivity properties.

b: The *Reissner-Nordstrom solution* (HE:156-161, TCE:172,174) is spherically symmetric, being the analogue of the Schwarzschild solution but with a non-zero electromagnetic field due to an overall electric charge. It has much more complex global properties than the Schwarzschild solution (there are many asymptotically flat regions in this case).

c: The *Gödel universe* (HE:168-170) is a rotating fluid-filled universe that is homogeneous in space and time. It exhibits causal anomalies (there exists no global time coordinate, and there are closed timelike lines through every point). It serves as a cautionary example against taking for granted causal properties, or the nature of time in a rotating cosmological solution.

d: The *Taub-NUT spacetime* (HE:170-178, TCE:124) (NUT stands for Newman, Unti, and Tamburino; the relation between the separately discovered Taub and NUT spacetimes was elucidated by Misner) is a locally rotationally symmetric vacuum space with two Killing horizons which are also causality horizons (they separate stationary regions where there are causal violations, from spatially homogeneous regions without causal violations). It has the remarkable property of having inequivalent analytic extensions across these horizons, which are geodesically incomplete despite being surfaces of homogeneity.

e: Whole classes of solutions can be obtained by *generating function techniques*, for example involving Bäcklund transformations. These demonstrate subtle symmetries of the EFE and unexpected relations between apparently quite different solutions (Hoenselaers and Dietz, 1984).

2.2 Properties Arising from Exact Symmetries

The exact solutions listed above (and others not mentioned here) yield interesting insights when their global properties are investigated (see particularly Penrose (1968) as well as HE and TCE). The following themes have emerged as worth examining in order to help understand the nature of a solution.

Maximal extension : One must ensure that the solution has been extended as far as it can (there can in general be apparent edges to space-time, due to bad coordinate choice rather than to any real space-time boundary).

Then one can look at the *global topology* of the space-time, considering for example orientability properties (which determine if a global arrow of time can exist, and if spinors can be globally defined). In general a maximal extension will have an unusual topology (e.g. it may have many asymptotically flat regions) and may look quite different in terms of different spatial slicings (determining the time coordinate in which the solution is expressed). Furthermore the *symmetries* of the full solution may be unexpected (for example, the Schwarzschild solution is spatially inhomogeneous inside the Killing horizon).

The space-time boundary : there will usually be a part of the space-time boundary at infinity, with a particular asymptotic and causal structure. There will also in general be a part of the boundary at a finite distance, that is, characterised by *geodesic incompleteness*. Thus one examines whether the geodesics in the maximally extended space-time are complete or not. Where they are incomplete, one runs into a space-time singularity. This may be spacelike, timelike, or null; isotropic or anisotropic (cigar or pancake, for example); scalar, non-scalar, or locally extensible; and may be associated with chaotic behaviour. In each case one may attempt various forms of extension through the singularity, or a definition of a singular boundary for the spacetime (for example, using Penrose' causal boundary or Schmidt's b-boundary). The point here is to try to understand in what ways space-time has a beginning or comes to an end, and what the physical implications may be.

Causal properties are significant, in particular the boundaries of causal domains; these can often be examined in terms of causal diagrams showing the conformal boundary of the space-time, and the nature of causal limits such as various kinds of horizons.
These include *causality horizons*, limiting the regions where causal violations occur; *Cauchy horizons*, limiting the domain of dependence of initial data, and related to geodesic connectivity and the existence of cosmic time; *particle horizons*, limiting communication in cosmology, and so putting bounds on the possibilities of physical causation; *event horizons*, limiting the possibility of communication to infinity, and so forming the basis of black hole definitions; and *Killing horizons*, where symmetries change their nature from timelike to spacelike, often associated either with singularities or with horizon bifurcations (the latter being associated with quantum production of Hawking and Unruh radiation).

2.3 Relevance of Exact Solutions
While exact solutions of the full non-linear field equations reveal important properties of those equations, they are of course idealised because of their exact symmetries (a lot of degrees of freedom have been 'frozen out' in order to obtain solvable models). The immediate issue then is, to what extent do these solutions reflect accurately the important physical properties of more realistic solutions, that do not have special symmetries ? This leads to attempts to show either that
* particular exact solution properties are generic (for example, that almost all space-times have singular boundaries),

or that

* generic solutions approach the exact solutions in appropriate limits (for example, the 'no-hair' theorems for black holes and inflationary universes).

Certain *uniqueness theorems* are helpful here, providing partial evidence of the physical significance of particular exact solutions. The prime example is *Birkhoff's theorem* (HE:369-372), showing that an exact spherically symmetric vacuum solution is always equivalent to part of the Schwarzschild solution, and so has an extra exact symmetry (in the exterior region, it is static); this confirms the conclusion that no monopole gravitational radiation is possible in General Relativity. Similarly there are powerful theorems (HE:308-347, De Witt and de Witt 1973, Mazur 1987) showing the Schwarzschild solution in the non-rotating case and the Kerr solution in the rotating case are the unique black hole solutions, provided we assume an exact symmetry (the solution is static or stationary) together with specific asymptotic flatness and horizon conditions. One starts by proving the solution is spherically or axially symmetric, respectively; thus showing the spacetime is invariant under a larger symmetry group than initially assumed.

The kind of problem that arises is that real rotating stars have external fields with multipole structures which are completely different from the Kerr solution. The idea is that when such stars collapse they radiate off these multipoles, and eventually settle down asymptotically to a stationary state which (by the uniqueness theorems) is the Kerr solution. Thus while these uniqueness theorems are a step in the right direction, they need supplementation by examination of generic exact solutions where possible, and where this does not suffice, by examination of approximate solutions.

2.4 Generic Exact Properties

The search for generic exact properties and the stability of particular solutions should in principle be based on systematic study of the space of space-times, choice of a suitable measure on that space determining how generic any specific solution is. This has proved technically difficult, so attention has focused on more specific areas.

A: Generic matter properties have been studied in depth, in particular the nature of fluid solutions, and their relation to a kinetic theory description (Ellis, Ehlers in Sachs 1971). This has led to detailed study of congruences of timelike (HE:78-85,96-101) and null (HE:86-88,101-102) curves, their nature being characterised by various kinematic quantities; it is possible to obtain exact equations governing the general behaviour of these kinematic quantities. Three specific issues have come to light through such studies:

Firstly, the complexity of rotating (vorticity) solutions, where the congruence of curves is not orthogonal to a family of hypersurfaces in space-time. Although there are well defined vorticity propagation laws, these solutions are an order of magnitude more difficult to investigate than non-rotating solutions.

Second, the importance and significance of shear of congruences: almost all congruences of curves do distort, and shear-free solutions are very special in character in both the timelike and null cases (the latter leading to the Goldberg-Sachs theorem for vacuum space-times). In general null shear conveys gravitational 'news', this idea being made explicit in the case of asymptotically flat space-times.

Thirdly, provided certain energy conditions are satisfied, both timelike and null non-rotating families of curves tend to be focused by any gravitational field. This is the basis of the simplest singularity theorems for cosmology, and of the properties of closed trapped surfaces and the event horizon that are so important in the study of black holes.

Quite different in technique but similar in spirit is the work on exact equations of motion of massive bodies (Dixon in Ehlers 1979).

B: *Generic Causal structure* is well understood (Penrose 1972, HE:180-221, TCE:116-123), the foundation being study of the generic nature of the boundary of a causal domain (HE:186-188). This leads firstly to the general *definition of horizons* (TCE:121-123), and in particular of particle and event horizons in general space times; secondly, to understanding the nature of *the domain of development of a spacelike surface* and of *Cauchy surfaces* and *Cauchy Horizons* (HE:201-212), which are closely related to properties of geodesic connectivity (HE:212-217); thirdly, to the definition of various kinds of *causal violations* (HE:189-204), including in particular stable causality, and of causality horizons.

C: *Incompleteness theorems* of a generic nature have been proven by Penrose, Hawking, and Geroch (HE:261-275, TCE:129-138), applicable both to the case of gravitational collapse of stars (HE:299-308) and to the early universe (HE: 348-359). They show that in both these cases the occurrence of geodesic incompleteness is not a property of high symmetry models but rather is a generic property of the field equations, provided the matter obeys specific energy conditions. This is usually interpreted as showing that space-time singularities exist. While these theorems demonstrate the existence of such singularities, they do not tell us their nature, although a number of studies have classified singularities into different classes, and characterised their properties (Ellis and Schmidt 1975, Clarke 1992).

D: *Existence and uniqueness theorems* Complementary to the singularity theorems are generic existence and uniqueness theorems for solutions of the Einstein field equations (HE: 226-255, Muller zum Hagen and Seiffert in Ehlers 1979, Friedrich 1990, Klainerman in GR13), based on the spacelike or null initial value formulation of General Relativity. Firstly one can show that there are generic solutions to the four initial value equations on an initial surface. Then given initial data which satisfies these equations, they can be shown to lead to existence of unique exact solutions in some domain around the initial surface (six propagation equations taking the data off the surface and into the space-time). The theorems for data on a null surface can be related directly to observable cosmological data (Ellis et al, 1985).

E: Black hole properties   A whole series of theorems can be proven about black hole properties, characterising the nature of the event and apparent horizons (HE:308-323, Hawking in de Witt and de Witt 1973). These assume specific behaviour at infinity, defining a family of asymptotically flat space-times, together with causal properties that specify a black-hole situation. However no exact symmetries are assumed; thus the space-times considered are generic within this class.

F: Positive mass theorems   These show that the mass at infinity of a generic asymptotically flat space-time is necessarily positive (Flaherty 1984), thus providing the foundation for showing that flat space-time is stable. A series of related investigations examine the relation between gravitational masses defined at spatial infinity, timelike infinity, and null infinity (Ehlers 1979).

2.5 Different Approaches
This series of exact results is an impressive achievement of the past decades. I suggest the basis of this upsurge is two particular features (spearheaded particularly by Roger Penrose and Stephen Hawking). Firstly,
* an emphasis on geometric rather than analytic thinking has been the driving force behind many of the achievements, not surprising in view of the geometric nature of General Relativity as a theory of gravitation.
Given an idea obtained from geometric insight, there has then been a delicate interplay between analytic and geometric methods of proof in establishing correct results of an unfamiliar kind. In doing so,
* there has been an emphasis on bounds and inequalities rather than equalities, in many cases (for example, use of inequalities to define closed trapped surfaces and to specify energy conditions (HE:88-96), resulting in the area theorems associated with the broad qualitative nature of black hole solutions and singularity theorems).
These are related to each other; for example the singularity theorems are of an essentially geometrical nature because the space-time does not have to satisfy the Einstein equations with specific matter sources, but rather the curvature has to have a certain sign (and this explains also why one cannot obtain detailed insight into the structure of the field equations this way). Similarly Penrose (1968) introduced a nice geometrical framework for defining the properties of null infinity; this enabled an examination, using the theory of partial differential equations, of how the different

geometrical features work together. It is this kind of emphasis that has underlain the ability to characterise and prove many generic properties of the equations, and so obtain results with realistic application to the real universe.

To a limited extent, the essential further feature of functional analysis (functions or sets of functions are considered as points in a more abstract space) has been of use in these developments (e.g. in the proofs of existence and uniqueness theorems). The disappointment perhaps has been that the 'space of spacetimes' has turned out to be rather intractable and difficult to deal with, and so has not yielded as many insights as one might have hoped.

3 INEXACT SOLUTIONS

While the generic exact results just discussed enable a broad understanding of many important properties, detailed analysis of complex situations and geometries demands introduction of approximate methods, necessarily leading to various types of inexact solutions. Some are very useful, but some may be misleading.

3.1 Approximate Properties: Perturbation Equations

A large literature has developed on approximate solutions of the field equations, based on approximate description of geometries with sufficient degrees of freedom to allow detailed representation of realistic space-times. These are essential in understanding many detailed properties of the gravitational field.

A: Solar system   Detailed description of the solar system (including specification of the position and masses of all the planets and hundreds of asteroids) is needed in relativistic celestial mechanics (Damour in HI), underlying space tests of General Relativity, where astonishing precision has been attained in laser and radar ranging of the Moon, Mars, and various spacecraft. The needed approximation scheme for the Schwarzschild solution has been developed in detail as the PPN (parametrized post-Newtonian) formalism (Will 1981). It is the basis of relativistic tests in the solar system, in particular predicting accurately the motion of planets.

B: Gravitational radiation   The motion of interacting bodies in an asymptotically flat background has been modelled in great detail during investigation of the emission of gravitational radiation in collisional interactions and by orbiting binaries, and in particular examining the effect on binary orbital evolution of gravitational radiation emission, a controversial subject where approximation methods have been hotly debated. Provided we believe the standard results (Smarr 1979, Will 1981, Thorne in HI, Schutz in GR13), the observations of such evolution in the case of the binary pulsar can be claimed to be the most accurate measurement made in physics (Penrose in GR13). This theory is of course related to the development of gravitational wave detectors, currently reaching the stage where we can begin to expect positive results (see GR13). Related work considers mass definitions at infinity, conserved quantities, and mass loss (related to the Bondi 'news function'), see Ehlers (1979).

C: Cosmology   The analysis of perturbations in cosmology underlies our best attempts at theories of growth of large scale structure in the universe and of galaxy formation (Rees in HI), that have been pursued for many decades without conclusive results; investigation of observational relations in perturbed cosmologies, particularly of CMBR anisotropies (Sachs and Wolfe 1967); and models of gravitational lensing in a cosmological setting (see Ehlers, Schneider and Falco 1992; Ehlers in GR13).
It is in these studies in particular that one runs across the problem of the lack of a unique background space in perturbation analysis, and so the problem of gauge freedom, leading to the occurrence of non-physical gauge modes (The problem is also present, but in less acute form, in the analysis of asymptotically flat space-times, where the asymptotically flat background is relatively rigid in its relation to the real space-time.). One can proceed by using special gauges, provided one keeps extremely clear track of the remaining gauge freedom and only calculates physical quantities; the perils of this approach are apparent in the confusions in the literature. The alternative is to use explicitly gauge-invariant methods of analysis, initially introduced in a major paper by Bardeen (1980), and subsequently followed by more geometrically based methods (e.g. Ellis and Bruni 1989).

D: Linearisation stability   A major issue underlying all such perturbation analyses is that of linearisation stability, that is, is the linearised solution a good approximation to full solutions of the field equations ? In general the answer is that locally everything is all right, but problems arise globally if there are symmetries in the background space, which will be the case almost always (Fischer and Marsden in Ehlers 1979, and in Hawking and Israel 1979). In any case, this needs checking in any particular application of perturbation analysis (see for example D'Eath (1976) for the cosmological situation).

E: Newtonian limit of GR   A particular issue of importance is the definition of a Newtonian limit for particular physical situations, and proof that the field equations imply such a limit is in fact attained. This is fundamental in that, assuming General Relativity is the correct classical theory of gravity, Newtonian Theory holds as a valid description when and only when such a limit exists. The point then is that this issue cannot be adequately examined through linearisation studies, precisely because we use Newtonian theory in many situations where non-linear effects occur; indeed a non-linear analysis is required to justify common astrophysical usage (where for example density contrasts can be very high, and we certainly take into account the non-linear effects resulting from gravitational clumping affecting the motion of matter which results in the gravitational clumping.)

A theory of how to define and justify such a limit has been developed by Ehlers and Lottermoser (1992), but intriguing questions remain unanswered. An example: consider a pressure-free fluid ('dust') that is moving in a shear-free way. In General Relativity theory, it is an exact result that such a fluid cannot both expand and rotate (Ellis 1967, Collins 1986); however this is possible in Newtonian Theory. Thus there is in this situation more freedom in the Newtonian solution than in the Relativistic one. Furthermore, this has interesting implications: in Newtonian theory one can find shear-free cosmological solutions that expand and rotate, which are without an initial singularity (following the model back in time, the rotation causes a 'bounce' before the singularity occurs). These would provide counterexamples to the singularity theorems of General Relativity, were such relativistic solutions possible; but they are not (Ellis in Sachs 1971).

F: Numerical solutions Because of the discretisation involved in numerical work, these are based on a form of approximation of the field equations that does not involve linearisation (but in some forms is related to the Regge calculus approximation). This is an area that has developed rapidly in recent times (Evans Hobill and Finn 1989, Matzner 1991, Evans and Teukolsky in GR13, d'Inverno 1992).

3.2 Inexact Definition
Some solutions are problematic because they are inexactly defined in relation to the problem they are supposed to illuminate. This can happen in a number of ways.

A: Vague Definition Problems occur where the situation considered is not specified uniquely enough, leading to ill-defined results.

A particular case is the issue of *Inflation, the generality of Ω, and the Inflation no-hair theorem*. Claims that the inflationary universe idea necessarily leads to Ω at the present day being extremely close to unity, are wrong; there are inflationary models where this is not so (Ellis 1991, Ellis Lyth and Mijic 1991). Given these counter-examples, inflationists move the goal-posts and claim that these models are not compatible with current microwave anisotropy results. This being disproved, the goal posts are moved again and the claim is made that these models are improbable. However no proper measure is used on the space of cosmological models when these claims are made, so they are without a proper basis; in fact they arise by one method or another of restricting the phase space of possibilities to a subset of all inflationary models, where the usual claims are true.

Similar issues arise in the case of the 'no hair' theorem for inflation: the claims made are often vague, and in fact do not show that almost all anisotropies will be smoothed out by inflation; indeed enough anisotropy has the possibility of preventing inflation happening at all. Underlying this malaise in both cases is the unsolved issue of a suitable measure on the space of cosmological models that will unambiguously specify what is probable and what is not. Without such a measure, conflicting claims will be unresolvable.

Other issues where we do not yet have an unambiguous and generally accepted definition are the old theme of *Mach's principle*, which continues to elude a precise and commonly agreed definition; and the concept of *gravitational entropy*, studied in depth in the case of black holes but not satisfactorily resolved in more general situations. A proper definition is of interest because it determines the correctness or not of the Penrose Weyl-tensor conjecture, proposing smoothness of the start of the universe (Penrose in Hawking and Israel 1980). If correct, this again provides reasons to be cautious about the inflationary universe idea.

B: Relevance of definition Another form of problem arises where the formulation of exact or approximate solutions is reasonably precise, but the correspondence of the solutions to the reality they are intended to describe is not fully satisfactory. Thus they are inexact in terms of understanding the physical question at hand.

A case is *cosmic censorship*, where plausible a physical conjecture having been made, spherically symmetric counter-examples were found (TCE:167-189); so reasons were found why these were 'unphysical', in effect leading to a reformulation of the conjecture. The goal-posts then being moved, again apparent counter examples have been claimed (Teukolsky in GR13). The conjecture may be right in a physical sense, but we have not yet found the right formulation which is counter-example free. When one has the phenomenon of successive counterexamples followed by moving goal-posts, one should worry a little if the conclusion is really true (In the case of the singularity theorems, the issue of the definition of singularity was resolved by a singularity with geodesic incompleteness, resulting in many powerful theorems; the cost is that the nature of the singularity is not always what we might have expected.).

A further case of this kind (which appears to worry no one but the present author) is the issue of *isolated systems and asymptotic flatness*. At present we study isolated systems in terms of asymptotic flatness conditions at infinity, which are clearly not satisfied in the real universe. Now the supposition is that this does not matter; if so, why cannot we rigorously prove this?
For example the present definitions of black holes, and resulting theorems, are highly non-local in character, but use a wrong global description. However real astrophys-

ical situations are local. One should be able to reformulate black hole definitions in a local form, if they are to describe the real physical universe. One way to attempt this is through use of finite spheres surrounding the collapsing object, providing an effective infinity for the problem that is a world tube at a finite distance (Ellis 1984); and replacing equalities at infinity by inequalities imposed at a finite distance, thus enabling the real physical situation to be described (An example of the kind of approach envisaged is the local study of definitions and properties of electromagnetic radiation by Hogan and Ellis 1991). This would not only enable astrophysical usable definition of black holes, but also investigation of the interaction between the universe outside and a local system; this is not possible with present formulations of asymptotic flatness. In particular it could provide a sound basis for study of the localisation property involved in the idea of an 'isolated system' in the expanding universe, underlying the fact we can understand local physical behaviour largely independent of the properties of the larger universe.

Again, there are interesting unresolved problems relating to the issue of *coarse graining* in GR approximations, and *the effect of averaging on field equations*. The issue is that we may actually be using the wrong field equations in cosmological applications if we do not allow for such effects (Ellis 1984). An increasing effort in this direction is leading to a series of interesting results, for example the suggestion that such coarse-graining effects could be significant in terms of affecting the age of the universe (Bildhauer and Futamase 1991, Zotov and Stoeger 1991); however the relation of different approaches to each other (and to the issue of gauge choice) has yet to be clarified.

A final example is the widespread use of *the perfect fluid idea* in understanding astrophysical objects with generic kinematics, while kinetic theory studies of the foundations of perfect fluids suggest this approximation may only be valid in shear-free situations (Ellis Matravers and Treciokas 1983).

3.3 Incompatible Approximations
Sometimes inexactness takes the form of assuming approximations that cannot exist, or incompatible approximations.

An example of the first is repeated attempts to study the properties and motion of *point particles* by a δ-function expansion about a (timelike) world line. The problem is that this does not take into account the nature of the Schwarzschild solution: inside the horizon there is no singular timelike world line (in a static space-time region), rather the singularity has a spacelike structure (in a spatially homogeneous but time-dependent region). The hoped for δ-function expansion is incompatible with the nature of spherical 'particles' in general relativity.

An example of the second is the issue of *infinitely thin gravitating cosmic strings*. A conical singularity approximation is widely used to determine the effect of such strings on bending of light (the gravitational lensing by the strings) and on motion of matter (the formation of sheets of matter as the 'wake' behind a moving string). A quite different approximation is used (a field theory solution on a fixed background, Vilenkin in HI) to examine the bending and intersection of such strings. Now if one takes the first viewpoint seriously, these models are strictly incompatible: it is impossible for a string that has a conical singularity structure to bend wildly (Clarke Vickers and Ellis 1990) because its central world line spans a geodesic 2-space in space time (Unruh et al 1989). Loops of such a nature can only exist either on the scale of the universe, or if very large gravitational fields exist, such as at the throat of the Schwarzschild solution (because closed loops with a conical singularity structure implies there are *spacelike* geodesics that are closed; this is only possible in very high gravitational fields). Again the two descriptions give quite different viewpoints on what happens when cosmic strings intersect.

The point here is that two different and incompatible descriptions of the nature of strings are being used in different studies. This may be acceptable and indeed appropriate for many purposes, but it certainly requires exceptionally careful handling; particularly if one in effect tries to use two incompatible descriptions simultaneously in examination of a single physical problem.

4 CONCLUSION

The progress that has been made is great; however there is still quite a bit to be done. The examples and theorems briefly indicated here show how careful one must be in dealing with the non-linearities of General Relativity, and specifically the lack of a fixed background space-time as a framework for physics.

In many cases particle physicists still treat the universe as a flat space-time, with gravity represented by a symmetric tensor field. This cannot account for much of the structure outlined above, and can be very misleading as to what is and is not possible. Indeed in understanding General Relativity one cannot just go ahead and blindly use analytic tools; one needs geometric and physical insight as well. The exact solutions and methods that have been devised can give such understanding, even where results are initially quite counter-intuitive.

I thank B. G. Schmidt, C. Hellaby, M. A. H. MacCallum, and J. C. Miller for useful comments, and the FRD (South Africa) and MURST (Italy) for financial support.

REFERENCES

B. Allen and J. Simon (1992): *Nature* **357**, 19.

J. Bardeen (1980): *Phys Rev* **D 22**, 1881.

S. Bildhauer and T. Futamase (1991): *Gen Rel Grav* **23**, 1251.

R. H. Boyer (1969): *Proc Roy Soc* **A311** 245.

H. Buchdahl (1959), *Phys Rev* **116** 1027.

C. J. S. Clarke (1992): *The analysis of spacetime singularities* Cambridge University Press, Cambridge [to appear].

C. J. S. Clarke, G. F. R. Ellis and J. A. Vickers (1990): *Class Qu Grav* **7**, 1.

C. B. Collins (1986): *Canadian Journ Phys* **64**, 191.

P. D'Eath (1976): *Ann. Phys. (N.Y.)* **98**, 237.

R. A. d'Inverno (1992) (Ed.): *Approaches to numerical relativity.* Cambridge University Press, Cambridge [to appear].

C. de Witt and B. de Witt (eds) (1973): *Les Astres Occlus.* Gordon and Breach, New York.

J. Ehlers (ed) (1979): *Isolated Gravitating Systems in General Relativity.* Enrico Fermi Course LXVII. North Holland, Amsterdam.

J. Ehlers and M. Lottermoser (1992): *Ann Inst Henri Poincare* [to appear].

J. Ehlers, P. Schneider and J. Falco (1992): *Gravitational lenses.* Springer, Berlin.

G. F. R. Ellis (1984). In *General Relativity and Gravitation*, Ed. B. Bertotti et al. Reidel, Amsterdam, 215.

G. F. R. Ellis (1987). In *Vth Brazilian School on Cosmology and Gravitation*, Ed. M. Novello. World Scientific, Singapore, 83.

G. F. R. Ellis (1991). In *Gravitation, Proceedings of Banff Summer Research Institute on Gravitation*, Ed. R. Mann and P. Wesson. World Scientific, Singapore, 3.

G. F. R. Ellis and M. Bruni (1989): *Phys Rev* **D40**, 1804.

G. F. R. Ellis, D.H. Lyth, and M.B. Mijic (1991): *Phys Lett* **B271**, 52.

G. F. R. Ellis, S. D. Nel, W. Stoeger, R. Maartens, and A. P. Whitman (1985): *Phys Reports* **124** 315.

G. F. R. Ellis and B. G. Schmidt (1975): *Gen Rel Grav* **8**, 915.

G. F. R. Ellis, R. Treciokas, and D. R. Matravers (1983): *Annals of Physics* (NY) **150**, 487.

C. R. Evans, L. S. Finn and D. W. Hobill (Eds) (1989): *Frontiers in numerical relativity.* Cambridge University Press, Cambridge.

F. J. Flaherty (ed) (1984): *Asymptotic Behaviour of Mass and Spacetime Geometry.* Springer Lecture Notes in Physics 202. Springer, Berlin.

A. H. Friedrich (1990). In *General Relativity and Gravitation* Ed. E. N. Ashby, D..F. Bartlett and W. Wyss. Cambridge University Press, Cambridge, 41.

G. W. Gibbons, S. W. Hawking and S. T. C. Siklos (eds) (1983): *The very early universe.* Cambridge University Press, Cambridge.

J. B. Griffiths (1991): *Colliding Plane Waves in General relativity*. Oxford University Press, Oxford.

M. Harwit (1992): *Astrophysical Journal* **392** 394.

S. W. Hawking and G. F. R. Ellis (1973): *The Large Scale Structure of Space-Time*. Cambridge University Press, Cambridge.

S. W. Hawking and W. Israel (eds) (1979): *General Relativity: An Einstein Centenary Survey*. Cambridge University Press, Cambridge.

S. W. Hawking and W. Israel (eds) (1987): *300 years of Gravitation*. Cambridge University Press, Cambridge.

A. Held (Ed.) (1980): *General Relativity and Gravitation: One hundred years after the birth of Albert Einstein* (Volume 1 and Volume 2). Plenum Press, New York.

C. Hoenselaers and W. Dietz, eds. (1984) *Solutions of Einstein's equations: techniques and results*. Lecture Notes in Physics, 205. Springer Verlag, Berlin

P. A. Hogan and G. F. R. Ellis, (1991): *Ann Phys* **210**, 178.

W. Israel (1966): *Nuovo Cimento* 44, 1; 48, 463.

B. S. Kay and R. M. Wald (1991): *Physics reports* **207** 49.

D. Kramer, H. Stephani, M. MacCallum and E. Herlt (1980): *Exact Solutions of Einstein's Field Equations* Cambridge University Press, Cambridge.

R. A. Matzner (1991): *Ann. N.Y. Acad. Sci.* **631**, 1.

P. O. Mazur (1987): In *General Relativity and Gravitation*, ed. M.A.H. MacCallum, Cambridge University Press, Cambridge, 130.

R. Pavelle and P. S. Wang (1985), *J.Symb. Comp* **1**, 97.

R. Penrose (1965). *Rev Mod Phys* **37**, 215.

R. Penrose (1968). In *Battelle Rencontres*. Ed C. de Witt and J. A. Wheeler. Benjamin, New York.

R. Penrose (1972): *Techniques of Differential Topology in General relativity*. SIAM Regional Conference Series in Applied Mathematics. SIAM, Philadelphia.

T. Regge (1961): *Nuovo Cimento* **19** 558.

R. K. Sachs and A. Wolfe (1967): *Astrophys Journ* **73**, 147.

R. K. Sachs (Ed.), (1971): *General Relativity and Cosmology*, Proc Int School of Physics "Enrico Fermi", Course XLVII. Academic Press, New York.

L. Smarr (ed), (1979): *Sources of Gravitational Radiation*. Cambridge University Press, Cambridge.

F. J. Tipler, C. J. S. Clarke, and G. F. R. Ellis (1980). In Held (1980), Vol. 2.

W. G. Unruh, G. Hayward, W. Israel and D. McManus (1989): *Phys Rev Lett* **62**, 2897.

R. Wald (1984): *General Relativity*. University of Chicago Press, Chicago.

S. Weinberg (1972) *Gravitation and cosmology*. Wiley, New York.

C. M. Will (1981): *Theory and experiment in gravitational physics*. Cambridge University Press, Cambridge.

R. M. Williams and P. A. Tuckey (1992): *Class Qu Grav* **9** 1409.
N. Zotov and W. Stoeger (1991): *Class Qu Grav* **9**, 1023.

Inertial Forces in General Relativity

MAREK A. ABRAMOWICZ

> *It is therefore not just an idle game to exercise our ability to analyse familiar concepts, and to demonstrate the conditions on which their justification and usefulness depend, and the way in which these developed, little by little.*
>
> Albert Einstein (1916)

1 INTRODUCTION

Some expert relativists believe that the very concept of gravitational, centrifugal and other inertial forces is fundamentally alien to general relativity. However, this concept may be quite useful in particular applications. For example, our recent investigation of centrifugal effects in a strong, time-independent, gravitational field[1] has revealed some unexpected and rather important links between dynamics described in terms of inertial forces, and the geometry of three-dimensional space. In particular, our work demonstrated the dynamical importance of the *optical reference geometry* of space $\tilde{\gamma}_{ik}$ which is connected to the directly projected geometry of space γ_{ik} by the conformal rescaling,

$$\tilde{\gamma}_{ik} = e^{2\Phi}\gamma_{ik}. \tag{1}$$

In time-independent gravitational fields (stationary spacetimes) *stationary observers* have their 4–velocity n^i parallel to a timelike Killing vector η^i which exists due to the time symmetry. It follows from the Killing equation, $\nabla_{(i}\eta_{k)} = 0$, that the 4-acceleration of static observers may be expressed by the gradient of a scalar function $\Phi = (1/2)\ln(-\eta^i\eta_i)$,

$$n^i\nabla_i n_k = -\nabla_k \Phi. \tag{2}$$

---

[1] A series of about ten papers which I wrote in collaboration with Jean-Pierre Lasota, Brandon Carter, A.R. Prasanna, John Miller, Jiří Bičák, Zdenek Stuchlík, Norbert Wex and Ewa Szuszkiewicz. See Abramowicz (1992) for references.

We shall see later that from the principle of equivalence and equation (2) one may deduce that the scalar function Φ, the same one which also appears in the definition of the optical reference geometry (1), has a physical interpretation as the gravitational potential. Thus, the gravitational force acting in a static spacetime on a test particle with the rest mass m should be defined as,

$$G_k = m\nabla_k \Phi = m\tilde{\nabla}_k \tilde{\Phi}. \tag{3}$$

A tilde indicates that the corresponding quantity is defined not in the directly projected three space with geometry $\gamma_{ik} = g_{ik} + n_i n_k$, but in the optical reference geometry $\tilde{\gamma}_{ik} = e^{2\Phi}\gamma_{ik}$.

If the particle moves in space along a curve having unit tangent vector τ^i, then the centrifugal force acting on the particle in its instantaneously corotating reference frame should be defined, in strict analogy with the conventional Newtonian definition (Frenet's formula), by the covariant expression,

$$Z_k = -m\tilde{v}^2\tilde{\tau}^i\tilde{\nabla}_i\tilde{\tau}_k = \frac{m\tilde{v}^2}{\tilde{R}}\tilde{\lambda}_k. \tag{4}$$

Here $\tilde{v}$ is the speed along the curve measured with respect to a static observer, $\tilde{R}$ is the geodesic curvature radius of the curve, and $\tilde{\lambda}_k$ is the first (outward pointing) normal to the curve.

Note two important points: (1) Light rays in a space corresponding to a static spacetime follow geodesic lines in the optical reference geometry, $\tilde{\tau}^i\tilde{\nabla}_i\tilde{\tau}_k = 0$. Thus, centrifugal force always repels in the local outward direction $\tilde{\lambda}_k$, defined with respect to the light trajectories in space and it vanishes for particles which move along the paths of light rays. (2) Some strongly curved three dimensional spaces close to the centre of a very compact star or a black hole may be turned inside out in the sense that the local outward direction, $\tilde{\lambda}_k$, may point in the (globally defined) direction towards the centre of the star. When this happens the centrifugal force attracts towards the centre of a circular motion.

These two points provide a simple and clear geometrical explanation for several paradoxical examples of the reversal of the action of centrifugal effects in a strong gravitational field which puzzled experts for quite some time:
(a) Abramowicz and Lasota (1974) noticed that identical test rockets orbiting along the path of the circular photon ray $r = 3GM/c^2$ around a Schwarzschild black hole

(with mass M) all have to use the same rocket thrust $T = mc^4/6GM$ in order to stay on the orbit, irrespective of their orbital speed v. This behaviour looks quite paradoxical because Newtonian theory (which governs our intuition) predicts the need for a *speed dependent* thrust, $T = (v_K^2 - v^2)/r$, where $v_K = (GM/r)^{1/2}$ is the Keplerian orbital speed. According to our approach, however, the effect is not paradoxical at all: it occurs because the rockets move along a straight line in the optical geometry and thus they experience no centrifugal force. The thrust only balances the (speed independent) gravitational force which is the same for all of the orbiting rockets.

(b) Chandrasekhar and Miller (1974) and Miller (1977) discovered that the ellipticity of quasi-stationarily contracting relativistic Maclaurin spheroids with fixed mass M and total angular momentum J *decreases* with decreasing mean radius R when $R < 5GM/c^2$. This is in contrast to Newtonian theory which predicts that when a rotating body shrinks conserving angular momentum, it always becomes progressively more flattened. The full calculation by Chandrasekhar and Miller required numerical integration of Einstein's field equation. On the other hand, an approximate calculation based on our approach, which offers an improved physical understanding of the behaviour of inertial forces, was carried out entirely analytically by myself and Miller. Not only did it give, for the first time, the correct physical explanation for the phenomenon, but it also reproduced results of the full calculation with impressive accuracy. Abramowicz and Wagoner (1976) found a similar effect: rotation *increases* internal pressure of a sufficiently compact body (having $R < 2.5GM/c^2$).

(c) Anderson and Lemos (1988) found that close to a Schwarzschild black hole, viscous stresses in thin accretion disks transport angular momentum *inwards*. This is contrary to a well-known result from the classical theory of thin accretion disks that viscous stresses always transport angular momentum outwards (Lynden-Bell and Pringle, 1974).

(d) Abramowicz and Prasanna (1989) demonstrated that close enough to a Schwarzschild black hole (for $r < 3GM/c^2$), the Rayleigh stability criterion is reversed: stable equilibria correspond to angular momentum *decreasing* outwards, whereas the conventional Rayleigh criterion demands that the angular momentum must *increase* outwards for stability.

(e) According to Newtonian theory, a gyroscope moving round a circle should always precess in the opposite sense to its circular motion in order to point in a fixed direction in space: gyroscopes on clockwise orbits precess anticlockwise and *vice versa*. However, very close to a Schwarzschild black hole orbiting gyroscopes precess in the same sense as their orbital motion: gyroscopes on clockwise orbits precess clockwise (de Felice 1991, Abramowicz 1992).

The seemingly paradoxical behaviour in the last three examples, (c), (d), (e), is explained in our approach as the dynamical consequence of the turning inside-out of the three space which is deeply connected with the behaviour of several geometrical and dynamical quantities. For example, the direction of the gradient of the radius of gyration was found to be identical with the local outward direction relevant for discussion of the dynamical effects of rotation. The level surfaces of the radius of gyration are the von Zeipel cylinders which are found also to correspond to the equipotential surfaces of the effective potential for photon motion. These strong (and unexpected) connections between different physical and geometrical quantities are obvious and clear in the optical reference geometry, which in a sense is just a special way of making particularly useful maps of a curved space. The difficulty with mapping in this case is quite similar to that in the case of conventional cartography, where it is impossible to accurately represent the spherical surface of the Earth on a flat plane without some kind of distortion. Several types of projection are used in cartography to minimize the distortion of the features of interest, while some other features may at the same time be distorted beyond recognition. The choice of a particular projection depends on the purpose for which the map is constructed. For example, the well-known Mercator projection exaggerates polar regions to an enormous extent, but is invaluable to navigators because it shows all lines of constant direction as straight lines. Exactly the same is true for the optical geometry which distorts the "true" geometry of space by a conformal rescaling, but is invaluable for studying dynamics because it introduces the concept of dynamically straight lines in space. It turns out that this helps in isolating particular and complicated technicalities from the basic geometrical and physical issues. For example, in the optical geometry it is easy to understand, without any calculations, why gyroscopes orbiting a Schwarzschild black hole on the $r = 3GM/c^2$ orbit do not precess: they follow a straight line! A long and rather technical calculation (based on the Killing vector and Fermi-Walker transport) is necessary for deriving the same conclusion in the directly projected metric. While the physical and geometrical meaning of the result concerning non-precessing gyroscopes can be fully understood in terms of the optical geometry even by non-experts, only those who have a rather solid background in general relativity can follow the technical calculation. This shows the practical value of the optical reference geometry approach.

Although the practical values of our approach have been widely recognized, its very foundations have been questioned by some experts. They have argued that the use of the Killing equation makes our approach a very special one, and that it therefore has no generic significance in Einstein's general relativity.

I shall answer this criticism by putting the results of our work in a wider perspective: I propose here a covariant and unique definition of inertial forces valid for all spacetimes, including ones which have no symmetries at all. This general definition has not been discussed before. It is based on the remark that the Killing equation was not necessary for introducing the optical reference geometry and that the much weaker equation (2) is sufficient. Equation (2) is covariant and has, as we shall see later, a local solution in *any* spacetime. Thus, the optical reference geometry and inertial forces could be defined in any spacetime in a mathematically satisfactory way. However, I believe that mathematical correctness, although crucial, is not by itself sufficient for a physically acceptable definition of inertial forces. Such a definition should be experimentally verifiable and agree with what is usually understood by this term. For this reason, in the next Section I give a short account of how our understanding of inertial forces has historically been evolving and maturing, and what are considered today to be the most basic attributes of the gravitational and centrifugal forces.

2 GRAVITATIONAL AND CENTRIFUGAL FORCES

The gravitational and centrifugal forces are the two most familiar forces of Nature, known to everybody from direct personal experience. Every day we work against attraction of gravity when walking up hills or climbing stairs, and every day we feel the repulsion of centrifugal force in trains, cars and planes when they make turns. The attractive tendency of gravity and the repulsive tendency of centrifugal forces were known and studied already in antiquity but, before the publication of Christiaan Huygens' *Horologium Oscillatorium* (Paris, 1673) and Isaac Newton's *Philosophiae Naturalis Principia Mathematica* (London, 1687) at the end of the seventeenth century, these studies were only qualitative. It was Newton's monumental achievement to demonstrate that the balance between gravitational attraction and centrifugal repulsion explains the three laws of planetary motion which had been discovered empirically earlier in the seventeenth century by Johan Kepler and described in the book *De Harmonice Mundi* (Ausburg, 1619).

2.1 Gravitational force

Under the influence of gravity, all bodies fall with the same acceleration. This was first discovered by Galileo Galilei and described in his unpublished work *De Motu*, written between 1589 and 1592, where he reported results of his experiments on dropping various objects from the Leaning Tower of Pisa: *The variation of speed in air between balls of gold, lead, copper, porphyry, and other heavy materials is so slight that in a fall of 100 cubits a ball of gold would surely not outstrip one of copper by as much as four fingers. Having observed this, I came to the conclusion that in a medium totally void of resistance all bodies would fall with the same speed.*

In the modern formulation of Newton's theory, the gravitational force **G** acting on a particle with mass m is derived from the gradient of gravitational potential Φ,

$$G = m\nabla\Phi. \tag{5}$$

Because the gravitational potential does not depend on the properties of the test bodies (in particular it does not depend on their speed) all bodies fall with the same acceleration,

$$g = \frac{1}{m}G = \nabla\Phi. \tag{6}$$

Albert Einstein realized in 1908 that, in his own words, *the independence of the gravitational acceleration from the nature of the falling substance, may be expressed as follows: In a gravitational field (of small spatial extension) things behave as they do in a space free of gravitation...* And in another place he wrote: *We arrive at a very satisfactory interpretation of this law of experience, if we assume that the systems K and K' are physically exactly equivalent, that is if we assume that we may just as well regard the system K as being in a space free from gravitational fields, if we then regard K as uniformly accelerated. This assumption of exact physical equivalence makes it impossible for us to speak of the absolute acceleration of the system of reference, just as the usual theory of relativity forbids us to talk of the absolute velocity of a system; and it makes the equal falling of all bodies in a gravitational field seem a matter of course.* Thus, equal acceleration of all falling bodies came from the acceleration of the observer standing on the ground: the push of the ground under his feet makes him accelerate, *i.e.* move along a non-geodesic line in the spacetime. Therefore, non-accelerated, freely-falling bodies observed in *his* accelerated frame appear for *him* to be falling in the same way because they all have acceleration opposite to his. Thus, if the 4–velocity of the observer on the ground is n^i, the universal gravitational acceleration of falling bodies is $g_k = -n^i\nabla_i n_k$, and the gravitational force,

$$G_k = -mn^i\nabla_i n_k. \tag{7}$$

In the geometrical description of gravity (finally formulated by Einstein in November and December 1915 in a series of articles which he presented to the Prussian Academy of Sciences in Berlin) the concept of gravitational force is no longer necessary. A heavy body, such as the Earth, modifies the geometry of 4–dimensional spacetime around itself in such a way that the shortest lines in this geometry (geodesics) are, in the three dimensional space, the curved trajectories of attracted bodies. Free-falling bodies move on geodesics in spacetime and they feel no inertial forces, in particular no attraction of gravity — they are weightless. In the freely-falling reference frames (*local inertial frames*) every free object moves in a straight line with uniform velocity.

2.2 Centrifugal force

In Newton's view (forcefully explained in *Principia* in terms of the famous gedanken experiment with a rotating pail of water) the centrifugal force appears in reference frames which rotate with respect to *absolute space*. If the angular velocity of the rotating frame measured with respect to the absolute space is Ω, and the distance of a particle (with mass m) from the axis of rotation of the frame is $\mathbf{r}$, then the centrifugal force acting on this particle *in the rotating frame* is given by

$$\mathbf{Z} = m\Omega \times (\mathbf{r} \times \Omega). \tag{8}$$

Let us consider a fixed curve $\mathbf{x} = \mathbf{x}(S)$ and a particle moving with a prescribed speed $v = v(S)$ along it. The distance along the curve is denoted by S. The reference frame instantaneously corotating with the particle is given by the Frenet formulae which define the three mutually orthogonal unit vectors – the tangent vector τ, the first (outward) normal λ and the second normal Λ:

$$\tau = \frac{d\mathbf{x}}{dS}, \tag{9a}$$

$$\lambda = -R(\tau \cdot \nabla)\tau, \tag{9b}$$

$$\Lambda = \lambda \times \tau. \tag{9c}$$

R is referred to as the geodesic curvature radius (or first curvature). One should note an important minus sign in formula (9b). The instantaneous angular velocity Ω of the instantaneously corotating reference frame and the instantaneous distance $\mathbf{r}$ of the particle from the instantaneous axis of the frame are given by

$$\Omega = \left(\frac{v}{R}\right)\Lambda, \tag{10a}$$

$$\mathbf{r} = R\lambda. \tag{10b}$$

From the last two formulae and from Newton's definition of centrifugal force (8) one finds the expression for the centrifugal force in the instantaneously corotating reference frame $\mathbf{Z} = -mv^2(\tau \cdot \nabla)\tau = m[v^2/R]\lambda$. In coordinate notation this equation takes the form,

$$Z_k = -m\tilde{v}^2\tilde{\tau}^i\tilde{\nabla}_i\tilde{\tau}_k = m\frac{\tilde{v}^2}{\tilde{R}}\tilde{\lambda}_k, \tag{11}$$

which is also valid for a curved space[2]. Here $\tilde{v}$ is the speed of the particle along a curve in space, $\tilde{\tau}_k$ is the unit vector tangent to the curve, $\tilde{\nabla}_i$ is the covariant derivative in the space, $\tilde{R}$ is the geodesic curvature radius of the curve, and $\tilde{\lambda}_k$ is the first normal to it. A geodesic line in space is defined by $\tilde{\tau}^i\tilde{\nabla}_i\tilde{\tau}_k = 0$ and so, for this, $1/\tilde{R} = 0$. Thus, from (11) it follows that the centrifugal force is non-zero (in the instantaneously corotating reference frame of a particle) if and only if the particle moves along a curve which differs from a geodesic line in space.

2.3 Relativity of motion

Most of the spectacular development of theoretical astronomy in the eighteenth and nineteenth centuries was directly based on analysing how the gravitational, centrifugal, Coriolis and other forces balance in particular situations involving heavenly bodies. However, despite the unquestionable achievements of the theory, many physicists and philosophers, including Christiaan Huygens and Gottfried von Leibniz in Newton's time, and later Ernst Mach and Albert Einstein (to name just a few), were deeply disturbed by the ideas of absolute space and absolute motion which were fundamental for Newton's theory. Taking the Cartesian viewpoint that motion is not absolute but relative, Huygens strongly rejected Newton's concept of absolute rotation. When he read *Principia*, he reacted violently: *One can in no wise conceive of the true and natural motion of the body as differing from its [state of] rest... In circular motion as in free and straight motion, there is nothing except relativity.*

The theory of general relativity has finally rejected the concepts of absolute motion and absolute space, but it has not fully answered a very closely related question about the origin of inertia. This question is still discussed today. Dennis Sciama's important contribution to the problem was given in his often quoted article published in *Monthly Notices of the Royal astronomical Society* (Sciama, 1953). He also described a historical account of the subject in his book, *The Unity of the Universe* (Sciama, 1959) and his *Scientific American* article "Inertia" (Sciama, 1957).

---

[2] It was not Newton, but Huygens, who first gave a formula for the centrifugal force equivalent to $\mathbf{Z} = m[v^2/R]\lambda$. He did this in his treatise *De vi centrifuga* written in 1659 but only published in 1703 after his death. In 1669 Huygens established his claim to authorship of the formula by sending it, in anagrammatic form, to Oldenburg, who was then secretary of the Royal Society. The solution of the anagrams were given in Huygens' book *Horologium Oscillatorium* which was published fourteen years before the publication of Newton's *Principia*.

3 INERTIAL FORCES IN EINSTEIN'S RELATIVITY

The short historical discussion given in the previous Section stresses the most important attributes of the gravitational and centrifugal forces:

Reference frame: Motion in space is not absolute but only relative: it should always
be related to some reference frame. Locally, physics is described by the same laws
(formulae) in all reference frames (Einstein's principle of general covariance). However, in some special reference frames description of some particular aspects of physics
may look simpler than in other frames. A distinction between the different kinds of
inertial force may only be made in specific reference frames.

Gravitational force: All bodies fall with the same acceleration in a gravitational field
independently of their substance and nature (Galileo). The same effect of universal
acceleration independent of the nature of the moving bodies is observed in an accelerated frame of reference, and indeed gravity is locally equivalent to the kinematic
effect of the acceleration of the reference frame (Einstein's principle of equivalence).
Gravitational force can be derived from a potential (Newton, Poisson).

Centrifugal force: Centrifugal force acts only in a rotating frame of reference (Newton). The centrifugal force is present in the instantaneously corotating frame of a
particle moving along a trajectory in space if and only if the trajectory differs from
a geodesic line in space. The force per unit mass is equal to the square of the speed
of the particle divided by the curvature radius of the trajectory and it points in the
direction of the trajectory's first outside normal (Huygens, Frenet).

In this Section, I will show that these properties suggest how the different kinds of
inertial force should be defined in general relativity.

3.1 Basic equations and a special reference frame

Here, I will introduce equations which determine both a special reference frame and
the "gravitational potential" Φ (note the $-+++$ signature):

$$n^i n_i = -1, \qquad n^i \nabla_i n_k = -\nabla_k \Phi. \tag{12}$$

In the special reference frame an observer's 4–velocity is n^i. From equations (12) it
follows that

$$n^i \nabla_i \Phi = 0, \tag{13}$$

which means that the special observers n^i register no change in the gravitational
potential as their proper time passes by: the observers have fixed positions with

respect to the "gravitational field". It is exactly this special property which makes them useful for distinguishing between different types of inertial force.

As far as I know, equations (12) have not been discussed before in Einstein's general relativity. However, they have several remarkable mathematical properties[3]. Obviously, equations (12) are a set of *five* equations (one algebraic and four first order, linear partial differential) for *five* unknown functions (four components of the observer's velocity n^i and one scalar function Φ corresponding to the gravitational potential). Thus, when the intial conditions are given, (12) has always at least a local solution in any spacetime with a known metric g_{ik}.

Equations (12) are identically satisfied in two important special cases: when n^i corresponds to a stationary observer in a stationary spacetime, and when n^i corresponds to a freely-falling (geodesic) observer in a general spacetime. In the first case n^i is parallel to a Killing vector η^i and $\Phi = -(1/2)\ln(-\eta^i\eta_i)$. In the second case $\Phi =$ const.

One may consider a metric $\tilde{g}_{ik} = \exp(2\Phi)g_{ik}$ which is conformal to the original spacetime metric g_{ik}. The covariant derivative in the conformal spacetime can be expressed as

$$\tilde{n}^i\tilde{\nabla}_i\tilde{n}_k = n^i\nabla_in_k + \nabla_k\Phi + n_kn^i\nabla_i\Phi. \qquad (14)$$

From this equation and (13) one concludes that equations (12) always determine a geodesic in the conformal spacetime. However, not all geodesics in the conformal spacetime correspond to a solution of (12) because of the term $n_kn^i\nabla_i\Phi$ which is zero when (12) is fulfilled, but is non-zero in general.

3.2 Acceleration formula and inertial forces
Let us now consider a particle with 4–velocity u^i and rest mass m. The 4–momentum of the particle is $P^i = mu^i$ and its 4–acceleration is defined as $a_k = u^i\nabla_iu_k$. With $F_k = dP_k/ds$ being the 4–force acting on the particle, its motion is governed by

---

[3] I am grateful to Brandon Carter for pointing out to me several of these properties and also some possible different applications of equations (12). These equations introduce a special standard of rest and therefore they may be useful for discussing the concept of "uniform motion" which is relevant for understanding under which general circumstances a moving charge will radiate or not radiate as seen by some preferred observers. They also introduce a quite general notion of *time*, and thus may be useful in discussing issues in quantum gravity.

$F_k = ma_k$. The physical three velocity of the particle, $\tilde{v}^k$, in the local instantaneous three dimensional space of the observer n^i, is given by

$$\tilde{v}^k = \gamma v^k = u^i h^k{}_i = \gamma v \tau^k. \tag{15}$$

Here $\gamma \equiv n^i u_i$ is the Lorentz γ factor, v is the Lorentz speed of the particle ($\gamma = 1/\sqrt{1-v^2}$), and τ^k a unit spacelike vector parallel to the three velocity of the particle in the three dimensional space. Note that $|\tilde{v}| \le \infty$, while $|v| \le 1$ and that in the units used here, the speed of light $c = 1$. Using these quantities, the 4–velocity of the particle can be written as

$$u^i = \gamma(n^i + v\tau^i), \tag{16}$$

from which it is easy to derive the following formula for the acceleration (in which the terms are arranged in order according to the powers of the speed and its derivative $(v^0, v^1, v^2, \dot{v})$),

$$\begin{aligned}
a_k = &- \gamma^2 \nabla_k \Phi \\
&+ \gamma^2 v(n^i \nabla_i \tau_k + \tau^i \nabla_i n_k) \\
&+ \gamma^2 v^2 \tau^i \nabla_i \tau_k \\
&+ (v\gamma)^{\cdot}\, \tau_k \\
&+ \dot{\gamma} n_k.
\end{aligned} \tag{17}$$

Such an arrangement is both unique and covariant.

A considerable amount of simple but tedious algebra is needed in order to write the acceleration formula (17) in an alternative form which is particularly convenient for discussing inertial forces. First of all, one locally introduces the directly projected three dimensional space and its conformally adjusted version, the *optical reference geometry*. The special observers define three dimensional (spacelike) quantities by projecting the spacetime 4–dimensional quantities, using the projection tensor $h^i{}_k = \delta^i{}_k + n^i n_k$, into their local instantaneous three dimensional space which has the metric $\gamma_{ik} = g_{ik} + n_i n_k$. In addition to the directly projected metric γ_{ik} one may consider the conformally adjusted optical reference geometry which has the metric

$$\tilde{\gamma}_{ik} = e^{2\Phi} \gamma_{ik}. \tag{18}$$

The optical reference geometry was first introduced by Abramowicz, Carter and Lasota (1989) for stationary spacetimes. One also uses the identity $\gamma^2 g_k = (1+v^2\gamma^2)g_k$, introduces the the vector $\eta^i = e^{-\Phi}n^i$, and the scalar[4] $E = \eta^i u_i$, and writes $\tilde{v} = \gamma v$. Then, in general,

$$(\gamma v)^\cdot = \dot{E}ve^\Phi + \gamma\dot{v} + \gamma^2 v^2(\tau^i\nabla_i\Phi). \tag{19}$$

One also defines $\tilde{\nabla}_i$ as being the covariant derivative in the optical reference geometry, and

$$\tilde{\tau}^i = e^{-\Phi}\tau^i, \quad \tilde{\tau}_i = e^\Phi\tau_i, \tag{20}$$

as being the contravariant and covariant components of the spacelike unit vector in that geometry, and proves that [cf equation (14)],

$$\tilde{\tau}^i\tilde{\nabla}_i\tilde{\tau}_k = \tau^i\nabla_i\tau_k + \nabla_k\Phi + \tau_k\tau^i\nabla_i\Phi. \tag{21}$$

Using all of these equations, one can finally write the acceleration formula in the desired form:

$$\begin{aligned}
a_k = {} & -\nabla_k\Phi \\
& + \gamma^2 v(n^i\nabla_i\tau_k + \tau^i\nabla_i n_k) \\
& + \tilde{v}^2\tilde{\tau}^i\tilde{\nabla}_i\tilde{\tau}_k \\
& + (\dot{E}ve^\Phi + \gamma\dot{v})\tau_k \\
& + \dot{\gamma}n_k.
\end{aligned} \tag{22}$$

The equation of motion for the particle in its own comoving frame may be projected into the three space, and it can then be written in the form $F_k - ma_k = 0$ which suggests that the particle is not accelerated and that the real force F_k is balanced by the sum of the inertial forces — gravitational G_k, centrifugal Z_k, Coriolis C_k, and Euler E_k,

---

[4] In the special case when n^i corresponds to a stationary observer in a stationary spacetime, η^i is the timelike Killing vector, and E is the (conserved) energy of the particle.

$$F_k + G_k + Z_k + C_k + E_k = 0. \tag{23}$$

The real force, which can be directly measured experimentally, is

$$\text{Real force}: \quad F_k = \frac{dP_k}{ds}. \tag{24}$$

By comparing (22) with (23) and by using the physical arguments discussed in the previous Section, one concludes that the following definitions of the different kinds of inertial forces should be adopted,

$$\text{Gravitational force}: \quad G_k = m\nabla_k \Phi, \tag{25}$$

$$\text{Coriolis force}: \quad C_k = -m\gamma^2 v(n^i \nabla_i \tau_k + \tau^i \nabla_i n_k), \tag{26}$$

$$\text{Centrifugal force}: \quad Z_k = -m\tilde{v}^2 \tilde{\tau}^i \tilde{\nabla}_i \tilde{\tau}_k, \tag{27}$$

$$\text{Euler force}: \quad E_k = -m(\dot{E}ve^\Phi + \gamma \dot{v})\tau_k. \tag{28}$$

The last term in the acceleration formula, $\dot{\gamma} n_k$, becomes zero when projected into the three space, and therefore does not correspond to any force in space.

Note that the different inertial forces are proportional to different powers of the speed with respect to the special observer. Because both the speed and the real force are directly measurable experimentally, one can measure also the different inertial forces (see Abramowicz 1992 for a detailed discussion of this point.)

3.3 Effective inertial mass?
Some critics of our approach argue that there is a different way to define inertial forces. They suggest that one may introduce the concept of an "effective" inertial mass,

$$m^* = \frac{m}{\sqrt{1 - v^2}}, \tag{29}$$

and use a special coordinate frame in which the time coordinate t is given by $n^i = \delta^i{}_{(t)}$. In this frame $dt/ds = \gamma = 1/\sqrt{1 - v^2}$. They use the last formula together with formula (17) in order to write,

$$\frac{dP_k}{dt} \equiv F_k^* = m^* \left[-g_k + v(n^i \nabla_i \tau_k + \tau^i \nabla_i n_k) + v^2 \tau^i \nabla_i \tau_k + \gamma^{-2} (v\gamma)^{\cdot} \tau_k + \gamma^{-2} \dot\gamma n_k \right].$$
$$(30)$$

At this point they argue that F_k^* can be identified with the three force and, consequently, that the gravitational force should be defined as

$$G_k^* = -\frac{m}{\sqrt{1-v^2}} n^i \nabla_i n_k = m^* g_k. \qquad (31)$$

They argue that this definition is also consistent with the equivalence principle, because the gravitational acceleration (equal to the gravitational force G_k^* divided by the effective inertial mass m^*) is the same for all bodies, $G_k^*/m^* = g_k = -n^i \nabla_i n_k$.

In my opinion the above interpretation is not physically satisfactory because the physical three force should *not* be defined — as this interpretation assumes — as the *coordinate time* derivative of the momentum, $F_k^* = dP_k/dt$. Instead, it should be defined as the 4–force projected into the three space. The 4–force is the *proper time* derivative of the 4–momentum, $F_k = dP_k/ds$.

It is relevant to quote here Christian Møller who wrote on a very closely connected point in the second edition of his book *Theory of Relativity* (Oxford, 1972): *In the older literature (and in the first edition of this book) distinction* [between F_k^* and F_k] *was not clearly made.* [...] *In the matter of definitions we have, of course, a certain freedom, and a quantity may be very well defined in several different ways. However, unless a clear distinction is made between the different definitions there is a risk of creating confusion. Also, not all definitions are equally practical.* [...] *it is not always practical or even reasonable to define the force as the time derivative of the momentum.* After saying this, Møller demonstrated in terms of a specific example that adopting the definition of F_k^* as the real physical force can imply an incorrect consequence: that a force is needed for maintenance of a constant velocity motion in an empty space free of gravity. He then concluded: *To say,* [...] *that F_k^* is a real physical force* [...] *is coming close to leaving the solid ground of Galilean experimental physics and returning to the philosophy of Aristotle.*

Thus, it seems that the idea of the "effective" inertial mass (popular in the older literature) cannot be used for introducing a physically satisfactory definition of gravitational force and of the other inertial forces.

4 CIRCULAR MOTION

In this Section, I will show that in the important special case of observers moving on circles in a stationary and axially symmetric spacetime, the basic equations (12) have a whole class of different, non-trivial and physically interesting solutions.

In a stationary, axially symmetric spacetime, two Killing vector fields exist: the timelike vector η_α, which has open trajectories, and the spacelike vector ξ_α with closed trajectories. These two vectors commute, $\xi^i \nabla_i \eta^k = \eta^i \nabla_i \xi^k$, but they are not orthogonal in general, and so

$$\omega - \frac{(\xi^i \eta_i)}{(\xi^i \xi_i)}, \tag{32}$$

is in general non-zero.

An observer moves on a circle if and only if his 4–velocity is a linear combination of the two Killing vectors,

$$n^i = e^{A_0}(\eta^i + \Omega_0 \xi^i), \tag{33}$$

where Ω_0 (in general not a constant) is the angular velocity of the observer (with respect to static inertial observers at infinity) and

$$A_0 = -\frac{1}{2} \ln\left[-(\eta^i \eta_i) - 2\Omega_0(\eta^i \xi_i) - \Omega_0^2(\xi^i \xi_i)\right]. \tag{34}$$

One can define the specific angular momentum of the observer by the formula

$$L_0 = -\frac{n^i \xi_i}{n^k \eta_k}. \tag{35}$$

The 4–acceleration of the observer is

$$n^i \nabla_i n_k = -\nabla_k A_0 + \frac{L_0 \nabla_k \Omega_0}{1 - L_0 \Omega_0}. \tag{36}$$

Therefore, the acceleration is the gradient of a scalar function if and only if the angular velocity and the angular momentum fulfill the von Zeipel condition (Abramowicz, 1974):

$$F(\Omega_0, L_0) = 0. \tag{37}$$

Indeed, when (37) holds, there is a unique connection between the angular momentum and angular velocity, $L_0 = L_0(\Omega_0)$ and one may write,

$$B_0(\Omega_0) = \int \frac{L_0(\Omega_0)d\Omega_0}{1 - L_0(\Omega_0)\Omega_0}, \quad \Phi = (A_0 + B_0), \quad n^i \nabla_i n_k = -\nabla_k \Phi, \tag{38}$$

which means that equation (12) is fulfilled. Therefore, for observers moving on circles, condition (37) is both necessary and sufficient for introducing the optical reference geometry and defining the inertial forces according to the formalism described in the previous Section. For any assumed form of the von Zeipel condition, $F(\Omega_0, L_0) = 0$, one may *directly* calculate n^i and Φ using formulae (38), (33) and (34). It is also convenient to introduce the *radius of gyration*. In Newtonian theory the radius of gyration $\tilde{r}$ is defined in connection with the motion of rigid bodies as the radius of the circular path on which a point-like particle having the same mass and angular velocity as the rigid body would also have the same angular momentum. This suggests introducing the concept of radius of gyration of a *particle* in general relativity by means of the definition (Miller, 1992):

$$\tilde{r} \equiv \sqrt{\frac{L_0}{\Omega_0}} \tag{39}$$

A particle moving on a circular orbit has 4-velocity

$$u^i = \gamma(n^i + v\tau^i) = e^A(\eta^i + \Omega \xi^i). \tag{40}$$

One concludes from this formula (after a small amount of easy algebra), that the gravitational, centrifugal and Coriolis forces are given by

$$G_k = m\nabla_k \Phi, \tag{41}$$

$$Z_k = m\frac{\tilde{v}^2}{\tilde{r}}\nabla_k \tilde{r}, \tag{42}$$

$$C_k = -m\tilde{v}\sqrt{1 + \tilde{v}^2}\nabla_k\omega, \tag{43}$$

and that the Euler force always vanishes. I show elsewhere how these forces change when one relates fixed circular motion of a particle to different special observers moving with different Ω_0 on a fixed circle. As different special observers are accelerated with respect to each other, the inertial forces do not transform according to the Lorentz transformations!

There are two particularly important special observers; the stationary observer N^i, characterized by $\Omega_0 = 0$, and the locally non-rotating observer H^i, characterized by $L_0 = 0$:

$$N^i = e^{\Psi}\eta^i, \quad H^i = e^{\Phi}\tilde{\eta}^i, \tag{44}$$

where $\tilde{\eta}^i = \eta^i + \omega\xi^i$ and

$$\Psi = \frac{1}{2}\ln\left[-(\eta^i\eta_i)\right], \quad \Phi = \frac{1}{2}\ln\left[-(\eta^i\eta_i) + \omega^2(\xi^i\xi_i)\right]. \tag{45}$$

The accelerations of the stationary and locally non-rotating observers are explicitly given by

$$N^i\nabla_i N_k = -\nabla_k\Psi, \quad H^i\nabla_i H_k = -\nabla_k\Phi. \tag{46}$$

The stationary observers have their 4–velocity parallel to the timelike Killing vector, but their trajectories are not hypersurface orthogonal, except for the special case of a strictly static spacetime (in this case $\omega = 0$). On the other hand, the locally non-rotating observers have hypersurface orthogonal trajectories, but their velocity is not parallel to a Killing vector, except for a strictly static spacetime.

The angular velocity of a particle with respect to the locally non-rotating observer is given by $\tilde{\Omega} = \Omega - \omega$, and the specific angular momentum of the particle $\tilde{L}$ and its orbital speed $\tilde{v}$ may be conveniently written in a form identical to the Newtonian counterpart:

$$\tilde{L} = \tilde{r}^2\tilde{\Omega}, \quad \tilde{v} = \tilde{r}\tilde{\Omega}. \tag{47}$$

For circular motion at the speed of light, $(u^i u_i = 0)$, the angular velocity is

$$\tilde{\Omega} = \pm \frac{1}{\tilde{r}}. \qquad (48)$$

The acceleration associated with the circular motion can be written in the form

$$a_k = -\nabla_k \Phi - \frac{\tilde{v}^2}{\tilde{r}} \nabla_k \tilde{r} + \tilde{v}\sqrt{1 + \tilde{v}^2} \nabla_k \omega. \qquad (49)$$

A gyroscope moving along the circular orbit would precess with respect to the direction tangent to the orbit, τ^i. The precession rate is given by the vector Ω_i:

$$\frac{\Omega_i}{\tilde{\Omega}} = \frac{1}{2}\tilde{r}e^{3(A-\Phi)}\left[\nabla_i \ln \tilde{r}^2 - (1 + \tilde{v}^2)\frac{\nabla_i \omega}{\tilde{\Omega}}\right]. \qquad (50)$$

5 CONCLUSIONS

In this article I have proposed a general–relativistic definition of different kinds of inertial forces. The definition is valid in any spacetime and is based on the introduction of a particular timelike congruence of special observers which, in a certain sense discussed in the article, are at rest with respect to the gravitational field. The congruence is defined by the the basic equation

$$n^i \nabla_i n_k = -\nabla_k \Phi. \qquad (51)$$

Equation (51) has several remarkable mathematical properties and for this reason it could find other important applications in general relativity in addition to that discussed here. It would therefore be interesting to find and study all of the solutions of these equations (in addition to the particular class of solutions connected with observers in circular motion found and described here) starting with the simple spacetimes of Minkowski, Schwarzschild, and Kerr.

REFERENCES

Abramowicz M.A., 1974, *Acta Astr.*, **24**, 45.
Abramowicz M.A., 1992, *Mon. Not. R. astr. Soc.*, **256**, 710.
Abramowicz M.A. and Lasota J.-P., 1974, *Acta Physica Polonica* **B5**, 327.
Abramowicz M.A., Carter B., and Lasota J.-P., 1988, *Gen. Relat. Grav.*, **20**, 1173.
Abramowicz M.A. and Miller J.C., 1989, *Mon. Not. R. astr. Soc.* **245**, 729.

Abramowicz M.A., Miller J.C. and Stuchlík Z., 1992, *Phys. Rev. D.*, in press.
Abramowicz M.A. and Prasanna A.R., 1989, *Mon. Not. R. astr. Soc.*, **245**, 270.
Abramowicz M.A. and Wagoner R.V., 1976, *Astrophys.J.*, **204**, 896.
Anderson M.R. and Lemos J.P.S., 1988, *Mon. Not. R. astr. Soc.*, **233**, 489.
Chandrasekhar S. and Miller J.C., 1974, *Mon. Not. R. astr. Soc.*, **167**, 63.
De Felice F., 1991, *Mon. Not. R. astr. Soc.*, **252**, 197,
Lynden-Bell D. and Pringle J.E., 1974, *Mon. Not. R. astr. Soc.*, **168**, 603.
Miller J.C., 1977, *Mon. Not. R. astr. Soc.*, **179**, 483.
Miller J.C., 1992, in *Proceedings of the meeting "Classical and numerical relativity", Southampton, December 1991*, ed. C.J.S. Clarke and R. d'Inverno, Cambridge University Press.
Sciama D.W., 1953, *Mon. Not. R. astr. Soc.*, **113**, 34.
Sciama D.W., 1959, *The Unity of the Universe*, Faber & Faber, London.
Sciama D.W., 1957, *Scientific American*, **196** (February), 99.

Relativistic Radiation Hydrodynamics:
A Covariant Theory of Flux-Limiters

A. MARCELLO ANILE & VITTORIO ROMANO

A fully covariant approach to transfer phenomena by using flux-limiters is presented. Explicit formulas for the radiation flux and radiation stress tensor are given for a wide class of physical situations.

1 INTRODUCTION

In several areas of cosmology and astrophysics the transfer of radiation through high-speed moving media plays a crucial role (accretion flow into black holes, X-ray bursts on a neutron star, supernova collapse, jets in radio sources, galaxy formation, phase transition in the early universe). If one wants to take into account all the effects associated with these transport processes, the full relativistic transport equation must be used.

Early discussion of radiative viscosity was performed by several authors in a non covariant formulation (Jeans 1925, Rosseland 1926, Vogt 1928, Milne 1929), but the appropriate transfer equation for the case of special relativity was given in a classical paper by Thomas (1930). A manifestly covariant form of the transfer equation was obtained by Hazelhurst and Sargent (1959), by using a geometrical formalism. Finally Lindquist (1966) performed the extension to the general relativistic situation and Mihalas (1983) analyzed in depth the order of magnitude of the various terms which appear in the transport equation.

From a mathematical point of view, the transport equation is an integro–differential equation and the task to solve it is in general very hard. In particular when the radiative transfer equation is coupled to fluid motion (radiation hydrodynamics) it is beyond present day computational possibilities to solve the full coupled equations directly by numerical methods. Instead one has to resort to analytical approximations in order to obtain a set of reduced differential equations to couple with the fluid equations.

In the diffusion approximation, by developing in powers of the photon mean free path, Thomas (1930) obtained analytical expressions for the heat flux and radiative viscosity. The resulting equations for the coupled system of fluid and radiation can be put in the form of the Eckart theory of relativistic dissipative fluids by a suitable definition of the equilibrium temperature (Weinberg, 1979). Hence such a formulation is beset by the well known causality and instability inconsistencies plaguing the Eckart theory. However this theory gives good results when the medium is optically thick, but one can expect it to break down in transparent regions.

The Grad method of moments as an alternative to the diffusion approximation has also been considered for the radiative transfer equation. In a relativistic setting a first treatment of the moment formalism for the radiative transfer equation was given by Anderson and Spiegel (1972), who showed that the classical Eddington approximation in the relativistic case is inadequate even if all scales of interest are reasonably opaque because it is unable to account for radiative stresses (Hsieh and Spiegel, 1976). A complete and systematic treatment of the moment formalism for the radiative transfer equation was given by Thorne (1981), who considered all the important scattering, absorption and emission processes. However, difficulties arise also with the Grad method. Since the collision terms in general contain the distribution function it follows that the source terms are very complicated to handle when the scattering is anisotropic (Schweizer, 1988). Moreover, in the equation of order m, the (m+1)-th moment is present and therefore the closure problem arises. Then one needs to express the (m+1)-th moment as a function of the lower moments in order to obtain a closed system.

One of the most popular closures is the 14-moment Grad approximation, consisting in neglecting all moments of higher order than the quadratic one in the harmonic expansion of the photon distribution function. Udey and Israel (1982) and Schweizer (1988) showed that in this case one can put the constitutive equations for radiative flux and shear in the same form as the usual laws of causal linear irreversible thermodynamics (Israel and Stewart 1979). However, although the latter thermodynamical theory is not beset by the inconsistencies of the Eckart one, it presents the very serious drawbacks of not being able to describe a regular shock structure even for moderately strong shock waves (Israel, 1988).

Several different closures have appeared in the literature. For example, Turolla and Nobili (1988) suggested analytical expressions of the higher order moments as functions of the preceeding ones by physical considerations while Levermore (1984) looked for a closure at the second order, giving the radiation stress tensor as a function of the radiation energy and flux by means of a variable Eddington factor.

Lately, A.M.Anile, S.Pennisi and M.Sammartino (1991, 1992) have improved on Levermore's approach casting the theory in a covariant form in the context of extended thermodynamics. The results obtained in this way are physically well founded, present nice mathematical features (hyperbolicity, causality) and allow a relatively simple numerical implementation. However, a full investigation of the structural properties of the theory (e.g. shock structure) is still lacking.

Here, following the ideas developed in the articles of A.M. Anile and M. Sammartino (1989), V. Romano (1991), A.M. Anile and V. Romano (1992), A. Bonanno and V. Romano (1992), we present another method in order to deal with radiation phenomena of great interest in several areas of astrophysics: a covariant flux-limiter approach. Although such an approach does not lead to a mathematical description in terms of a hyperbolic system of equations but to non–linear parabolic equations, it still enjoys a weak form of relativistic causality (as will be seen in the sequel) and allows a relatively simple mathematical description for numerical purposes. In section 2 the classical formulation of flux-limiters is presented and in section 3 the covariant generalization is given and conclusions are drawn.

2 CLASSICAL FLUX-LIMITED DIFFUSION THEORY

When radiation propagates into an extremely opaque medium, the mean free path of the photons, λ_p, is negligible compared to the characteristic length l of the material $\lambda_p/l \ll 1$) and a state of near thermal equilibrium is achieved. Then, if one expands the transport equation at the various orders (see Mihalas, 1984) in the parameters λ_p/l, v/c, $\lambda_p v/lc$, λ_p^2/l^2, v^2/c^2, a satisfactory diffusion theory is obtained (v represents the velocity of the medium). In particular at the order $\lambda_p v/lc$, by using the Eckart decomposition of the radiation energy-momentum tensor, one obtains for the radiative flux a Fourier-like law

$$\underline{F} = -\kappa(T)(\nabla T + T\underline{a}), \tag{1}$$

where $\underline{a}$ is the acceleration of the medium (hereafter we use units such that c=1), T is the absolute temperature, $\kappa(T)$ is the thermal conductivity and $\underline{F}$ is the radiative flux.

Because of its simplicity and numerical applicability, the diffusion theory is often used in both opaque and transparent regions, although a complete justification exists only when $\lambda_p/l \ll 1$.

When the material, into which the radiation is propagating, is optically thin the flux predicted by the diffusion theory becomes very large and it may exceed the energy density times the velocity of light,

$$|F_{diffusion}| > E. \tag{2}$$

This implies, moreover, that the effective speed of energy propagation V_E exceeds the velocity of light

$$V_E \simeq \frac{\lambda_p}{3l} \tag{3}$$

The first result is inconsistent with the same definition of $\underline{F}$ (see eq. (8)) and the relation (3), of course, contravenes relativistic causality.

To overcome these difficulties Wilson (in an unpublished work) first suggested to modify eq.(1) in such a way as to give the standard diffusion results in the high opacity medium, but simulating free streaming in the transparent regions, for example changing (1) into an expression of the form

$$\underline{F} = -\frac{\nabla E}{3/\lambda_p + |\nabla E/E|} \tag{4}$$

Levermore and Pomraning in a fundamental article (1981) laid down the theoretical foundations for a formal justification of a flux–limited diffusion theory, giving, in an inertial frame, a flux–limited expression for the energy flux as a function of the energy–density gradient.

If we neglect effects such as polarization, dispersion and coherence, the integrated transport equation for the specific intensity $I(\underline{r}, \underline{\Omega}, t)$ in a grey stationary medium is

$$\frac{\partial I}{\partial t} + \underline{\Omega} \cdot \nabla I + \sigma I = \frac{1}{4\pi}(\sigma_a B + \sigma_s E), \tag{5}$$

where $\sigma_a(\underline{r}, t)$ is the absorption coefficient, $\sigma_s(\underline{r}, t)$ is the scattering coefficient, $\sigma = \sigma_a + \sigma_s$ is the total interaction coefficient, $B(\underline{r}, t)$ is the blackbody energy density and $E(\underline{r}, t)$ is the radiation energy density as defined by

$$E(\underline{r}, t) = \int_\Omega I d\Omega. \tag{6}$$

Integrating (5) all over the solid angle, one obtains

$$\frac{\partial E}{\partial t} + \nabla \cdot \underline{F} = \sigma_a(B - E), \tag{7}$$

with $\underline{F}(\underline{r}, t)$ the radiative flux,

$$\underline{F}(\underline{r}, t) = \int_\Omega I \underline{\Omega} d\Omega \tag{8}$$

(We observe that eq. (8) implies $|F| \leq E$).

Now, if we introduce the normalized specific intensity, defined by

$$I = E\psi, \tag{9}$$

with

$$\int_\Omega \psi d\Omega = 1, \tag{10}$$

eq. (5) reads

$$\left(\frac{\partial\psi}{\partial t} + \underline{\Omega}\cdot\nabla\psi\right)E + (\underline{\Omega}\cdot\nabla E - \underline{f}\cdot\nabla E - E\nabla\cdot\underline{f} + \sigma_s E + \sigma_a B)\psi$$

$$= \frac{1}{4\pi}(\sigma_a B + \sigma_s E) \tag{11}$$

The function ψ is known in two limiting cases. In the isotropic diffusion limit

$$\psi = \frac{1}{4\pi}\left(1 - \underline{\Omega}\cdot\frac{\nabla E}{\sigma E}\right), \tag{12}$$

with $|\nabla E/\sigma E| \ll 1$, while in the extreme streaming limit $(\sigma_s = B = 0)$

$$\psi = \delta(\underline{\Omega}\cdot\underline{n}), \tag{13}$$

where $\underline{n}$ is a unit spatial vector in the direction of $\underline{F}$.

Since in both these cases ψ is seen to be slowly varying, then Levermore and Pomraning made the ansatz that ψ is a slowly varying function along the characteristics of eq. (11), setting

$$\frac{\partial\psi}{\partial t} + \underline{\Omega}\cdot\nabla\psi = 0 \tag{14}$$

If we introduce the normalized radiative flux, defined by

$$\underline{F} = E\underline{f}, \tag{15}$$

$$\underline{f} = \int_\Omega \psi d\Omega, \tag{16}$$

and if we use eq. (10) and the relation

$$\nabla\cdot\underline{f} = 0 \tag{17}$$

(which follows from eqs. (10) and (14)), we obtain, by using eqs. (11) and (14)

$$\psi = \frac{1}{4\pi}\frac{1}{1 + \underline{f}\cdot\underline{R} - \underline{\Omega}\cdot\underline{R}}, \tag{18}$$

where

$$\underline{R} = -\frac{\nabla E}{\sigma \omega E} \tag{19}$$

and ω is the effective albedo given by

$$\omega = \frac{\sigma_a B + \sigma_s E}{\sigma E}. \tag{20}$$

Writing $\underline{f} = \lambda(R)\underline{R}$, it is possible to invert equation (18), obtaining

$$\lambda(R) = \frac{1}{R}\left(\coth R - \frac{1}{R}\right). \tag{21}$$

It represents the Levermore flux–limiter which was originally derived (Levermore 1979) by employing an approach similar to the Chapman–Enskog method in the kinetic theory of gases. By means of (21), one has the following diffusion law

$$\underline{F} = -\frac{D_F}{\sigma}\nabla E, \tag{22}$$

where

$$D_F = \frac{\lambda(R)}{\omega}. \tag{23}$$

A comparison with analytic and numerical solutions, shows that eq. (22) gives very accurate results for a wide class of radiative transfer phenomena.

A.M. Anile and M. Sammartino (1989) performed a detailed investigation of the propagation speed of disturbances predicted by a flux–limited diffusion theory. At first glance one might think that diffusion theories lead to parabolic equations and one usually associates infinite propagation speed with equations of such a type. However, the situation is not so simple because flux–limited diffusion theories lead to a class of non–linear parabolic equations which enjoy peculiar properties with respect to propagating pulses. In fact for a class of initial data (which is sufficiently broad to encompass most application in astrophysics and cosmology) the pulse propagating speeds are always bounded by the light speed when the flux–limiters are used.

For a small signal perturbation, assuming the unperturbed state is homogeneous, let $E = B + \delta E$, with $|\delta E| \ll 1$. Linearizing eq. (7) gives

$$\frac{\partial \delta E}{\partial t} - \frac{1}{\sigma}D_F^o \nabla^2 \delta E = -\sigma \delta E, \tag{24}$$

where $D_F^o = \frac{1}{3}$.

Eq. (24) is the standard diffusion equation whose solutions have an acausal behavior.

The root of this result is in the assumption that both δE and its derivative are very small. However for harmonic waves of the type

$$\delta E \propto e^{i(\underline{K}\cdot\underline{z}-\omega t)} \tag{25}$$

for large wave numbers or frequency the derivatives are no longer infinitesimal. Then we look for solutions of eq. (24) in the form of asymptotic waves

$$E(\underline{x},t) = B + \varepsilon\hat{E}(\zeta,\underline{x},t), \tag{26}$$

with $\zeta = \eta\varphi(\underline{x},t)$ (φ being the phase function) and η and ε dimensionless parameters. For $\eta \to \infty$ we have large frequencies waves, for $\varepsilon \to 0$ we have small amplitude.

Observing that $\lambda(R) \propto 1/R$ when $R \to \infty$, in the case of large frequency ($\varepsilon\eta \gg 1$) eq. (24) at the order $\varepsilon\eta$ leads to

$$\left(\frac{\partial\hat{E}}{\partial\zeta}\right)(\varphi_t^2 - \varphi_i\varphi_i) = 0. \tag{27}$$

It follows that

$$\varphi_t^2 - \sum_{j=1}^{3}\varphi_j\varphi_j = 0, \tag{28}$$

which shows that the limiting propagation speed is the velocity of light. One can obtain similar results by studying the propagation of discontinuity waves.

3 COVARIANT FORMULATION OF FLUX-LIMITERS

As pointed out in the introduction, in several astrophysical applications (e.g. stellar core collapse) one usually deals with radiation propagating through a highly inhomogeneous and non–stationary medium and, moreover, relativistic effects must be taken into account. Therefore, a general relativistic formulation of the Levermore and Pomraning approach is warranted. In this section we present a general relativistic flux–limited diffusion theory, which holds for arbitrary inhomogeneous and non–stationary media, limiting ourselves to the case of a grey medium.

However, this is not a serious limitation and the theory can be easily developed, along similar lines, in order to yield a multigroup flux–limited diffusion approximation. Let us consider the relativistic, explicitly covariant, integrated form of the radiative transfer equation (Anderson and Spiegel, 1972)

$$(u^\mu + l^\mu)\left[\nabla_\mu I + 4Il^\sigma(\nabla_\mu u_\sigma) + (\nabla_\mu u_\sigma)l^\rho l^\sigma\frac{\partial I}{\partial l^\rho} + (\nabla_\mu u_\sigma)u^\rho l^\sigma\frac{\partial I}{\partial l^\rho}\right.$$

$$\left.-(\nabla_\mu u^\rho)\frac{\partial I}{\partial l^\rho}\right] = \rho\left[\varepsilon - \int_0^\infty k_{\nu_0}f\nu_0^3 d\nu_0\right], \tag{29}$$

where $I = I(u^\mu, l^\mu)$ is the integrated intensity,

$$I = \int_0^\infty f\nu_0^3 d\nu_0 \tag{30}$$

(f being the invariant distribution function), ν_0 is the rest–frame frequency as measured by the observer whose 4–velocity is u^μ; l^μ is an unit vector in observer's rest frame,

$$l^\mu l_\mu = 1,$$
$$l^\mu u_\mu = 0; \tag{31}$$

and

$$\epsilon(\underline{l}) = \int_0^\infty \epsilon_{\nu_0} \nu_0 \tag{32}$$

is the integrated emissivity, $k_{\nu_0}(\underline{l})$ is the absorbtion coefficient and ρ is the rest mass density.

Let $J, H^\mu, K^{\mu\nu}$ denote the radiation energy density, energy–flux and stress tensor as measured in the observer's rest frame,

$$J = \frac{1}{4\pi} \int_\Omega I d\Omega \qquad H^\mu = \frac{1}{4\pi} \int_\Omega l^\mu I d\Omega \qquad K^{\mu\nu} = \frac{1}{4\pi} \int_\Omega l^\mu l^\nu I d\Omega, \tag{33}$$

where $d\Omega$ is the element of solid angle $d\Omega = \delta(u_\mu l^\mu)\delta(l^\alpha l_\alpha - 1)d^4 l$. The energy–momentum tensor $T^{\mu\nu}$ of radiation is

$$T^{\mu\nu} = 4\pi \left[J u^\mu u^\nu + u^\mu H^\nu + u^\nu H^\mu + K^{\mu\nu} \right] \tag{34}$$

and satisfies the conservation law

$$\nabla_\nu T^{\mu\nu} = \rho u^\mu \left(\int_\Omega \int_0^\infty \epsilon_{\nu_0} d\nu_0 d\Omega - \int_\Omega \int_0^\infty k_{\nu_0} f\nu_0^3 d\nu_0 d\Omega \right)$$
$$- \rho \int_\Omega \int_0^\infty f\nu_0^3 k_{\nu_0} l^\mu d\nu_0 d\Omega, \tag{35}$$

wherefrom we obtain the energy–conservation equation by contracting with u_μ:

$$u_\mu \nabla_\nu T^{\mu\nu} = -\rho \left(\int_\Omega \int_0^\infty \epsilon_{\nu_0} d\nu_0 d\Omega - \int_\Omega \int_0^\infty k_{\nu_0} f\nu_0^3 d\nu_0 d\Omega \right). \tag{36}$$

In terms of the normalized intensity $\psi = I/J$ eq. (29) can be rewritten as

$$\psi(u^\mu + l^\mu)\nabla_\mu J + J(u^\mu + l^\mu)\nabla_\mu \psi + 4J\psi l^\sigma (a_\sigma + l^\mu \nabla_\mu u_\sigma)$$
$$+ J\left(a_\sigma l^\sigma l^\rho - a^\rho + l^\mu l^\sigma l^\rho \nabla_\mu u_\sigma + l^\mu l^\sigma u^\rho \nabla_\mu u_\sigma + a_\sigma l^\rho l^\sigma - l^\mu \nabla_\mu u^\rho \right) \frac{\partial \psi}{\partial l^\rho}$$
$$= \rho \left(\epsilon - \int_0^\infty k_{\nu_0} f\nu_0^3 d\nu_0 \right). \tag{37}$$

By using the relation (see Anile and Romano 1992)

$$u^\rho \frac{\partial \psi}{\partial l^\rho} = 0 \tag{38}$$

and the standard kinematic decomposition

$$\nabla_\mu u_\sigma = \frac{1}{3}\Theta h_{\mu\sigma} + \sigma_{\mu\sigma} + \omega_{\mu\sigma} - a_\sigma u_\mu, \tag{39}$$

where $\Theta, \sigma_{\mu\sigma}, \omega_{\mu\sigma}$ are, respectively, the expansion, shear and rotation and $h_{\mu\sigma}$ is the projection tensor, equation (37) becomes

$$\psi(u^\mu + l^\mu)\nabla_\mu J + J(u^\mu + l^\mu)\nabla_\mu\psi + 4J\psi\left(l^\sigma a_\sigma + \frac{\Theta}{3} + \sigma_{\mu\sigma}l^\mu l^\sigma\right)$$

$$+ J\left(a_\sigma l^\sigma l^\rho - a_\rho + \sigma_{\mu\sigma}l^\mu l^\sigma l^\rho - l^\mu\sigma_{\mu\rho} - l^\mu\omega_{\mu\rho}\right)\frac{\partial\psi}{\partial l^\rho}$$

$$= \rho\left(\varepsilon - \int_0^\infty k_{\nu_0} f\nu_0^3 d\nu_0\right). \tag{40}$$

As discussed by Anile and Sammartino (1989), the appropriate generalization of the Levermore and Pomraning ansatz (1981) is that ψ is slowly varying along the bicharacteristics of (40), i.e.

$$(u^\mu + l^\mu)\nabla_\mu\psi + \left(a_\sigma l^\sigma l^\rho - a_\rho + \sigma_{\mu\sigma}l^\mu l^\sigma l^\rho - l^\mu\sigma_{\mu\rho} - l^\mu\omega_{\mu\rho}\right)\frac{\partial\psi}{\partial l^\rho} = 0. \tag{41}$$

By using (41), equation (40) becomes

$$\psi\left[(u^\mu + l^\mu)\nabla_\mu J + 4J\left(l^\sigma a_\sigma + \frac{\Theta}{3} + \sigma_{\mu\sigma}l^\mu l^\sigma\right)\right] = \rho\left(\varepsilon - \int_0^\infty k_{\nu_0} f\nu_0^3 d\nu_0\right). \tag{42}$$

Let f^μ denote the normalized flux ($H^\mu = Jf^\mu$), with $f^\mu = \frac{1}{4\pi}\int_\Omega l^\mu\psi d\Omega$. We have

$$\nabla_\mu H^\mu \simeq f^\mu\nabla_\mu + 3a_\mu H^\mu + 3K^{\mu\nu}\sigma_{\mu\nu}. \tag{43}$$

Equation (38) and (41) give

$$u^\mu\nabla_\mu J = -\frac{4}{3}J\theta - f^\mu\nabla_\mu J - 4a_\mu Jf^\mu - 4K^{\mu\nu}\sigma_{\mu\nu}$$

$$+ \rho\left(\int_\Omega\int_0^\infty \varepsilon_{\nu_0} d\nu_0 d\Omega - \int_\Omega\int_0^\infty k_{\nu_0} f\nu_0^3 d\nu_0 d\Omega\right). \tag{44}$$

By using (16) in (14), we finally obtain

$$\psi = \frac{S}{S_0 + (l^\mu - f^\mu)R_\mu + 4J\sigma_{\mu\nu}(l^\mu l^\nu - K^{\mu\nu}/J)}, \tag{45}$$

where

$$R_\alpha = h_\alpha^\beta \left(\nabla_\beta J + 4 J a_\beta \right), \tag{46}$$

$$S_0 = \rho \left(\int_\Omega \int_0^\infty \varepsilon_{\nu_0} \nu_0^3 d\nu_0 d\Omega - \int_\Omega k_{\nu_0} f \nu_0^3 d\nu_0 d\Omega \right) \tag{47}$$

and

$$S = \rho \left(\varepsilon - \int_0^\infty k_{\nu_0} f \nu_0^3 d\nu_0 \right) \tag{48}$$

is the source term.

Substituting (45) into (33), one should be able to invert eqs. (33), obtaining $K^{\mu\nu}$ and H^μ, but in general a host of highly non-linear terms describing the coupling of radiation and shear would appear. Thus in order to obtain explicit expressions for $K^{\mu\nu}$ and H^μ, we develop S in spherical harmonics and linearize for small shear.

Let us consider first the case of isotropic source,

$$S = S_0. \tag{49}$$

Then, from the exact representation theorems (S. Pennisi 1986, S. Pennisi and M. Trovato, 1987) keeping only terms linear in $\sigma^{\mu\nu}$, we obtain

$$f^\mu = (\lambda_0 + \lambda_{0,1} \sigma^{\tau\rho} \tilde{R}_\tau \tilde{R}_\rho) \tilde{R}^\mu + \sigma\sigma^{\mu\nu} \tilde{R}_\nu, \tag{50}$$

$$K^{\mu\nu} = \frac{1}{3}(\alpha_0 + \alpha_{0,1} \sigma^{\tau\rho} \tilde{R}_\tau \tilde{R}_\rho) J h^{\mu\nu} + \beta \sigma^{\mu\nu} + \gamma \sigma^{\mu\nu} \tilde{R}_\rho \tilde{R}^\nu$$
$$+ (\delta_0 + \delta_{0,1} \sigma^{\tau\rho} \tilde{R}_\tau \tilde{R}_\rho) \tilde{R}^\mu \tilde{R}^\nu, \tag{51}$$

with λ_0, $\lambda_{0,1}$, σ, α_0, $\alpha_{0,1}$, β, γ, δ_0 and $\delta_{0,1}$ functions of J and $\tilde{R}^\mu \tilde{R}_\mu$ and

$$\tilde{R} = \frac{R}{S}. \tag{52}$$

Equations (33) with the help of equations (50) and (51), after linearizing for small shears, give (for details see Anile and Romano, 1992)

$$\alpha_0 = \frac{3}{2} \left[1 + \coth \tilde{R} \left(\frac{1}{\tilde{R}} - \coth \tilde{R} \right) \right], \tag{53}$$

$$\delta_0 = \frac{J}{2\tilde{R}^2} \left[3 \coth \tilde{R} \left(\coth \tilde{R} - \frac{1}{\tilde{R}} - 1 \right) \right], \tag{54}$$

$$\beta = -J\tilde{J} \left[\frac{1 - 4\coth^2 \tilde{R} + \coth^4 \tilde{R}}{\tilde{R}^2 (\coth^2 \tilde{R} - 1)} \right.$$
$$\left. + \frac{9\coth^2 \tilde{R} - 5}{3\tilde{R}^2} + \frac{4\coth^2 \tilde{R}(1 - \coth^4 \tilde{R})}{\tilde{R}} \right], \tag{55}$$

$$\gamma = \frac{J\tilde{J}}{\tilde{R}^2}\left[\frac{1 - 12\coth^2\tilde{R} + 5\coth^4\tilde{R}}{\tilde{R}^2(\coth^2\tilde{R} - 1)}\right.$$
$$\left. - \frac{13}{3\tilde{R}^2} + \frac{15\coth^2\tilde{R}}{\tilde{R}^2} - \frac{12\coth\tilde{R} - 20\coth^3\tilde{R}}{\tilde{R}}\right], \qquad (56)$$

$$\sigma = -\frac{\tilde{J}}{\tilde{R}}\left[\frac{2}{\tilde{R}} + \frac{4\coth\tilde{R}}{\tilde{R}^2} + \frac{6\coth^2\tilde{R}}{\tilde{R}} - \frac{2\coth\tilde{R}(1 - \coth^2\tilde{R})}{\tilde{R}^2(\coth^2\tilde{R} - 1)}\right], \qquad (57)$$

$$\lambda_0 = -\frac{1}{\tilde{R}}(\coth\tilde{R} - \frac{1}{\tilde{R}}), \qquad (58)$$

$$\lambda_{0,1} = \frac{2\tilde{J}}{\tilde{R}^2}\left[3\coth\tilde{R}(\coth\tilde{R} - \frac{1}{\tilde{R}}) - 1\right](\frac{\coth^2\tilde{R} - 1}{\tilde{R}\coth\tilde{R}} - \frac{1}{\tilde{R}^2})$$
$$+ \frac{J(\coth^2\tilde{R} - 1)}{\tilde{R}^2\coth\tilde{R}}(\frac{2}{\tilde{R}} + \frac{18\coth\tilde{R}}{\tilde{R}^2} - \frac{18\coth\tilde{R}}{\tilde{R}}), \qquad (59)$$

$$\alpha_{0,1} = \frac{3}{2}(\lambda_{0,1}\tilde{R}^2 + \frac{4\tilde{J}\delta_0}{J} + \sigma)\left[\frac{1 - 2\coth^2\tilde{R}}{\tilde{R}^2(\coth^2\tilde{R} - 1)} - \frac{1}{\tilde{R}^2} + \frac{2\coth\tilde{R}}{\tilde{R}}\right]$$
$$+ \frac{3\tilde{J}}{\tilde{R}^2}\left[\frac{1 - 6\coth^2\tilde{R} + 2\coth^4\tilde{R}}{\tilde{R}^2(\coth^2\tilde{R} - 1)} + \frac{6\coth^2\tilde{R}}{\tilde{R}^2}\right.$$
$$\left. + \frac{6\coth\tilde{R} - 8\coth^3\tilde{R}}{\tilde{R}} - \frac{7}{3\tilde{R}^2}\right], \qquad (60)$$

$$\delta_{0,1} = \frac{(1 - \alpha_0 - \alpha_{0,1}\sigma^{\mu\nu}\tilde{R}_\mu\tilde{R}_\nu)J}{\tilde{R}^\tau\tilde{R}_\tau\sigma^{\mu\nu}\tilde{R}_\nu\tilde{R}_\nu} - \frac{\gamma\sigma^{\mu\nu}\tilde{R}_\mu\tilde{R}_\nu + \delta_0\tilde{R}^\tau\tilde{R}_\tau}{\tilde{R}^\tau\tilde{R}_\tau\sigma^{\mu\nu}\tilde{R}_\nu\tilde{R}_\nu}, \qquad (61)$$

with

$$\tilde{J} = \frac{J}{S}. \qquad (62)$$

Now we consider the case of anisotropic source functions (see Bonanno and Romano, 1992). Then, in order to invert eqs. (33) we develop the source term in spherical harmonics

$$S = \sum_{k=0}^{\infty} \frac{(2k + 1)!!}{4\pi k!}(\int_\Omega Sl^{\alpha 1}\ldots l^{\alpha k}\,d\Omega)^{SPTF}l_{\alpha 1}\ldots l_{\alpha k}. \qquad (63)$$

However, in several astrophysical and cosmological applications we deal with small anisotropic source terms, so the approximation

$$S = S_0 + S_\mu l^\mu + o^2, \qquad (64)$$

can be used, with $|S_0^{-1}S^\mu| \ll 1$, where S_0 is defined in (19) and S_μ is the first moment of the source term,

$$S^\mu = \frac{3}{4\pi}\int_\Omega Sl^\mu\,d\Omega \qquad (65)$$

(see Thorne, 1980).

The function ψ becomes

$$\psi = \frac{1 + S_0^{-1}S_\mu l^\mu}{1 + (l^\mu - f^\mu)\tilde{R}_\mu + 4\tilde{J}\sigma_{\mu\nu}(l^\mu l^\nu - J^{-1}K^{\mu\nu})}, \tag{66}$$

where

$$R_\alpha = \frac{1}{S_0}h_\alpha^\beta(\nabla_\beta J + 4Ja_\beta), \tag{67}$$

$$\tilde{J} = \frac{J}{S_0}. \tag{68}$$

If we suppose that the shear is of the same order of $\tilde{S} = S/S_0$, we can neglect terms as $\sigma_{\mu\nu}\tilde{S}^\mu\tilde{S}^\nu$ or $\sigma_{\mu\nu}\tilde{S}^\mu\tilde{R}^\nu$, obtaining from the representation formulas and relations (33)

$$f^\mu = \lambda\tilde{R}^\mu + \zeta\frac{S^\mu}{S_0} + \sigma\sigma^{\mu\nu}\tilde{R}_\nu, \tag{69}$$

$$K^{\mu\nu} = \frac{1}{3}\alpha Jh^{\mu\nu} + \beta\sigma^{\mu\nu} + \delta\tilde{R}^{(\mu}\tilde{R}^{\nu)} + \xi\frac{S^{(\mu}\tilde{R}^{\nu)}}{S_0} + \gamma\sigma^{(\mu\tau}\tilde{R}_\tau\tilde{R}^{\nu)}, \tag{70}$$

where

$$\alpha = \alpha_0 + \sigma_{\mu\nu}\tilde{R}^\mu\tilde{R}^\nu\alpha_{0,1} + \alpha_1\tilde{S}^\tau R_\tau, \tag{71}$$

$$\delta = \delta_0 + \sigma_{\mu\nu}\tilde{R}^\mu\tilde{R}^\nu\delta_{0,1} + J\delta_1\tilde{S}^\tau R_\tau, \tag{72}$$

$$\lambda = \lambda_0 + \sigma_{\mu\nu}\tilde{R}^\mu\tilde{R}^\nu\lambda_{0,1} + \lambda_1\tilde{S}^\tau R_\tau, \tag{73}$$

$$\lambda_1 = \frac{1}{\tilde{R}}\left[(\coth^2\tilde{R} - 1)(\coth\tilde{R} - \frac{1}{2\tilde{R}}) - \frac{\coth\tilde{R}}{2\tilde{R}^2}\right], \tag{74}$$

$$\zeta = \frac{1}{2}\left[\frac{\coth\tilde{R}}{\tilde{R}} - \coth^2\tilde{R} + 1\right], \tag{75}$$

$$\alpha_0 = \frac{3}{2}\left[1 + \coth\tilde{R}(\frac{1}{\tilde{R}} - \coth\tilde{R})\right], \tag{76}$$

$$\alpha_1 = 3\left[\frac{\xi}{2J} - \frac{\zeta\lambda_0}{\tilde{R}}\right] - 3\lambda_0\lambda_1\tilde{R}^2, \tag{77}$$

$$\delta_0 = \frac{J}{2\tilde{R}^2}\left[3\coth\tilde{R}(\coth\tilde{R} - \frac{1}{\tilde{R}}) - 1\right], \tag{78}$$

$$\delta_1 = \frac{1}{\tilde{R}^4}\zeta + \frac{1}{\tilde{R}^3}\left[\frac{1}{\tilde{R}} - \lambda_0\tilde{R}\coth^2\tilde{R}\right] - \frac{\alpha_1}{3\tilde{R}^2} - \xi\frac{1}{J\tilde{R}^2}, \tag{79}$$

$$\xi = \frac{J}{\tilde{R}}\left[\frac{2}{3\tilde{R}} - \coth\tilde{R} + \frac{\coth^2\tilde{R}}{\tilde{R}} - \coth^3\tilde{R}\right], \tag{80}$$

with $\tilde{R}^2 = \tilde{R}^\mu\tilde{R}_\mu$.

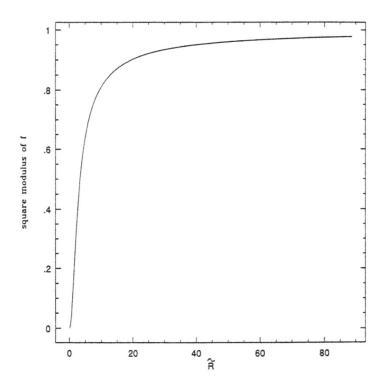

Figure 1 shows the graph of the square modulus of f for $|\widetilde{S}_\mu|$ of order 10^{-1} in the shear free case with anisotropic source. We find similar behavior in the presence of shear too.

The previous formalism is not applicable when S_0 is zero but it is easy to overcome this problem by following a different procedure. Here for sake of brevity we do not report this alternative approach (the interested reader is referred to A. Bonanno and V. Romano, 1992).

We have presented a theory of radiation hydrodynamics in which the radiation energy flux and stresses are given by non–linear constitutive functions. The resulting theory is fully covariant and flux–limited and should be used in problems in astrophysics and cosmology (e.g. stellar core collapse or evolution of primordial fluctuations) instead of resorting to *ad hoc* heuristic methods.

REFERENCES

Anderson J.L. and Spiegel E.A. (1972), *Ap. J.*, **171**, 127.
Anile A.M. *Relativistic Fluids and Magnetofluids*, Cambridge University Press, 1989.
Anile A.M. and Romano V. (1992), *Ap. J*, **386**, 325.
Anile A.M. and Sammartino M. (1989), *Annales de Physique*, **14**, 1.
Anile A.M., Pennisi S. and Sammartino M. (1991), *J Math. Phys.*, **32**, 544.

Anile A.M., Pennisi S. and Sammartino M. (1992), *Ann. Inst. H. Poin.*, **56** , 49.

Bonanno A. and Romano V. (1992), *Ap. J* to appear.

Bruenn S.W. (1985), *Ap. J.*, **58**, 771.

Hazelhurst J. and Sargent W.L.W. (1959), *Ap.J.*, **130**, 276.

Hsieh S.H. and Spiegel E.A. (1976), *Ap.J.*, **207**, 244.

Israel W. in *Relativistic Fluid Dynamics*, 1987, A. Anile and Y. Choquet-Bruhat editors, Lecture Notes in Mathematics, Springer-Verlag, p.152

Israel W. and Stewart J.M. (1979), *Ann. Phys.*, **118**, 341.

Jeans J.H. (1925), *Mon.Not.R.astr.Soc.*, **85**, 917.

Levermore C.D. (1984), *J.Q.S.R.T.*, **31**, 149.

Levermore C.D (1979), report UCID-18229, Livermore: Lawrence Livermore Laboratory.

Levermore C.D. and Pomraning G.C. (1981), *Ap. J.*, **248**, 321.

Lindquist R.W. (1966), *Ann. Phys.*, **37**, 341.

Mezzacappa A. and Matzner R. (1989), *Ap. J.*, **343**, 853.

Mihalas D. (1983), *Ap. J*, **266**, 242.

Mihalas D. and Mihalas B.W. (1984), *Foundation of Radiation Hydrodynamics*.

Pennisi S. (1986), *Suppl. B. UMI, Fisica Matematica*, **5**, 191.

Pennisi S. and Trovato M. (1987), *Int. J. Engng Sci*, **25**, 1059.

Romano V. (1991), *Le Matematiche*, Vol. XLVI, 1.

Rosseland S. (1926), *Ap J.*, **63**, 342.

Schweizer M.A. (1988), *Annals of Physics*, **183**, 80.

Synge J.L. (1934), *Trans.R.Soc.Canada*, **28**, 439.

Synge J.L. (1957), *The Relativistic Gas*, North-Holland.

Thomas L.H. (1930), *Q.J. Math*, **1**, 239.

Thorne K.S.(1981), *Mon.Not.R.astr.Soc.*, **194**, 439.

Terasawa W. and Sato K.(1989), *Phys. Rev. D*, **39**, 2893.

Terasawa W. and Sato K.(1990), *Ap.J.*, **326**, L47.

Turolla R. and Nobili L. (1988), *Mon.Not.R.astr.Soc.*, **235**, 1273.

Udey N. and Israel W. (1982), *Mon.Not.R.astr.Soc.*, **199**, 1137.

Vogt H. (1928), *Astr. Nachr.*, **232**, 1.

Weinberg S. (1979), *Ap J.*, **168**, 175.

Relativistic Gravitational Collapse

JOHN C. MILLER

1 INTRODUCTION

I started as a research student with Dennis Sciama in 1971 at the beginning of his time at Oxford, after he had transferred there from Cambridge, and was subsequently a post-doc with his groups in Oxford and Trieste. It is a great pleasure to have the opportunity of contributing to this book.

In the renaissance of general relativity and cosmology, which is our subject here, one of the central themes has been the study of relativistic gravitational collapse, black holes and neutron stars. At the beginning of my research work, Dennis emphasized to me the role which was going to be played in this by numerical computing and he pointed me in that direction despite some initial reluctance on my part. Applying general relativity to real problems in the real world is a complicated business but gradually it has entered the mainstream of astrophysics to the extent that it now no longer seems to be an exotic curiosity but has come of age as an equal member of the collection of physical theories which are brought into service in attempting to explain how things work. Computing has played a key role in this, making it possible to move beyond theoretical models which have been simplified to the point where analytical techniques are sufficient for studying them, to the development of more detailed models which probe more deeply into the consequences of the theory and come closer to contact with possible observations.

This article does not attempt to be an exhaustive review of the subject of relativistic gravitational collapse but, rather, is a personal overview of what has been done and of what I think may be the most important directions for future work. In Section 2, I discuss the motivation for thinking about continued collapse and black holes in the first place; Section 3 is concerned with studies of non-rotating spherical collapse; Section 4 discusses non-spherical collapse (including rotation) and Section 5 is the conclusion.

2 THE MOTIVATION FOR THINKING ABOUT BLACK HOLES

The history of this idea can be traced back to the paper by John Michell (published in 1784) which, on the basis of straightforward considerations within Newtonian theory, pointed out that the minimum escape velocity for particles propelled outwards from the surface of a gravitating object could reach the speed of light (and beyond) if the object were sufficiently compact. If light is thought of as being composed of particles, then this has the implication that a sufficiently compact object could not be seen by distant observers (although its presence could still be detected as a result of the influence of its gravitational field on neighbouring objects). Michell's paper is a truly remarkable one but a fully satisfactory discussion of the phenomenon did not become possible until the present century with the arrival of general relativity.

The first relativistic calculation of gravitational collapse leading to black hole formation was carried out by Oppenheimer & Snyder in 1939 and made use of the very considerable simplifications which come from considering zero rotation, spherical symmetry, uniform density (at any given time) and zero pressure. From Birkhoff's theorem (Birkhoff 1923) it follows that the geometry outside a non-rotating, spherically symmetric collapsing object is just the ordinary vacuum Schwarzschild geometry (Schwarzschild 1916) and, for the situation being considered, the interior is described by the closed Friedmann cosmological solution (Friedmann 1922). The two geometries can be joined together continuously at the surface of the collapsing object and evolve in exactly the same way as if they had not been joined with each element of the surface following a geodesic path in the Schwarzschild space-time. From the known properties of the Schwarzschild and Friedmann solutions it is then possible to build up an overall picture of the collapse and of black hole formation. Since this picture is part of the phenomenology of non-rotating spherical collapse, discussion of it will be delayed until the next section.

An important further question though is whether conditions ever arise *in practice* which lead to continued gravitational collapse and black hole formation. In 1957, John Wheeler and his collaborators embarked on an extensive study (see the review by Harrison et al. 1965) of the "issue of the final state", namely: what would be the fate of a star which passed through all of the stages of thermonuclear evolution leading to the minimum possible energy per baryon and then cooled to zero temperature? (This is clearly an idealization which would only be approximated in nature but studying it is nevertheless a very useful exercise.) What emerged is that there exist just two regimes of cold stellar configurations in stable equilibrium, corresponding to *white dwarfs* (which form a sequence joining continuously with a sequence of planet-like objects at low masses) and *neutron stars*.

The white dwarfs (masses $\sim 1\,M_\odot$, radii $\sim 10^{-2}\,R_\odot$, mean densities $\sim 10^6\,g\,cm^{-3}$) are supported by electron degeneracy pressure and have the property that in the absence of rotation there is a *maximum mass* which they can have for any particular chemical composition (Chandrasekhar 1931, 1935; Landau 1932). After subsequent further refinements, present values for the limiting mass are in the range from $1.0\,M_\odot$ to $1.4\,M_\odot$ depending on the chemical composition. Rapid rotation raises the limit however (Roxburgh 1965) and extreme differential rotation can remove it entirely (Ostriker, Bodenheimer & Lynden-Bell 1966). It seems though that observed white dwarfs are rotating quite slowly.

The name "neutron star" is rather a misnomer since these objects are by no means completely composed of neutrons. In fact, they are normally thought to consist of a mixture of neutrons, protons, electrons, muons, hyperons and neutron-rich nuclei with the pressure support coming from the rather complicated interactions among this collection of particles (not just from neutron degeneracy as was once proposed). Typical neutron stars have masses of $\sim 1\,M_\odot$, radii of $\sim 10\,km$ $(10^{-5}\,R_\odot)$ and mean densities $\sim 10^{15}\,g\,cm^{-3}$. From measurements made for pulsars in binary systems (pulsars are thought to be rotating neutron stars), it seems that $1.4\,M_\odot$ is a "canonical" mass. Neutron star models were first calculated by Oppenheimer & Volkoff (1939) who demonstrated the existence of a *maximum* mass also for non-rotating objects of this sort. Current values for this maximum are mainly in the range from $1.4\,M_\odot$ to $2.5\,M_\odot$. As for white dwarfs, rapid rotation would lead to an increase in the limit (Friedman, Ipser & Parker 1986) but, up to the present time, no pulsar has been observed with a rotation speed which is large enough to make a very significant difference.

The important conclusion coming from this is that there exists a maximum mass above which a non-rotating, burned-out stellar core could not be held in equilibrium by pressure forces and so would collapse to form a black hole. Any observed compact object whose mass seems to be above the limit is then a strong candidate for being a black hole. However, there are two important questions which need to be asked about this picture. Firstly, how likely is it that the processes of stellar evolution would lead to burned-out cores with masses higher than the limits just quoted? Secondly, how secure are we about the *values* for these limits?

Regarding the first question, it used to be emphasized that there are a very large number of stars with masses well in excess of the limit and that if any significant fraction of these ended up as black holes then stellar mass black holes should be very common (see, for example, Peebles 1972). However, subsequent work on stellar evolution showed that even high mass stars develop quite small cores by the time that they have reached the end of their thermonuclear evolution (Arnett 1974) and so the question arose of whether *any* stellar cores would be sufficiently massive to form black holes. (Of course, one could always point to the fact that neutron stars might sometimes increase their mass by accretion until they passed the limit but this would not be a compelling mechanism for forming *large numbers* of black holes.) The current view from stellar evolution (see Brown, Bruenn & Wheeler 1992) is that black holes *may*, after all, be formed fairly frequently by iron core collapses although not as often as was once thought.

Regarding the second question, it has to be said that there is extreme uncertainty about the nature and properties of matter above nuclear density which is reflected in the range of values quoted for the neutron star maximum mass. Since this is a very important quantity, it makes sense to ask what would be the highest possible value consistent with some general physical principles rather than concentrating only on particular equations of state. Rhoades & Ruffini (1974) made a calculation of this sort (based on assuming the correctness of GR, microscopic stability, causality and knowledge of the equation of state *up to* around nuclear density) and found a maximum of $\lesssim 3.2\,M_\odot$. (Hartle – 1978 – argued that the causality condition used was not appropriate because of the dispersive nature of the medium. Removing this constraint, he obtained $5\,M_\odot$.) Any sufficiently compact object with a measured mass greater than these limits would then seem to be a strong candidate for being a black hole.

Recently, a loophole has been opened in this kind of argument by Bahcall, Lynn & Selipsky (1990) who pointed to the lack of experimental constraints on the nature of high density bulk matter even below nuclear density. Using an effective field theory approach for representing the strong force, they demonstrated that it is possible to find a coherent solution in which the strong force is able to confine bulk *hadronic* matter (i.e. protons, neutrons, etc.) at densities significantly *below nuclear* without any presently available experimental evidence being contradicted. They constructed corresponding stellar models which they called "Q stars". (There is also another variant of Q stars involving matter at higher densities but we will not be concerned with this here.) Because of the modification to the equation of state at densities below nuclear, the Rhoades-Ruffini and Hartle mass limits do not apply and indeed the calculated Q star models have masses going up to several hundred $M_\odot$.

Q stars have some similarity with *strange stars* (Witten 1984) which are models resulting from the hypothesis that the absolute ground state of baryonic matter could be a mixture of up, down and strange quarks *even at zero pressure*. A strange star (composed of such matter) would behave like a single massive nucleon ($A \sim 10^{57}$) with the strong force aiding gravity. It is to be stressed, however, that the idea of strange stars (which have a mass limit in the standard neutron star range) is reasonably conventional in the sense that the particles being confined are quarks and the density is above nuclear; we are used to the idea of the strong force giving confinement under these circumstances. Witten's hypothesis may or may not be correct but it does not fall outside the range of everyday ideas. In contrast, having ordinary protons and neutrons being confined by the strong force in the same way is certainly a radical suggestion. However, until this possibility has been definitively ruled out, arguments made on the basis of a neutron star maximum mass of $\lesssim 3.2\,M_\odot$ or $5\,M_\odot$ must remain somewhat tentative even if we may take the view that they are most probably correct.

We have been concentrating on the possibilities for having black holes formed as a result of the normal evolution of stars and we will continue to concentrate on this for the remainder of the article but it should be mentioned that there are also two other contexts in which black hole formation is discussed:

(i) In the early universe where the density was very high everywhere and black holes of any mass could have been formed;

(ii) By the collapse of galactic nuclei, dense star clusters or supermassive stars.

For a black hole to be produced, it is necessary for a quantity of matter to become compacted within a suitably small volume but the *degree* of compactness required depends on the mass involved. If one considers the total mass M as being evenly distributed through the region within the black hole event horizon then the density would be $\propto 1/M^2$ (if there is no rotation). While for a solar mass black hole, this density is considerably higher than that of nuclear matter, for $10^8\,M_\odot$ it is comparable to that of water. Forming larger black holes is a less extreme process than forming small ones.

3 NON-ROTATING SPHERICALLY SYMMETRIC COLLAPSE

As mentioned in the previous section, the first relativistic calculations of gravitational collapse leading to black hole formation were the ones made analytically by Oppenheimer & Snyder (1939) for objects with zero pressure and uniform density at any given time (as well as no rotation and spherical symmetry). Updated presentations of their discussion were given by Beckedorff (1962) and Misner (1969). It is still possible to make calculations analytically if the restriction to constant density is removed (Liang 1974) but inclusion of non-zero pressure requires numerical computation. The first extensive computations with non-zero pressure were carried out by

May & White (1966). We will here give an overview of the picture of non-rotating spherical collapse resulting from all of these studies. (A more detailed review of work prior to 1980 relating to this Section and to the next one was given by Miller & Sciama (1980).) The description given here is for cases where pressure forces (if present) are never sufficient to halt the collapse, even temporarily. A more complete picture would include the possibility of a bounce occurring which might then be followed by a second collapse phase (see Shapiro & Teukolsky 1980).

Consider first the collapse as seen by a distant observer. In the early stages this observer sees the infall velocity increasing but then it reaches a maximum and decreases again tending asymptotically to zero as the object reaches its Schwarzschild radius ($R = 2GM/c^2$) where it appears to become "frozen". During the collapse, light emitted radially outwards from the surface becomes progressively more redshifted and the redshift tends asymptotically to infinity as the Schwarzschild radius is approached.

The radial (and nearly radial) light rays are the best source of information about the collapse. However, in the late stages as the redshift of radial photons tends to infinity, the total luminosity is dominated by photons which were deposited in unstable circular orbits when the surface of the object passed through $R = 3GM/c^2$ and have then leaked out to infinity (see Ames & Thorne, 1968). The distant observer then sees the object as an apparent disc which is brightest and bluest at the rim (where the "leaked" photons come from) and is darker and redder towards the centre (where the "direct" photons are coming from). The observed intensity and redshift (z) of light from the rim are time independent (with $z = 2$) but the width of the time-independent zone decreases exponentially with time.

The picture of the collapse is very different for a hypothetical observer standing on the surface of the collapsing matter. To him the infall velocity appears to continue increasing. Nothing special happens as he passes the Schwarzschild radius but as the collapse proceeds he experiences enormous tidal forces and, after a finite interval of proper time, he is pulled into a singularity where the space-time curvature diverges (at least when viewed classically). Once the "event horizon" at $R = 2GM/c^2$ has been crossed, any physical particle or light ray emitted from the surface of the object is trapped within the black hole and cannot reach infinity. Collapse to a singularity is then inevitable. A black hole *can* emit radiation as a result of a quantum process (Hawking 1975) but this energy appears in an incoherent form and does not carry with it information from inside the event horizon.

The possibility of having two such very different pictures of the same collapse is an indication of a fundamental difference between general relativity and Newtonian theory. Whereas in Newtonian theory the scale of time is universal, this is not the case in general relativity. There space-time is a fully four-dimensional entity with each of the four dimensions having similar status. If one wants to make a calculation starting from some initial conditions at an initial time and then evolving forward from these, it is necessary to specify how the space-time is to be sliced (divided into a stack of hypersurfaces on each one of which the time coordinate is defined to be constant). This is not a unique process and which procedure is best to use depends on the nature of the problem being considered. Once a slicing criterion has been chosen, the calculation proceeds by advancing the solution from one time-slice to the next, solving coupled Einstein equations and fluid equations (if the matter is to be treated as a fluid). Methods based on this approach are referred to as "3+1" methods (since the space-time is split into 3 spatial dimensions and 1 temporal dimension). An alternative class of methods use 2 null coordinates and 2 spatial ones and so are referred to as "2 + 2" methods.

The phenomenology described above for spherical collapse can all be inferred from a pressure-free calculation; the inclusion of pressure changes the speed of infall of the surface of the object and also the solution internal to the collapsing matter. In their calculations, May & White used a lagrangian technique with a "cosmic" time coordinate (which makes the metric diagonal). This type of time slicing is suitable for studying the production of a singularity but has the disadvantage that the computation must be terminated as soon as the singularity forms since the singular state cannot be handled numerically. A particular feature seen in their calculations is the formation of a region within the collapsing matter where the circumferential radial coordinate R (equal to $1/2\pi$ times the proper circumference of a circle centred at the symmetry centre of the object) is a *decreasing* function of proper distance measured radially outwards from the centre. When the singularity first forms, R goes to zero and the density becomes infinite at a finite proper distance away from the centre of symmetry. This behaviour has been discussed in many places (see, for example, the review by Miller & Sciama quoted earlier). Regions where R is a decreasing function of radial proper distance have a number of strange properties. For example, matter within them gives a *negative* contribution to the mass of the whole object; a negative pressure gradient (with respect to radial proper distance) *accelerates* collapse rather than opposing it as would normally be the case. It can be shown that these regions always occur within Schwarzschild surfaces (where $R = 2GM/c^2$ with M here being the mass internal to the spherical shell under consideration).

For a physically realistic centrally condensed initial model, the collapse is very non-homologous and the singularity is formed at a cosmic time when the outer material is still far outside the Schwarzschild radius. In order to follow the later stages of collapse of this outer material, some alternative slicing is needed. For problems with spherical symmetry, one convenient choice is to use "observer time" (Hernandez & Misner 1966) where the time coordinate of any event is taken to be the time at which a radial outgoing light ray emanating from the event would be received by a distant observer (ignoring possible scattering or absorption). Use of such a time coordinate has the consequence that all events which could in principle be seen by a distant observer fall within the scope of the calculation while events which could not be seen by such an observer (including singularity formation) do not. Calculations made with this slicing (Miller & Motta 1989) give a particularly satisfactory "outside" view of the collapse: one can see the central matter reaching the "frozen out" condition first (with the redshift diverging) and then the outer matter falling down onto it. The Hernandez & Misner type of formulation is suitable for spherical collapse but not for more general situations. More widely applicable conditions, giving similar behaviour in avoiding singularities and allowing continued calculation of the outer solution, include "maximal slicing" (Estabrook et al. (1973)), "hypergeometric slicing" (Nakamura (1981)) and "polar slicing" (Bardeen & Piran (1983)).

4 NON-SPHERICAL COLLAPSE

The main features of general collapse which go beyond those discussed in the previous section, can be summarised as follows:

(i) Whenever a collapse proceeds to a stage where a "trapped surface" is formed (a surface from which light cannot escape outwards), it is then inevitable that a singularity will be produced irrespective of the geometry of the infall (Hawking & Penrose 1970).

(ii) Rotating relativistic objects encounter the phenomenon of *dragging of the inertial frames*. In general relativity, the behaviour of matter affects the nature of space-time and a rotating object tends to drag the local space round with it. A local zero angular momentum particle then has non-zero angular velocity with respect to infinity.

(iii) In the later stages of a rotating collapse, *ergo-regions* form within which particles can have *negative energy* with respect to infinity (although the locally-measured energy is always positive).

(iv) For a collapse which is not spherically symmetric, the external space-time is dynamical with gravitational radiation being emitted as a result of the motion of the collapsing object.

(v) As long as there is no significant net electric charge, the final end-point of continued collapse with non-zero rotation will be a *Kerr black hole* (one for which the space-time external to the event horizon is represented by the Kerr metric – Kerr (1963)). The nature of this final state depends only on the parameters

$$m = \frac{GM}{c^2} \quad \text{and} \quad a = \frac{J}{cM}$$

where J is the total angular momentum of the object which collapsed to form the black hole. All information about the external field of the black hole is represented by these two parameters; any further information about the structure of the object which collapsed is lost when the black hole forms. Any equilibrium rotating black hole has its external field represented by the Kerr metric (see Robinson 1975) and the Kerr metric only represents a black hole solution when

$$\frac{a}{m} \leq 1$$

The question of whether a collapse could ever give rise to a singularity not surrounded by an event horizon is still an open one. The *cosmic censorship hypothesis* (Penrose 1969) suggests that such naked singularities will never arise for a physical collapse starting from regular initial conditions. For axi-symmetry, it seems that collapse of a realistic rotating object with $a > m$ will be stopped by centrifugal support before black hole dimensions are reached (Miller & de Felice 1985).

Useful insight into rotating collapse can be obtained using the *slow rotation approximation* (Hartle 1967) in which structure equations are written within full general relativity but taking centrifugal effects to be everywhere small compared with gravitational ones. Chandrasekhar & Miller (1974) used this technique to construct a *quasi-stationary sequence* of equilibrium uniform-density models designed so as to mimic dynamical collapse (J and M being kept constant along the sequence). A striking result of this study was that whereas in Newtonian theory the centrifugal flattening of the object progressively increases as it becomes more compact, the GR calculation shows the flattening reaching a maximum when the radius of the object is around 2.5 times its Schwarzschild radius (a typical dimension for neutron stars) with further contraction leading to it becoming more spherical again. Newtonian theory gives errors of order 100% when calculating the ellipticity of objects having the degree of compactness of neutron stars and since rotational flattening is a prototype for all other rotational effects of a centrifugal nature (as opposed to those connected with frame dragging) this casts severe doubt on the use of Newtonian theory for any meaningful calculations of such effects for neutron stars. The cause of the reversal in behaviour of the ellipticity was first thought to be connected with dragging of the inertial frames but is now understood to be connected instead with the distortion of the von Zeipel cylinders which occurs for compact objects (Abramowicz, Miller & Stuchlík 1992). (The von Zeipel cylinders are surfaces on each of which specific angular momentum divided by angular velocity is a constant.) It has been found (Abramowicz & Miller 1990) that use of strong field correction factors (derived in connection with considerations of centrifugal forces in general relativity) allows one to reproduce the relativistic behaviour to surprisingly high accuracy by means of analytical calculations within a Newtonian framework (the relativistic calculations need to be numerical). This technique (which has been referred to as the CCN – "Centrifugally Corrected Newtonian" – approach) has also allowed deeper understanding of the processes involved and could perhaps be useful in the future for providing a simplified route to the calculation of properties of strong field sources.

The first relativistic computations of dynamical collapse of rapidly rotating objects were made by Nakamura and his collaborators (Nakamura 1981 et seq.; see Nakamura, Oohara & Kojima 1987 for a review) who obtained extremely interesting results for the behaviour of the matter undergoing collapse but, with their method, were not able to calculate the gravitational radiation emitted. The first computations able to do this as well were those of Stark & Piran (1986). Both of these series of calculations were for collapses where axial symmetry is maintained. The gravitational radiation calculated by Stark & Piran turns out to correspond very closely to that produced by the normal modes of the Kerr space-time towards which the exterior solution evolves. The efficiency of energy emission is quite low (the energy radiated is $\lesssim 7 \times 10^{-4}$ times the total mass-energy of the collapsing object for all of the cases examined) and the flux peaks at a frequency of $\sim 10 \times (1.4\,M_\odot/M)$ kHz.

There are many comments and points of discussion which arise from this work but we will focus here on just one further point which may be of particular importance. The initial models used by Stark & Piran were spherically symmetric polytropes which were set in uniform rotation and then destabilized by reducing the internal pressure. Some of these models continued to collapse down to form black holes while others bounced and then oscillated around an equilibrium state. It can be observed how initial models which are identical in every way except for the amount of rotation (a/m) then behave differently, those with the smaller amount of rotation forming black holes while those with the more rapid rotation bounce. Leaving aside the question of exact conditions for a bounce to occur (which depend on more detailed physics than was included in the calculations) we can say that there is a significant class of initial models for which collapse is halted directly because of the influence of rotation. (When the contribution from pressure is small, the minimum a/m of the initial model in order for collapse to be stopped is typically of order 1, a result first obtained by Nakamura and his collaborators which is interesting in the light of the cosmic censorship hypothesis.)

It is well-known that equilibrium objects for which rotation supplies a significant proportion of the force required to balance gravity are susceptible to *non-axisymmetric instabilities* (see Lindblom 1986). This provides an important motivation for future work. A possible picture is that a collapsing stellar core could have its collapse halted by a combination of internal pressure and centrifugal support. If the rotational speed is sufficient, a non-axisymmetric mode could then start to grow on a secular time-scale giving rise to gravitational radiation which would carry away both energy and angular momentum. As a result of this, together with a probable increase in the core mass coming from continuing infall of surrounding matter, the core might then be destabilised again and commence a second phase of collapse ending in black hole formation. Such a picture could alter the predictions for gravitational radiation emission. (Note that it is most likely that the non-axisymmetric structure would be produced on a secular time-scale during a "waiting phase" in the collapse as pre-collapse cores probably do not have enough angular momentum for dynamical instabilities to be excited which could grow significantly during phases of dynamical infall.)

5 CONCLUSIONS

A great deal has been achieved in the last half century or so towards gaining an understanding of relativistic gravitational collapse and black hole formation. At this point, the most interesting questions for further research seem to be ones which require fully three-dimensional calculations; in particular:

(i) Growth of non-axisymmetric modes in an initially axisymmetric stellar core and following of a possible subsequent collapse.

(ii) Coalescence of neutron star or black hole binaries (the mechanism which is likely to be responsible for the most powerful sources of gravitational radiation).

(iii) Examination of the status of the cosmic censorship hypothesis under the circumstances of a realistic non-symmetric collapse.

An important stimulus to this would be the arrival of the long-awaited first reliable detection of gravitational radiation from astrophysical sources. Gravitational waves would provide *the* way to get observational information about relativistic gravitational collapse and could also give important insights into the nature of compact objects. An interaction between theoretical calculations and observations will be fundamental in this. The most exciting prospect of all is perhaps the conceivable possibility of being able to obtain information about fundamental physics by observing astrophysical objects composed of matter under extreme conditions. The universe provides an environment where experiments are continually going on under conditions which we could not reproduce in a laboratory.

I gratefully acknowledge financial support from the Italian Ministero dell'Università e della Ricerca Scientifica e Tecnologica and from the Trieste Consorzio di Magnetoflu-idodinamica.

REFERENCES

Abramowicz, M.A. & Miller, J.C. (1990). *Mon. Not. R. astr. Soc.*, **245**, 729.

Abramowicz, M.A., Miller, J.C. & Stuchlík, Z. (1992). *Phys. Rev. D.*, in press.

Ames, W.L. & Thorne, K.S. (1968). *Ap. J.*, **151**, 659.

Arnett, W.D. (1974). In *Late Stages of Stellar Evolution*, eds. R.J. Taylor & J.E. Hesser (Proceedings of IAU Symposium No. 66, Reidel), p. 1.

Bahcall, S., Lynn, B.W. & Selipsky, S.B. (1990). *Ap. J.*, **362**, 251.

Bardeen, J.M. & Piran, T. (1983). *Phys. Rep.*, **96**, 205.

Beckedorff, D.L. (1962). *A.B. Senior thesis* (Princeton University).

Birkhoff, G.D. (1923). *Relativity and Modern Physics* (Harvard University Press), p. 253.

Brown, G.E., Bruenn, S.W. & Wheeler, J.C. (1992). *Comments Astrophys.*, **16**, 153.

Chandrasekhar, S. (1931). *Ap. J.*, **74**, 81.

Chandrasekhar, S. (1935). *Mon. Not. R. astr. Soc.*, **95**, 207.

Chandrasekhar, S. & Miller, J.C. (1974). *Mon. Not. R. astr. Soc.*, **167**, 63.

Estabrook, F., Wahlquist, H., Christensen, S., de Witt, B., Smarr, L. & Tsiang, E. (1973). *Phys. Rev. D*, **7**, 2814.

Friedman, J.L., Ipser, J.R. & Parker, L. (1986). *Ap. J.*, **304**, 115.

Friedmann, A.A. (1922). *Z. Phys.*, **10**, 377.

Harrison, B.K., Thorne, K.S., Wakano, M. & Wheeler, J.A. (1965). *Gravitation Theory and Gravitational Collapse* (University of Chicago Press).

Hartle, J.B. (1967). *Ap. J.*, **150**, 1005.

Hartle, J.B. (1978). *Physics Reports*, **46**, 201.

Hawking, S.W. (1975). *Commun. Math. Phys.*, **43**, 199.

Hawking, S.W. & Penrose, R. (1970). *Proc. Roy. Soc. Lond. A*, **314**, 529.

Hernandez, W.C. & Misner, C.W. (1966). *Ap. J.*, **143**, 452.

Kerr, R.P. (1963). *Phys. Rev. Lett.*, **11**, 237.

Landau, L.D. (1932). *Phys. Zs. Sowjetunion*, **1**, 285.

Liang, E.P.T. (1974). *Phys. Rev. D*, **10**, 447.

Lindblom, L. (1986). *Ap. J.*, **303**, 146.

May, M.M. & White, R.H. (1966). *Phys. Rev.*, **141**, 1232.

Michell, J. (1784). *Phil. Trans. Roy. Soc.*, **74**, 35.

Miller, J.C. & de Felice, F. (1985). *Ap. J.*, **298**, 474.

Miller, J.C. & Motta, S. (1989). *Class. Quantum Grav.*, **6**, 185.

Miller, J.C. & Sciama, D.W. (1980). In *General relativity and gravitation – one hundred years after the birth of Albert Einstein* ed. A. Held; Vol. II – Plenum) p. 359.

Misner, C.W. (1969). In *Astrophysics and General Relativity*, eds. M. Chrétien, S. Deser & J. Goldstein (Proceedings of the Brandeis University Summer Institute in Theoretical Physics, Gordon & Breach), Vol. I, p. 113.

Nakamura, T. (1981). *Prog. Theor. Phys.*, **65**, 1876.

Nakamura, T., Oohara, K. & Kojima, Y. (1987). *Prog. Theor. Phys. Suppl.*, **90**, 1.

Oppenheimer, J.R. & Snyder, H. (1939). *Phys. Rev.*, **56**, 455.

Oppenheimer, J.R. & Volkoff, G. (1939). *Phys. Rev.*, **55**, 374.

Ostriker, J.P., Bodenheimer, P. & Lynden-Bell, D. (1966). *Phys. Rev. Lett.*, **17**, 816.

Peebles, P.J.E. (1972). *Gen. Rel. Grav.*, **3**, 63.

Penrose, R. (1969). *Rev. Nuovo Cimento, Ser. 1*, **1**, 252.

Rhoades, C.E. & Ruffini, R. (1974). *Phys. Rev. Lett.*, **32**, 324.

Robinson, D. (1975). *Phys. Rev. Lett.*, **34**, 905.

Roxburgh, I.W. (1965). *Z. Astrophys.*, **62**, 134.

Schwarzschild, K. (1916). *Sitzber. Deut. Akad. Wiss. Berlin, Kl. Math.-Phys. Tech.*, **1**, 189.

Shapiro, S.L. & Teukolsky, S.A. (1980). *Ap. J.*, **235**, 199.

Stark, R.F. & Piran, T. (1986). In *Proceedings of the Fourth Marcel Grossman Meeting on General Relativity* (ed. R. Ruffini) p. 327.

Witten, E. (1984). *Phys. Rev. D*, **30**, 272.

The Cosmic Censorship Hypothesis

CHRISTOPHER J. S. CLARKE

1 INTRODUCTION

Those of us who had the privilege of being Dennis Sciama's students during what Hajicek has described as 'the Golden Age of General Relativity' can trace many of the current concerns of the subject back to the ideas which he fostered, either directly or indirectly, within his research group in Cambridge. This was the environment in which major contributions to most of the foundational ideas about singularities: from the controversies about the steady state and big bang theories; through the critique of the early Lifshitz-Khalatnikov arguments which at first suggested that the big bang singularity was not generic, leading to definitions of just what constituted a singularity; to the Hawking-Penrose singularity theorems themselves. The issue of cosmic censorship stemmed naturally from this work, and illustrates well the combination of rigorous mathematics with a firm hold on physical relevance which he established at that time. In this talk I shall try to give an outline of the historical work on cosmic censorship, focussing at the end on my own recent work on shell crossing singularities. I shall not be concerned with what George Ellis, in this meeting, has termed the position of the goal posts – the details of exactly what the target is; rather, I shall be arguing that we should in fact be playing a different game.

Once singularities were accepted as generic features of the universe, attention turned to their physical properties, and in particular, to their observational appearance. The basic concern in cosmic censorship is easily stated. We expect three sorts of singularities in general relativistic models of the universe: the big bang, the big crunch, and those due to collapsing stars. The first is hidden from us by the opacity of ionised matter; the second is hidden by the presumed unknowability of the future (of which more later); is it possible to observe the third, either with or without plunging suicidally into the collapse oneself? Exact solutions lent support to the idea that generically the singularity of collapse would be inside the event horizon, and hence unobservable. This led Penrose (1969) to ask: "does there exist a 'cosmic censor' who forbids the appearance of naked singularities, clothing each one in an absolute event horizon?" The affirmative answer to this is the basis of what is now known as the weak (or global) cosmic censorship hypothesis (WCCH).

The "absolute event horizon" in this initial statement of WCCH is the boundary of set of those points in space-time from which it is possible to escape to infinity. Both "infinity" and "singularity" are easily defined in terms of the generalised affine parameter on a curve (the measure of the length of a curve using a parallely propagated frame): we could say that a curve escaped to infinity if it was causal and of unbounded generalised affine parameter to the future (g.a.p. complete); while it comes from a singularity if it is past inextendible and g.a.p. incomplete to the past. Thus the above formulation amounts to

WCCH1 *Space-time does not contain any causal curve that is past inextendible and g.a.p. incomplete but future g.a.p. complete.*

There are, however, a number of problems with this formulation.
1. As it stands this it is violated by $k = 0$ Friedman cosmology; if we wish to include cosmological examples we must restrict to the part of space-time to the future of some partial Cauchy surface S that cuts off the past singularity.
2. Referring to all future-complete curves is very general and does not necessarily correspond to what one intuitively thinks of as 'infinity'. Penrose actually had in mind the situation where there is a well-defined $\jmath$, and the hypothesis is only to be applied to curves ending there.
3. There is evidence, reviewed below, that some very symmetrical space-times violate cosmic censorship, so that a genericness condition may be required.
4. It can be argued that the important issue is not the geometrical fact that some curves are incomplete, but the more physical fact that this incompleteness causes the space-time to be unpredictable from cauchy data; such unpredictability could come either from singularities or from causality violation (Newman, 1989b).
5 One needs to rule out singularities that are there from the beginning (primordial singularities), so that the Cauchy surface S referred to above needs to be regular in some sense. It must extend to infinity without any intermediate singularities, which is best expressed by requiring the outer part of the surface to generate an asymptotically flat space time.

To formulate these ideas, let us restrict attention to weakly asymptotically simple and empty space-times having a partial Cauchy surface S such that all past-directed generators of $\jmath^+$ enter and remain in the boundary of the future domain of dependence of S. We will also suppose that S is $\mathbb{R}^4$ (in the light of Newman (1989a) this is a real restriction, but not too drastic a one) with some map $\mathbb{R}^4 \to S$ specified in each case so that data on S can be pulled back to data on $\mathbb{R}^4$. Then we can (not entirely consistently) call a set of such things *generic* if the corresponding collection of data-sets on $\mathbb{R}^4$ is open and dense in some suitable topology. We also require that the domain of dependence of S should be the maximal Cauchy development of S, in order to exclude space-times with bits cut out of them.

Taking into account the other points above gives the following possible version of the hypothesis:

   WCCH2 *For a generic set of space-times and partial Cauchy surfaces as above, the whole of $\mathcal{J}^+$ lies in the boundary of the future domain of dependence of S.*

The last clause is what Hawking has termed the *future asymptotically predictable* condition. This formulation, while perhaps on the right lines, still lacks precision. We have not specified what equations are in force: vacuum, fields, fluids, or kinetic theory; we have not stated what constitutes a 'solution' (and hence a development of data) of these equations, in terms of its allowed differentiability; and there are subtle points revealed by Newman's work (1989a) about what constitutes an asymptotically flat space-time, in terms of causality restrictions. We shall return to all these later.

It has to be said that progress towards proving the above proposition has been limited. The difficulty, in essence, is that we are concerned with a global existence proof for the non-linear equations of general relativity, and such global proofs are notoriously hard to come by: there are some useful non-existence theorems (Rendall, 1992) and existence theorems in special cases (Chrusciel et al, 1989), but little beyond this. Strategies have therefore either concentrated on weaker forms of the hypothesis, or – a seeming paradox – on a stronger form: the strong (or local) cosmic censorship hypothesis (SCCH).

This form was introduced by Penrose (1974) on the grounds that by strengthening the statement it became much simpler, and the simpler statement might suggest correct strategies of proof more than the complex statement. Whereas the weak hypothesis, as given above, states that the development of data extends to the whole of $\mathcal{J}^+$, the strong hypothesis states that the data extends to the entire space-time: there are no regions, inside or outside the horizon, which are not predictable. This principle, moreover, extends to both past and future, so that:

   SCCH *Generically, space-time is globally hyperbolic.*

It is worth noting where the 'thunderbolts', the null singularities described by Stephen Hawking in this meeting, fit into this picture. They come in two forms, temporal and eschatological. The first are localised: they can get the other guy while missing you, and so are singularities which you can observe, and thus are naked. The eschatological ones, on the other hand, bring the whole of space-time to an end and so are not naked, in any sense. When Penrose formulated the strong hypothesis, he had precisely this outcome in mind for the universe.

We now turn to various lines of attack on these conjectures.

2 STRONG CURVATURE AND EXTENSION

One motivation for believing in the WCCH was the vague idea that a 'real' singularity should involve a strong focussing effect, and that this effect should also produce a horizon. This led to the strategy of trying to introduce a notion of a 'strong singularity', so chosen that

(a) all real singularities were strong singularities, because it should be possible to extend the space-time through anything weaker than these;

(b) strong singularities are always inside horizons.

The pursuit of this strategy led to a greatly improved understanding of the nature of singularities, but the problem has shown little sign of yielding to this approach, because of the gap between definitions of strong singularities that work for (a) and those that work for (b). In order to achieve goal (b) we need a definition of strength that is related to focussing, which is achieved by the following alternative definitions of *strong curvature*. They both refer to a timelike (resp. null) geodesic $\gamma : [0, a) \to M$, the singularity being approached as the affine parameter s tends to a, and are conveniently expressed in terms of Jacobi fields that vanish at a point $\gamma(b)$: we define $J_b(\gamma)$ for $b \in [0, a)$ to be the set of maps $Z : [b, a) \to TM$ such that

$$Z(s) \in T_{\gamma(s)}M, \quad Z(b) = 0, \quad \frac{D^2 Z}{ds^2} = R(\dot\gamma, Z)\dot\gamma, \quad \frac{DZ}{ds}\bigg|_b \cdot \dot\gamma(b) = 0.$$

For a timelike (resp. null) geodesic we use three (resp. two) such fields independent of each other (and of $\dot\gamma$). Their exterior product defines a spacelike volume (resp. area) element, whose magnitude at the affine parameter value s we denote by $V(s)$. The strong curvature conditions are then

SCSK *(Krolak, 1987) For all $b \in [0, a)$ and all independent fields* $\underset{1}{Z}, \underset{2}{Z}, \underset{3}{Z} \in J_b$ *(resp. $\underset{1}{Z}, \underset{2}{Z} \in J_b$) there exists $c \in [b, a)$ with $dV(c)/ds < 0$.*

SCST *(Tipler, 1977) For all $b \in [0, a)$ and all independent fields* $\underset{1}{Z}, \underset{2}{Z}, \underset{3}{Z} \in J_b$ *(resp. $\underset{1}{Z}, \underset{2}{Z} \in J_b$) we have $\liminf_{s \to a} V(s) = 0$.*

We shall retain the oldest terminology of *strong curvature* for SCST, but follow Newman (1986) in referring to SCSK as the *limiting focussing condition*.

With either of these definitions it seems equally difficult to prove (a) or (b) (which maybe suggests that the line is being drawn in the right place!) Some progress can be made in proving that the singularities corresponding to these definitions of strong curvature are censored (step (a)), but only with rather strong additional assumptions (Krolak, 1987, 1992) that are hard to justify a priori.

In the case of (b), in order to prove that weaker singularities can be removed by extension it is necessary to convert the conditions into properties of the Riemann tensor. Clarke and Krolak (1985) give separate necessary and sufficient conditions for SCST and SCSK in terms of the divergence of integrals of the Riemann, Ricci and Weyl tensors along the geodesic γ. When it comes to trying to extend the space-time, however, through a putative singularity, one has to cope with the tricky task of extending the metric in such a way that it is sufficiently differentiable to make sense as a solution of the field equations, knowing only bounds on the Riemann tensor, not on the metric itself. If one tries to construct a purely geometric extension, relying on the properties of geodesics and classical differential geometry, then one needs to maintain a high level of differentiability, and an extension seems only possible in cases where the Riemann tensor satisfies a Hölder condition (Clarke, 1982) which is far more well behaved than what is needed here. We shall see later, however, that by taking the dynamics of the evolution into account it could well be possible to advance further.

A different approach has been taken by Newman, using a different type of definition of strong curvature, called *persistent curvature*. For a null geodesic γ this is defined as

$$\sup_{I,\underset{1}{e},\underset{2}{e},m,n}\left\{\mathrm{mes}(I)^2\inf_{v\in I}\left|R_{ijkl}\underset{m}{e}^i\underset{n}{e}^k\dot\gamma^j\dot\gamma^l\right|_{\gamma(v)}\right\}$$

With this it is possible to prove useful censorship theorems (step (a)), and these provide the most solid general theorems available. However, this condition is non-local in nature and so ill suited to linking with step (b). Newman's theorems should therefore be regarded as providing information about what singularities might be naked, and what clothed, rather than as absolute censorship theorems. In this spirit, he has extended his work to those singularities that persist into the future, and has obtained a numerical bound on the maximum possible size of the persistent curvature in such cases. One important feature revealed by this analysis is that the results are obtained only under global asymptotic causal conditions on the space-time that are stronger than those normally imposed, while being in themselves very natural.

3 COUNTEREXAMPLES

Considerable progress has been made in finding counterexamples to stronger forms of the censorship hypotheses, thus exploring the nature of the conditions that need to be imposed if the proposition is to be true. All of the interesting examples are non-vacuum, which might suggest that the hypothesis is more likely to be true for vacuum solutions. Inevitably, the examples have high symmetry and this is widely thought to be crucial.

A major class of examples are solutions for dust: a perfect fluid with zero pressure, for which the conservation equation $T^{ij}_{;j} = 0$ implies that the dust moves on geodesics. The spherically symmetric solutions for this were found by Tolman (1934) and Bondi (1947). At some time specified as 'initial' the density of matter and its velocity, as functions of a radial coordinate (say the areal radius R), can be specified arbitrarily; the matter can be thought of as disposed in spherical 'shells' labelled by a comoving coordinate r, which can be taken as the value of R at the initial time. For a certain stable set of initial conditions it turns out that the analytic form of the solution breaks down at a stage where dR/dr becomes zero: infinitessimally neighbouring shells approach each other. This is termed a *shell crossing singularity*. We shall follow recent authors in reserving the term for the case where this happens only at isolated and non-zero values of R, calling the other cases *shell focussing singularities*. Shell crossing singularities were first found by Yodzis et al. (1973,4) and have been investigated extensively since. The shell focussing singularity that can be formed at the centre satisfy the limiting focussing condition (Newman, 1986) and even, provided the initial distribution is not smooth, the strong curvature condition (Gorini et al, 1989). With appropriate conditions the examples violate both strong and weak cosmic censorship.

Various questions are raised by these examples.
   (a) Is the violation of the hypothesis due to the symmetry, the singularity disappearing or becoming clothed when it is perturbed?
   (b) Is it in fact not a singularity; i.e. is an extension possible?
   (c) Is the problem due to the artificiality of the matter, being absent if a realistic pressure law is included?

Limited evidence is now accumulating that (a) and (b) may not provide a way out. Shapiro and Teukolsky (1991) have recently presented a numerical example of a cloud of non-colliding particles (Boltzman gas) with only axial symmetry which collapses to a spindle-like singularity that appears to be naked. Though we have a relaxation both of the symmetry and of the nature of the matter, the procedure has been criticised: Wald and Iyer (1991) have pointed out that nakedness has not been proven rigorously, and Rendal (1992) has drawn attention to analogous Newtonian situations where the initial conditions used by Shapiro and Teukolsky (zero temperature) are an exceptional case. Moreover, the work of Thorne reported by Miller at this meeting suggests that the inclusion of initial motion may remove the singularity.

Further evidence against (c) as the solution comes from the work of Ori and Piran (1988) who find naked singularities in a collapse of a perfect fluid with a barotropic equation of state. This solution has even more symmetry, being self-similar as well as spherically symmetric, but other examples (Lake, 1991) suggest that this is not the reason for the singularity.

4 EXTENDABILITY

Finally I want to consider option (b): that the anomalies in the Tolman-Bondi solution are in fact not singularities, but admit extensions. I shall outline the construction of an extension through a shell-crossing singularity below, fuller details appearing in Clarke and O'Donnell (1992). Since shell crossing singularities do not satisfy even the limiting focussing condition, it might be thought that this approach would fail with the stronger shell focussing singularities; Papapetrou and Hamoui (1967) have, however, constructed an extension through a simple shell focussing singularity which can be shown to satisfy Einstein's equations, suggesting that this approach may be generally applicable.

Since curvature polynomials diverge at a shell crossing singularity, it is clear that no C^2 extension to the metric can exist, and so we are concerned with a non-classical solution. The least radical approach is to adopt the formalism proposed by Geroch and Traschen (1987); the more radical approach, which might be needed for general shell focussing singularities, involves multiplying distributions (Vickers, 1992). For the former, we require that should exist an atlas in which:

   GT1. *The components g_{ij} and g^{ij} are locally bounded and locally integrable.*
   GT2. *There exist locally square integrable weak derivatives $g_{ij,k}$ of the metric components.*

It follows from this that the components G_{ij} of the Einstein tensor can be defined *as distributions* (or, if preferred, that the Einstein tensor multiplied by the volume form v can be defined as a tensor-valued distribution, which we refer to as the Einstein distribution and denote by vG).

The aim is then to find an extension in which the Einstein equations are satisfied distributionally. But first we need to understand the nature of the matter. The original solution had matter that was a pressure-free fluid, but it is clear that this is not the appropriate context in which to construct the extension. If we draw the space-time diagram of the analogous situation for Newtonian non-gravitating dust (Fig. 1) we see that a caustic is formed, after which there are three superimposed dust flows.

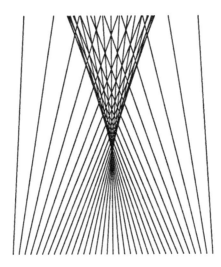

Figure 1. Space-time diagram of the world-lines of dust intersecting to form a caustic bounding the three-dust region.

This means that the initial Tolman-Bondi region needs to be regarded as a special case of a situation in which there are 1 or more superimposed flows of dust – a multi-dust solution. Such a situation is best represented by considering the tangent bundle. At each point in space-time there is defined a number of vectors representing the unit tangent vectors to the flows of dust there. This specifies a number of points in the tangent bundle, which vary continuously with position. In fact, it is easy to see that these define a four-dimensional surface N in the tangent bundle. The world lines of individual dust particles lift up to curves in this surface which are integral curves of the geodesic flow X there.

The projection map from N to M is, in the case we are considering here, a simple fold catastrophe. In standard catastrophe theory (which, as we shall see, is not strictly applicable here because of the low differentiability) it is possible to choose coordinates r, τ in N, which we could suggestively think of as a comoving shell label and proper time, respectively, and coordinates R, t in M (with the spherical coordinates θ, ϕ

suppressed) so that the map from N to M takes the form

$$f : (r, \tau) \mapsto (R, t)$$
$$t = \tau$$
$$R = r^3 - 3r\tau$$

In this case the caustic is given by

$$\frac{\partial R}{\partial r} = 0, \quad \text{i.e. } r = \pm t^{\frac{1}{2}} \text{ or } R = \mp 2t^{\frac{3}{2}}. \tag{1}$$

At the point $r = \pm t^{\frac{1}{2}}$ the dust crossing the caustic into the 3-dust region has r given by $r = \mp 2t^{\frac{1}{2}}$; we refer to this as the incoming flow.

If the mass increases uniformly with the label r then the density ρ of the dust is proportional to $\left(\frac{\partial R}{\partial r}\right)^{-1}$. Near a point on the caustic $r = t^{\frac{1}{2}}$ we introduce a coordinate $\xi = R + 2t^{\frac{3}{2}}$ measuring the distance in the R-direction from the caustic. We then find that the density of dust flows tangent to the caustic is given by

$$\rho^{-1} \propto \left|\frac{\partial R}{\partial r}\right| \approx 2\sqrt{3}\xi^{\frac{1}{2}}t^{\frac{1}{4}} \tag{2}$$

the approximation being asymptotic as $\xi \to 0$, while the density of the incoming flow is

$$\rho^{-1} \propto \left|\frac{\partial R}{\partial r}\right| = 9t.$$

A coordinate velocity of the flow can be defined as $\frac{\partial R}{\partial \tau}$, which for the tangent flows is

$$\frac{\partial R}{\partial \tau} \approx -3\left(\pm\frac{\xi^{\frac{1}{2}}}{\sqrt{3}t^{\frac{1}{4}}} + t^{\frac{1}{2}}\right) \tag{3}$$

(the $\pm$ here referring to the two flows tangent to $r = t^{\frac{1}{2}}$, not to the two different caustics), while the velocity of the incoming flow is given by

$$\frac{\partial R}{\partial \tau} = 6t^{\frac{1}{2}}.$$

Returning to our general discussion, the density will in general be specified by a bounded 3-form on the surface N. Any such surface, with its density, defines a multi-dust model. The caustics, as we have seen with the model just described, are the places where the projection π of N onto the space-time is singular, that is, where N is tangent to the fibres of the tangent bundle. If we require in addition that this projection map is *proper* (inverse images of compact sets are compact) then the density projects down to a distribution, from which we may construct an energy-momentum distribution to equate to the Einstein distribution, namely $T = \pi_*((u \wedge n \otimes X \otimes X)_D)$, where u is the 1-form corresponding to X and $_D$ denotes the interpretation of a form as a distribution.

Ideally we need to find an existence proof for a solution of this form; but so far this has not been possible – the equations are not of a form for which there exist ready-made existence proofs. What has been done, however, is to demonstrate the self-consistency of the approach. Namely, we have determined the form of the caustics that is required by the equations of motion, and we have shown that if a solution of this form can be found in the three-dust region, then, when this region is 'grafted' into the Tolman-Bondi metric, Einstein's equations are satisfied distributionally. The way is thus open to construct numerical solutions, for example, which has been successfully done.

The main points to emerge from the analysis are the following.
   (a) The surface N is not smooth. This is because it is determined by the geodesic equations, in which appear the connection of a metric which fails to be smooth at the caustic.
   (b) Despite this, the functional form of the caustic set, and the functional form of the density and velocity of the dust near the caustic, is to highest order the same as in the smooth case, provided that curvature coordinates are used.
   (c) The Geroch-Traschen conditions are not satisfied in curvature coordinates, but there exists a coordinate system (related to Gaussian coordinates by a non-linear transformation) in which they are satisfied.

We briefly describe these points in turn. In order to establish the geodesic equations we write the metric in curvature coordinates as

$$ds^2 = -e^\lambda dR^2 - R^2(d\theta^2 + \sin^2\theta d\phi^2) + e^\mu dt^2.$$

A convenient parametrisation for the geodesic equations is obtained by setting

$$\frac{dx^1}{ds} = qe^{-\frac{\lambda}{2}}, \qquad \frac{dx^4}{ds} = \sqrt{1+q^2}e^{-\frac{\mu}{2}} \tag{4}$$

when we find, using the field equations, the form

$$\frac{dq_1}{ds} = -\frac{\sqrt{1+q_1^2}}{2}e^{\frac{\lambda}{2}}\left\{\frac{\sqrt{1+q_1^2}}{R}(1-e^{-\lambda}) + \kappa R\rho_2 q_2\left(q_2\sqrt{1+q_1^2}-q_1\sqrt{1+q_2^2}\right)\right.$$

$$\left. + \kappa R\rho_3 q_3\left(q_3\sqrt{1+q_1^2}-q_1\sqrt{1+q_3^2}\right)\right\} \tag{5}$$

Here subscripts refer to the three dust flows: 1 approaching the caustic in the 3-dust region, 2 leaving it and 3 crossing the caustic into the region. The terms in ρ_1 cancel; each dust congruence is not affected by its own local density, but only by the densities of the congruences it is moving through and by its own integrated Coulomb field. Moreover, as the caustic is approached, $q_1 - q_2 \to 0$, while $\rho_1 \to \infty$ and $\rho_2 \to \infty$; thus it is possible for the terms in the geodesic equation, which essentially involve products of these, to remain bounded. This is crucial for the whole analysis: it means that the surface N is C^{1-} but not C^2.

Moving to (b), the form of the caustic, one can show that the equation of the caustic is $R = x(t)$, where

$$\frac{d^2 x}{dt^2} = \frac{x'}{2\beta^2}\left(\frac{d\mu}{d\sigma} - \frac{d\lambda}{d\sigma}\right) - \frac{(1 - e^{-\lambda})}{2R\beta^2} - \frac{\kappa H e^{\lambda/2} x' \sqrt{2} K R}{\beta^2 e^{-\lambda/2}} - \frac{K^2}{\beta^2} + \frac{A_0 e^{-(\lambda/2 + \mu)}}{\beta^4}$$

where σ is proper time on the caustic and $\beta = dt/d\sigma$, x' is the proper time derivative of x, and

$$A_0 = \frac{1}{2}\kappa R \rho_3 q_3 e^{\frac{\lambda + \mu}{2}} \beta \left(\beta e^{\mu/2} q_3 - \sqrt{1 + q_3^2 x' e^{\lambda/2}}\right).$$

To show that this yields the form (1) asymptotically, we first perform a translation of coordinates so that $x = 0$ occurs at the first point of the caustic, while t is replaced by a coordinate y also equal to zero here. We then investigate the similarity transformation

$$y \to \alpha y, \quad x \to \alpha^{3/2} x, \quad q \to \alpha^{1/2} q, \quad \rho \to \alpha^{-1} \rho, \quad s \to \alpha s, \quad \sigma \to \alpha \sigma$$

when $\alpha \to 0$. It is found that asymptotically all the equations take the form given by the standard model, including the equation for the caustic just cited, with the exception of the geodesic equation which reduces asymptotically to

$$\frac{dq_1}{ds} = -\frac{\kappa R_0}{2}\{\rho_2 q_2 (q_2 - q_1) + \rho_3 (q_3 - q_1)\}.$$

This indicates that there is a non-trivial gravitational effect at the caustic which alters the parameters of the asymptotic form, without altering its functional form.

Finally, we turn to (c), the choice of coordinates satisfying the Geroch-Traschen conditions. Here the basis of the method is the use of Gaussian coordinates based on one of the caustics; that is, from each point p in the region construct a geodesic γ_p from p to a given caustic, intersecting the caustic orthogonally at a point q, then let x^1 be the length of γ_p and x^4 be the proper time along the caustic to q. The metric then has the form

$$ds^2 = (dx^1)^2 + e^\beta \left[(d\theta)^2 + \sin^2\theta d\phi^2\right] - e^\gamma (dx^4)^2.$$

Introduce $u = e^{-\gamma/2}\beta_{,4}$, $\gamma_1 = \gamma_{,1}$ and write the field equations (Synge 1960) as the system of 4 ordinary differential equations in these new variables:

$$u_{,1} + \frac{1}{2}(2\gamma_1 + \beta_1)u = R_{14}e^{-\gamma/2} = R_{\hat{1}\hat{4}}$$

$$\beta_{1,1} - \frac{1}{2}u^2 + \frac{3}{2}\beta_1^2 - 2e^{-\beta} = R_{\hat{4}\hat{4}} + 2R_{\hat{2}\hat{2}} + R_{\hat{1}\hat{1}}$$

$$\beta_{,1} - \beta_1 = 0$$

$$\gamma_{1,1} + \beta_1^2 + \frac{1}{2}\gamma_1^2 = 2R_{\hat{1}\hat{1}}$$

where hats denote tetrad components. Then the standard theory of existence for ordinary systems shows that the integrability of the right hand sides, which follows from the integrability of ρ, implies that these variables are bounded. But the Christoffel symbols can be expressed in terms of these variables, and so these too are bounded. The geometry of the region implies, in fact, that the bounds are uniform in the region. This then enables us to deform the Gaussian coordinates in order to patch the region into the Tolman-Bondi part.

5 CONCLUSIONS

During the 23 years that have elapsed since the formulation of the cosmic censorship hypothesis, considerable evidence has built up that, as far as classical relativity is concerned, there will be places in the universe, in communication with us, where the conventional C^{2-} structure of space-time breaks down. The work of Newman, in particular, limits the extent of this breakdown: a stable 'singularity' (in this sense) visible from an 'infinity' where there is no causal breakdown cannot have too large a value of the persistent curvature, for example.

When it comes, however, to singularities that are so strong (or so topologically tangled) that space-time itself must come to an end – where one is inclined to appeal for rescue to the deus ex machina of quantum theory – then the possibility remains open that cosmic censorship, in one or other of its forms, may still hold. The possible line of proof, however, could well not involve the conventional classification of strong curvature and weak focussing: it would be useful to know, for example, whether the shell focussing 'singularity' of Papapetrou and Hamoui (1969), through which an extension is possible, is of strong curvature type. (This would appear to be the case at first glance.)

So I will conclude with a few questions indicative of possible ways ahead:

  (a) Can one prove a rigorous existence theorem for the extension just described?
  (b) Can the same method be used with the central shell focussing singularity?

(Both the above probably require a more robust handling of the distributions, using Vickers' method (1992) so as to avoid the delicacy of achieving the Geroch Traschen conditions).

   (c) Is the (extendible) Papapetrou-Hamoui shell focussing singularity of strong curvature type? If so (or if the strong curvature central singularities are extendible), what criterion might replace strong curvature so as to prevent extendibility?

   (d) Is there some way of characterising singularities that are so bad that no amount of generalisation of the field equations could produce an extension?

In short, we now need to switch attention from the strong curvature condition, which has probably outlived its usefulness, to examine the real relations between singularities, horizons and extendability, working in the first place with exact and numerical solutions to gain the understanding of these relations that is at present lacking.

REFERENCES

Bondi, H (1947) Spherically symmetrical models in general relativity *Mon. Not. Roy. Astron. Soc.* **107** 410—425

Chrusciel P T, Isenberg J, Moncrief V (1990) Strong Cosmic Censorship in Polarized Gowdy Spacetimes *Class. Quantum Grav.* **7** 1671—1680

Clarke, C J S (1982) Space-times of low differentiability and singularities *J. Math. Anal. Appl.* **88** 270—305

Clarke, C J S & Krolak A (1985) Conditions for the occurrence of strong curvature singularities *J. Geom. Phys.* **24** 127—143

Clarke, C J S & O'Donnell, N (1992) Dynamical extension through a space–time singularity, to appear in *Rendiconti di seminario matematico, Università e Politecnico di Torino*

Geroch, R and Traschen, J (1987) *Phys. Rev. D* **36** 1017–31

Gorini V, Grillo G, Pelizza M (1989) Cosmic Censorship and Tolman-Bondi Space-times *Phys. Lett. A* **135** 154—158

Krolak A (1992) Strong Curvature Singularities and Causal Simplicity *J. Math. Phys.* **33** 701—704

Krolak A (1987) Towards the Proof of the Cosmic Censorship Hypothesis in Cosmological Space-Times *J. Math. Phys.* **28** 138—141

Lake K (1991) Naked Singularities in Gravitational Collapse which is not Self-Similar *Phys. Rev. D* **43** 1416—1417

Newman, R P A C (1984) Persistent curvature and cosmic censorship *Gen. Rel. Grav.* **16** 1177—1187

Newman, R P A C (1986) Strengths of naked singularities in Tolman-Bondi space-times *Class. Quantum Grav.* **3** 527—539

Newman, R P A C (1989a) The global structure of simple space times. *Comm. Math.*

Phys. **123** 17—52

Newman, R P A C (1989b) Black holes without singularities *Gen. Rel. Grav.* **21** 981—995

Ori A & Piran T (1988) Self-Similar Spherical Gravitational Collapse and the Cosmic Censorship Hypothesis *Gen. Rel. Grav.* **20** 7—13

Papapetrou, A and Hamoui, A (1967) *Ann. Inst. H. Poincaré* **9** 343–364

Penrose R (1969) Gravitational collapse: the role of general relativity *Riv. del Nuovo Cimento* **1** (Numero Spec.) 252—276

Penrose R (1974) Gravitational collapse, in *IAU symposium 64 on Gravitational Radiation and Gravitational Collapse* Reidel, Dordrecht, 82—91 *Ann. N Y Acad. Sci.* **224** 125

Rendal A (1992) On the choice of matter in general relativity, in *Approaches to numerical relativity* ed R A d'Inverno, CUP

Shapiro S L, Teukolsky S A (1991) Formation of Naked Singularities - the Violation of Cosmic Censorship *Phys. Rev. Lett.* **66** 994—997

Synge, J L, (1960) *Relativity, the general theory*, North-Holland.

Tipler F (1977) Singularities in Conformally flat space-times *Phys. Lett. A* **64** 8

Tolman R C (1934) Effect of inhomogeneity on cosmological models *Proc. Nat. Acad. Sci.* **20** 169—176

Vickers, J A (1992) to appear in *Rendiconti del seminario matematico, Università e Politecnico di Torino*

Wald R M & Iyer V (1991) Trapped Surfaces in the Schwarzschild Geometry and Cosmic Censorship *Phys. Rev. D* **44**

Yodzis, P, Mueller zum Hagen, H and Seifert, H-J (1973,4) *Comm. Math. Phys.* **34** 135 and **37** 29.

The Kerr Metric: A Gateway to the Roots of Gravity?

FERNANDO de FELICE

1 INTRODUCTION

Since my first interaction with the Kerr metric, early in 1967, when Dennis Sciama suggested to me that I work on it, I was fascinated by the *magic* of that solution to reduce whatever mathematical expression to simple terms, and by the richness of the information it provided. After nearly 25 years of intense investigation of the Kerr metric carried out by almost all the relativists around the world, new properties continue to be discussed and perhaps deep information about the very nature of gravity is still to be brought to light.

There are basic questions about gravity which, in my opinion, still need to be answered. Some (and perhaps the most obvious ones) are:

  i) - Why do the properties of a physical system, like energy and momentum, bend the background geometry?

  ii) - How are energy and momentum actually transferred to the background geometry, leading to a non zero curvature?

  iii) - To what extent does energy and momentum of the background geometry contribute to these same properties of a physical system?

Answering these types of question is what I mean by going to the roots of gravity. Evidently, central to this issue is the concept of energy in general, for which we require, at the classical level at least, the fulfillment of the energy conditions. These appear to be satisfied by all classically defined matter having a generic T_ν^μ. The energy and momentum contribution to gravity also manifests itself as adding specifications to the background geometry, which then carries memory of the source. If the memory of the source is a manifestation that some energy and momentum are stored in the geometrical curvature, then we expect the curvature itself to generate more gravity with a further share of energy and momentum with itself. Quoting Deser (1970), we can state that the full non-linear Einstein equations are an infinite series in the deviation $h_{\mu\nu}$ of the metric $g_{\mu\nu}$ from its Minkowskian value $\eta_{\mu\nu}$. As a result of the self-interaction of the gravitational field, which arises from the unique property of gravity of being itself gravitationally charged, we have a sort of energy diffusion in the background geometry, which appears to be a reason why the gravitational energy is non localizable.

There are several definitions of quasi-local energy of a gravitational field (Hawking, 1968; Geroch, 1973; Penrose, 1982; Katz, Lynden-Bell and Israel, 1988; even local definitions are in Witten, 1981; Horowitz and Strötminger, 1983) but it seems that we are still far from reaching an agreement about what is the correct formulation. Despite the uncertainties on how to describe the energy and momentum content of the gravitational field, there is no doubt that it exists and allows the background geometry to behave by itself as a physical entity with dynamical properties.

It would then be helpful to find evidence of gravitational effects which are due to this energy and momentum content. Here I claim that the Kerr metric provides some of this evidence. In what follows Greek indices run from 0 to 3 and units will be such that $c = 1 = G$; c is the velocity of light, G the gravitational constant.

2 THE SPHERICALLY SYMMETRIC CASE
Let us first recall, for the sake of comparison, the Schwarzschild vacuum solution. This is generated by a source which has energy, stress but no net momentum; the space-time geometry outside the source, namely where $T^\nu_\mu = 0$, carries an amount of energy which, as a memory of the spherical symmetry of the source, is isotropically distributed. A scalar which can be naturally selected as a good indicator of this isotropy, is the first Kretschmann invariant:

$$K = R^{\mu\nu\rho\sigma} R_{\mu\nu\rho\sigma} = \frac{48M^2}{r^6} \qquad (1)$$

where M is the total mass-energy of the source in geometrized units. Among the various definitions of gravitational energy, let us select in this case that given by Katz, Lynden-Bell and Israel (1988). According to them the fraction of the total energy which is stored in any space-like section $\Sigma - (\Sigma \cap W)$, W being the domain of definition of T^ν_μ, is given by:

$$\widetilde{M}(\bar{r} > \bar{r}_0) = M - \int_{\Sigma \cap W} \sqrt{-g} T^{\mu\nu}_{\cdot} \xi_\nu d\Sigma_\mu \qquad (2)$$

where ξ is the time Killing one–form, $T_{\cdot}$ is the energy-momentum tensor of a matter-energy configuration confined in a thin shell of isotropic coordinate radius $\bar{r}_0$ and

defined so that the external space-time is described by the Schwarzschild metric and
the internal space-time is flat. Furthermore $d\Sigma_\mu$ is the surface element and g is the
determinant of the metric.

Written in isotropic coordinates the Schwarzschild metric reads:

$$ds^2 = -e^{2\bar{\nu}}dt^2 + e^{2\bar{\mu}}[d\bar{r}^2 + \bar{r}^2(d\theta^2 + \sin^2\theta d\phi^2)] \tag{3}$$

where:

$$e^{2\bar{\nu}} = \left(1 - \frac{M}{2\bar{r}}\right)^2 \left(1 + \frac{M}{2\bar{r}}\right)^{-2} ; \qquad e^{2\bar{\mu}} = \left(1 + \frac{M}{2\bar{r}}\right)^4 ; \tag{4}$$

$\bar{r}$ is related to the radial curvature coordinate r as:

$$\bar{r} = \frac{1}{2}(r - M + \sqrt{r(r - 2M)}). \tag{5}$$

Following Katz et al. (1988), one finds for the energy (2):

$$\widetilde{M}(\bar{r} > \bar{r}_0) = \frac{1}{2}\frac{GM^2}{\bar{r}_0} \equiv \widetilde{M}_0. \tag{6}$$

We can give operational significance to $\widetilde{M}_0$ by relating it to a physically measurable
quantity. In fact, the frequency shift of a photon which is emitted radially at $\bar{r}_0$ and
detected by an observer at infinity, can be written in the following way:

$$1 + z \equiv \frac{\omega_e}{\omega_\infty} = \frac{M + \widetilde{M}}{M - \widetilde{M}} \tag{7}$$

or alternatively:

$$\frac{\widetilde{M}}{M} = \frac{z}{2 + z}, \tag{8}$$

hence, if we knew M, we would have a direct measurement of $\widetilde{M}_0$.

An intriguing consequence of (6) and (8) is that $\widetilde{M}_0 \to M$ as $\bar{r}_0 \to M/2$ (the horizon in isotropic coordinates), hence one is lead to conclude that the total energy is all outside the black-hole. While this result appears consistent with the expectation that energy extraction from a classical black-hole can only occur from outside the hole, the question of how this final state of energy distribution is dynamically justified is rather an open question. Nevertheless, the only conclusion, compatible with the above definition of quasi-local energy, that the total energy inside a static black-hole is zero[1] implies, perhaps, that we are facing some kind of mutual annihilation of positive and negative energy terms inside the hole, but there is no way to identify these terms. Whatever the correct interpretation of equation (8) is, the most important consequence of definition (2) can be stated as follows: *gravitational collapse to a black-hole, in spherically symmetric space-times, is accompanied by a complete transfer of the source energy into the background geometry outside the hole.*

Since this result arises strictly as a consequence of the assumed definition of quasi-local energy, I will not try to defend it on the basis of a general validity of that definition, this being still a matter of debate (Nachmad-Achar and Schutz, 1987), but I will only point out the existence of relativistic effects which, in a completely different context and independently of the definition of the gravitational energy, appear to be consistent with what is implied by the above statement.

3 THE KERR METRIC

Let us then consider the Kerr metric (Kerr, 1963; Kerr & Schild, 1967). In Boyer and Lindquist (1967) coordinates it reads:

$$ds^2 = -\left(1 - \frac{2Mr}{\Sigma}\right) dt^2 - \frac{4Mar\sin^2\theta}{\Sigma} dt d\phi + \frac{A}{\Sigma}\sin^2\theta d\phi^2 + \frac{\Sigma}{\Delta}dr^2 + \Sigma d\theta^2 \quad (9)$$

where

$$\Sigma = r^2 + a^2\cos^2\theta$$
$$\Delta = r^2 + a^2 - 2Mr \quad (10)$$
$$A = (r^2 + a^2)^2 - a^2\Delta\sin^2\theta.$$

The gravitational field contributes non-locally to the total space-time energy M and angular momentum Ma; this means that the field itself carries a fraction of the total energy and momentum *stored* in the background curvature. The metric source is a

---

[1] This is also true for a charged Reissner-Nordström black-hole.

ring singularity which may be hidden by an event horizon when $a < M$ or naked if $a > M$. Indeed to identify a source other than this is a long standing problem. We all agree that the Kerr metric describes the space-time of a rotating black-hole, but I will extend considerations to the naked singularity solution since in that case the whole space-time would be accessible to asymptotic observation.

The memory of the angular momentum of the source is manifested by the gravitational dragging of the inertial frames. This is a general relativistic effect which is described by the angular frequency:

$$\omega = \frac{2amr}{A}.$$ (11)

Its behaviour shows the anisotropy in θ which is expected in a rotating system; namely the maximum dragging is on the equatorial plane ($\theta = \pi/2$) while the minimum is on the axis of symmetry ($\theta = 0$). We also deduce the rather anomalous behaviour, along the axis of symmetry, reaches a maximum at some $r > 0$ and then decreases to zero on the $r = 0$-disk. In a pictorial way, the effects of dragging is assimilated to that of a rotating viscous fluid; the best way to show this, is to study the properties of spatially circular orbits. These are characterized by a four-vector:

$$\ell^\alpha = e^\psi(k^\alpha + \Omega m^\alpha)$$ (12)

where k and m are respectively the time and axial Killing vectors of the metric, Ω is the angular frequency of revolution as it would be measured at flat infinity and e^ψ is the red-shift factor which reads:

$$
\begin{aligned}
e^\psi &= [-k^\alpha k_\alpha - 2\Omega k^\alpha m_\alpha - \Omega^2 m^\alpha m_\alpha]^{-\frac{1}{2}} \\
&= \left[1 - \frac{2Mr}{\Sigma}(1 - a\Omega \sin^2\theta)^2 - (r^2 + a^2)\Omega^2 \sin^2\theta\right]^{-\frac{1}{2}}
\end{aligned}
$$ (13)

The effects of dragging on (13) are different whether $a\Omega > 0$ or < 0. Condition $a\Omega > 0$ implies an effectively weaker source since the mass which enters the red-shift factor is reduced by $(1 - a\Omega \sin^2\theta)^2$. As expected, the maximum reduction takes place in the equatorial plane.

The observer's four-acceleration is given by the one-form:

$$\dot{l}_\alpha = -\nabla_\alpha \psi \qquad (14)$$

where the covariant derivative is with respect to metric (9). The thrust per unit mass needed to keep the orbit spatially circular is given by the chronometric invariant:

$$\tilde{f} = |\nabla_\alpha \psi \nabla^\alpha \psi|^{\frac{1}{2}} . \qquad (15)$$

As is well known, the principle of equivalence forbids one to distinguish between *true* gravity and inertial forces on the basis of measurements which are insensitive to the background curvature. This is certainly the case for the thrust whose expression (15) neatly illustrates the unavoidable ambiguity of any such splitting. Nevertheless, whenever required for a better conceptual understanding, it may be useful to have a criterion for identifying a gravitational component of the thrust as opposed to an inertial one; in our case the latter would be only centrifugal. From the point of view of the rotating observer, both gravitational and centrifugal components of the thrust must be rescaled, with respect to the values they have in the non relativistic limit, by the red-shift factor e^ψ which takes into account the actual rate of her or his proper-time. However since a force per unit mass has physically the dimensions of a length over the square of a time, the rotating observer will expect the red-shift factor to be counted twice (de Felice, 1991). Moreover, besides the red-shift factor, both components should equally suffer whatever geometric correction arises from the common background curvature; thus, once factorized, we shall identify as centrifugal force per unit mass that one of the remaining terms which does not depend on the source mass M.[2]

Written explicitly, the four-acceleration (14) reads:

$$\dot{l}_\alpha = e^{2\psi}\{ \left[\frac{M}{\Sigma}\left(1 - \frac{2a^2 \cos^2\theta}{\Sigma} \right)(1 - a\Omega \sin^2\theta)^2 - r\Omega^2 \sin^2\theta \right] \delta_r^\alpha$$
$$+ \frac{\sin 2\theta}{2\Sigma^2} \left[2Mra(1 - a\Omega \sin^2\theta)(2\Omega\Sigma - a(1 - a\Omega \sin^2\theta)) - \Omega^2\Sigma^2(r^2 + a^2) \right] \delta_\theta^\alpha \}.$$
$$(16)$$

Both components are clearly the result of a balance between two terms which, accord-

---

[2] This criterion is controversial. It differs from the one adopted by Abramowicz (1990) according to whom a gravitational force should be identified as being altogether independent of velocity.

ing to our criterion, are identified as gravitational and centrifugal. This is particularly evident in the equatorial plane where (16) reduces to:

$$\dot{\imath}_r = e^{2\psi} \left[\frac{M}{r^2}(1 - a\Omega)^2 - \Omega^2 r \right]. \tag{16}'$$

The radial component of the four-acceleration (16) is responsible for keeping the observer at a fixed r while the θ-component is a less transparent constraint needed to preserve the orbit on a surface with constant θ. Consistently with the asymptotic behaviour of (16), one is justified in interpreting as gravitational force per unit mass the term of the radial component of (16) which contains the corrected mass :

$$M' = M \left(\frac{1 - 2a^2 \cos^2\theta}{\Sigma} \right) (1 - a\Omega \sin^2\theta)^2 \qquad a\Omega < 1. \tag{17}$$

There are two correcting factors, both depending on the radial and the latitudinal coordinates. The first, namely:

$$f_1 \equiv 1 - \frac{2a^2 \cos^2\theta}{\Sigma}$$

does not depend on the frame and arises specifically from particle dynamics, (de Felice and Bradley, 1988). The second factor, namely:

$$f_2 \equiv (1 - a\Omega \sin^2\theta)^2$$

does depend on the frame; it will be interpreted here by invoking physical intuition. One basic point is that momentum contributes to gravity as much as energy, hence a rotating observer experiences in her or his rest frame, a partial suppression of the angular momentum of the source. This implies an effective suppression of the corresponding contribution to the gravitational field so justifying the mass reduction carried by f_2.

It is a well established property of the Kerr metric ring-singularity to be reachable only by trajectories which are either stably in the equatorial plane or tend to be equatorial in the limit on approaching the singularity (Carter, 1968; Israel, 1978). Confining our attention to strictly equatorial motion, we know (de Felice, 1968; de Felice and Usseglio-Tomasset, 1992) that corevolving circular orbits can only approach the singularity (we are in the naked singularity case) if $\Omega \to \omega \to a^{-1}$. In fact, from (9), we have that for any time-like (spatially) circular orbit, the angular velocity with respect to flat infinity $\Omega \equiv d\phi/dt$, satisfies the condition:

$$\omega - h < \Omega < \omega + h$$

where

$$h = \left(\frac{r^2}{A}\right)^2 \Delta.$$

Since $h \to 0$ when $r \to 0$, then $\Omega \to \omega$ as stated. When the observer approaches the singularity in the equatorial plane then she or he is forced to rotate with respect to infinity with the same angular velocity $1/a$ as the source, hence one expects that all the angular momentum of the source is being suppressed. The effect of this on the mass reduction is $M(1 - a\omega)^2 \to 0$. From (11) we find that $M(1 - a\omega)^2 \to O(r^2)$ as $r \to 0$ hence in the equatorial plane the divergence of l_r as $r \to 0$ is only due to the divergence of the red-shift factor, this being a symptom that the circular orbit approaches a null trajectory making the very concept of a measurement meaningless. The above considerations seem to indicate that the spinning singularity contributes to gravity *only* with its own angular momentum.

Before turning to the factor f_1, let us consider how the Kretschmann invariant behaves in the Kerr metric. From (9) and some algebra, we have:

$$K(r,\theta,a) = \frac{48M^2}{\Sigma^6}(r^2 - a^2\cos^2\theta)(a^2\cos^2\theta + r^2 + 4ar\cos\theta)(a^2\cos^2\theta + r^2 - 4ar\cos\theta).$$
$$(18)$$

The invariant K remains practically zero everywhere except nearby the ring singularity where it blows up in rapid oscillations about zero (see de Felice and Bradley, 1988). It is not clear what physical significance one should attribute to K and to its oscillations, but if we give to its square amplitude the role of tracing the amount of energy being stored in the geometry, then we see, consistently with the behaviour of dragging, that the energy is most concentrated on and nearby the equatorial plane as compared to nearby the axis of symmetry.

We can now analyze the factor f_1. We have already met it in the radial component of the acceleration of a circular orbit, but more generally it appears in the analysis of vortical geodesics (de Felice and Calvani, 1972); with respect to a family of stationary observers, the three-force tangential to the hyperboloids $\theta = $const. which contain the geodesics, is given by (de Felice et al. , 1975; de Felice and Bradley, 1988):

$$\tilde{f}_t = -\frac{\mu_0}{(1-v^2)^{\frac{1}{2}}} \frac{\Delta^{\frac{1}{2}} M}{\Sigma^{\frac{1}{2}}(\Sigma - 2Mr)} \left(1 - \frac{2a^2 \cos^2\theta}{\Sigma}\right) \left[1 - \frac{G-1}{G^2} \frac{4a^2 \sin^2\theta}{\Delta\Sigma}(\Sigma - 2Mr)\right]^{\frac{1}{2}}$$

$$(19)$$

where

$$G = \left(\frac{\Gamma+1}{\Gamma}\right)^{\frac{1}{2}} \quad ; \quad \Gamma = E^2 - 1$$

E being the total energy of the particle and μ_0 its rest mass. It is clear that $\tilde{f}_t$ reverses sign on the surface $r = a\cos\theta$, turning from attractive when $r > a\cos\theta$ to repulsive when $r < a\cos\theta$, independently of the particle energy. Hence the analysis of the geodetic motion leads to the following rotational paradox:
a test particle freely falling towards $r = 0$ feels a maximum repulsive effect on the axis and none in the equatorial plane.

This is opposite to what one expects from the most obvious explanation based on centrifugal effects. In fact a particle, freely falling along the axis of symmetry, once below the point with coordinate $r = a$, behaves as if it were not acted upon by a gravitational source (energy and momentum) concentrated in the ring singularity but rather by a source all smeared in the outside space with a latitudinal anisotropy inferred by the behaviour of the Kretschmann invariant, namely most concentrated on and nearby the equatorial plane and most diluted on and nearby the axis (de Felice, 1975). Evidently the only identifiable gravitational source of this type is the energy and momentum of the gravitational field itself.

4 CONCLUSION

One of the most important implications of General Relativity is that gravity, carrying energy and momentum, is itself gravitationally charged. This probably gives rise to relativistic effects which however are not easily discernible, mainly because of the difficulty of giving a unique and unambiguous definition of the gravitational energy and momentum. Nevertheless, in the limit of a curvature singularity, these effects may become strong enough to be identified and interpreted in physical terms. The

Kerr metric provides some of these effects which appear to be the result of a maximal storage of energy and momentum in the background geometry.

ACKNOWLEDGMENTS
Financial support by 40% Research Project Equazioni di Evoluzione e loro Applicazioni Fisico Matematiche and by the Italian Space Agency is gratefully acknowledged.

REFERENCES
Abramowicz M.A. 1990 *Mon. Not. R. astr. Soc.* **245**, 733
Boyer R.H. and Lindquist R.W. 1967 *J. Math. Phys.* **8**, 265
Carter B. 1968 *Phys. Rev.* **179**, 1559
de Felice F. 1968 *Il Nuovo Cimento* **57B**, 351
de Felice F. and Calvani M. 1972 *Il Nuovo Cimento* **10B**, 447
de Felice F. 1975 *Astron. & Astrophys.* **45**, 65
de Felice F., Calvani M. and Nobili L. 1975 *Il Nuovo Cimento* **26B**, 1
de Felice F. and Bradley M. 1988 *Class. Quantum Grav.* **5**, 1577
de Felice F. 1991 *Mon. Not. R. astr. Soc.* **252**, 197
de Felice F. and Usseglio-Tomasset S. 1991 *Class. Quantum Grav.* **8**, 1871
Deser S. 1970 *Gen. Rel. Grav.* **1**, 9
Geroch R. 1973 *Ann. NY Acad. Sci.* **224**, 108
Hawking S.W. 1968 *J. Math. Phys.* **9**, 598
Horowitz G.T. and Strominger A. 1983 *Phys. Rev.* **27**, 2793
Katz J., Lynden-Bell D. and Israel W. 1988 *Class. Quantum Grav.* **5**, 971
Kerr R.P. 1963 *Phys. Rev. Lett.* **11**, 237
Kerr R. P. and Schild A. 1967 *A new class of vacuum solutions of the Einstein field equations*; *Atti del Convegno su General Relativity* Ed. G. Barbera Firenze
Israel W. 1978 *Phys. Rev.* **2**, 641
Nachmad-Achar E. and Schutz B.1987 Gen. Rel. & Grav. **19**, 655
Penrose R. 1982 *Proc. R. Soc.* A**381**, 53
Witten E. 1981 *Comm. Math. Phys.* **80**, 381

Galactic Astronomy Since 1950

JAMES J. BINNEY

Galaxies are the building blocks of the Universe, and most of what we know about them has been discovered since Dennis Sciama became a research student. In the space available to me it is not possible to cover even in outline all significant developments during this period. So I have tried to concentrate on what seem to me to be the most important themes. My choice must surely be heavily influenced by personal taste and experience; I hope only that my prejudices are not too glaringly evident.

1 THE STRUCTURE OF THE MILKY WAY

Galactic astronomy in the 1950s was dominated by the discovery (Ewen & Purcell, 1951) of the 21 cm line predicted by H. C. van der Hulst in 1944. This made it possible for the first time to study the large-scale kinematics of the Milky Way. For the most part the 21 cm observations confirmed the picture of a disk in differential rotation developed by Oort more than twenty years before. However, there were surprises – most notably the discovery that the disk is warped rather than being perfectly flat (Burke, 1957; Kerr, 1957).

Extinction of stars by dust had first betrayed the existence of the interstellar medium (Trumpler 1930). From 1940 the ISM had been extensively studied by means of the narrow absorption lines, particularly those of sodium and calcium, that it introduces into the spectra of stars. However, in 1950 the overall picture of what the interstellar medium looked like, and how it functioned, was very hazy. In 1950 Spitzer and Savedoff concluded that atomic hydrogen forms two phases: clouds with densities $n_H \simeq 1\,\mathrm{cm}^{-3}$ and temperatures $T \simeq 60\,\mathrm{K}$, and the intercloud medium with $n \simeq 10^{-2}\,\mathrm{cm}^{-3}$ and $T \simeq 10^4\,\mathrm{K}$. Supernovae were known to blow hot cavities in the ISM, but it was only after the launch of the Copernicus satellite in 1973 that it became clear that much, possibly most, of interstellar space is filled with gas at temperatures in excess of $10^6\,\mathrm{K}$. With the discovery of this hot phase, the interstellar medium seemed more of a dynamic system, in which material was being constantly heated, cooled and violently stirred, than was the case in the picture of the 1950s and 60s.

During the 1970s it became possible to trace the predominantly molecular component of the interstellar medium through mm line emission of trace molecules, especially CO (Wilson Jefferts Penzias 1971). The launch of the Copernicus satellite in 1972, which could measure H_2 column densities directly from ultraviolet rotation-vibration lines, facilitated the calibration of CO data in terms of total molecular mass (Jenkins & Savage 1974). It was found that although $^{12}C^{16}O$ emission is usually optically thick, reliable estimates of molecular mass can be inferred by combining measurements of $^{12}C^{16}O$ with measurements of one of the rarer species $^{13}C^{16}O$ or $^{12}C^{18}O$. It turned out that roughly half the Galaxy's interstellar gas is in molecular rather than atomic form, the molecular gas being concentrated into a ring roughly 4 kpc in radius.

Progress in using radio line observations to refine our knowledge of the large-scale structure of the Disk has been disappointing. In the 1960s hopes were entertained that the Milky Way's spiral arms could be traced in the 21 cm data, but the work of Burton (1970) and others demonstrated that deviations from smooth galactic rotation were large enough to make distances to emitting clouds too uncertain to be useful in constructing detailed density plots. More recently, several studies have argued that the Galaxy is barred. A report that the entire Disk interior to the Sun is mildly elliptical (Blitz & Spergel 1991a) is highly controversial (e.g. Kuijken & Tremaine 1991), but a consensus does seem to be emerging that the inner Galaxy is dominated with a bar that is roughly 2 kpc long (e.g. Blitz et al. 1992).

From early on it was clear that non-negligible quantities of gas are seen at a few degrees from the galactic centre with velocities which do not occur in a model in which gas moves on circular orbits – these are the 'forbidden' velocities (Sinha 1979). One possible explanation for the occurrence of forbidden velocities is that gas is moving radially outwards from the centre, having been expelled by some energetic event in the nucleus. An alternative explanation was put forward in 1982, when Liszt & Burton (1980) successfully modelled the available 21 cm and CO data by material moving on elliptical orbits. In 1991 Binney et al. put this kinematical model on a dynamical basis by arguing that the CO data enable one to identify the pattern speed and orientation of the barred gravitational potential responsible for the elliptical Liszt & Burton's streamlines. Infrared data from balloon flights (Blitz & Spergel 1991b) and the COBE satellite (Hauser et al. 1990) provide direct evidence that the flattened bulge of the Milky Way is a barred peanut-shaped object.

Despite these recent successes, the inner couple of kiloparsecs of the Milky way are not at all well understood. For example, interior to 2 deg there is twice as much molecular gas at positive longitudes as at negative and nobody seems to know why. Also, from Sinha's work it has been thought that the gas at the centre moves in a

plane that is tipped with respect to the outer Plane by as much as 20 deg. This is a
hard phenomenon to explain. One possibility is that it is a symptom of the continued
infall of predominantly dark matter onto the Galaxy (Ostriker & Binney 1989).

Until the end of the 1970s attempts to deduce the large-scale structure of the Milky
Way followed the classical path that went back to the work of Kapteyn and others
at the turn of the century. This approach owes nothing to studies of other galaxies,
because Kapteyn and his contemporaries did not recognize that the Milky Way was
actually one of the 'spiral nebulae'. So they sought to deduce the shape and extent
of the Milky Way from star counts by inverting the integral equation which connects
the three-dimensional density distribution and star counts. By the late 1970s photometric studies of external galaxies to be described below had identified a few basic
'components' out of which all galaxies seem to be constructed, and it became natural
to fit the star counts to a model that consisted of a superposition of such components. This led to much more credible results because the very imperfect star-count
data (which themselves were mostly of alarming antiquity) were then required only
to yield a handful of fitting parameters rather than several full functions of three
independent variables; namely the space densities of stars of different spectral types.

Since lines of sight that lie close to the Plane are heavily obscured, it turns out that
some key parameters of galaxy models are not well determined by star counts in the
optical. Hence this programme of determining the Milky Way's vital statistics is
not likely to be satisfactorily completed until more extensive near-infrared data are
available. But the work of de Vaucouleurs & Pence (1978), Bahcall & Soneira (1980),
Gilmore & Reid (1983) and Kent (1991) considerably advanced our understanding of
the Milky Way. One important controversy generated by this work was whether the
Milky Way has a 'thick disk' in addition to the familiar thin one.

2 LARGE-SCALE STRUCTURE OF GALAXIES

The early 21 cm studies of external galaxies were severely hampered by poor spatial
resolution. So some of the most important work was that of Wannier & Wrixon
(1972) and Mathewson, Schwarz & Murray (1977) on the Magellanic clouds, which
subtend several degrees on the southern sky. This showed that in addition to a disk
of neutral hydrogen (HI) the Clouds are associated with a vast streamer of HI that
stretches half way around the sky: this has ever since been interpreted as evidence
that the Clouds are being tidally stripped by the Milky Way.

Until the completion of the large aperture-synthesis telescopes at Cambridge and
Westerbork around 1970, our knowledge of the rotation curves of external galaxies
derived mostly from measurements of emission lines by HII regions (e.g. Burbidge &

Burbidge 1975). These showed that the circular speed v_c generally rises quite steeply near the centre, and then flattens off. It seemed reasonable to assume that v_c falls off as $v_c \propto r^{-1/2}$ beyond the radius of the last observed point, but this fall-off was not actually detected. Schwarzschild (1954) pointed out that it was surprising that the anticipated fall-off was not detected in our neighbour, M31.

In the 1970s the new aperture-synthesis telescopes, especially that at Westerbork, revolutionized our understanding of spiral galaxies by probing orbiting gas at much greater distances than the old Hα studies had been able to do. Two key findings were:
(i) That warps like that of the Milky Way are extremely common – Bosma (1978) found that 8 of a sample of 25 galaxies are seriously warped beyond the optically conspicuous portion of the disk.
(ii) That the circular speed remained either rising or flat to the furthest radii probed, even though these were typically a factor two larger than those at which a roughly Keplerian decline should have set in.

The paper of van Albada *et al.* (1985) rammed this last point home with great thoroughness in the specific case of NGC 3198. In this galaxy the velocity field measured in the 21 cm line is precisely that expected for a flat disk in differential, circular rotation, and v_c is remarkably constant from $r = 6h_{100}$ kpc to $24h_{100}$ kpc. If light traced mass, the rotation curve should have peaked near $r = 4h_{100}$ kpc.

Prior to 1975, very few reliable data were available on stellar motions in elliptical galaxies or in spiral bulges: a few estimates of central velocity dispersions had been made, but these (e.g. Minkowski 1961) tended to be unreliable, and nothing was known about the motions of stars away from the nucleus. With the advent of reasonably linear detectors it became possible to obtain reliable velocity dispersions and mean-streaming velocities away from the nucleus. As the study of NGC 4697 by Bertola & Capaccioli (1975) and of twelve further galaxies by Illingworth (1977) first revealed, the streaming velocities in most giant elliptical galaxies are smaller than the straightforward dynamical models of Prendergast & Tomer (1970) and of Wilson (1975) had predicted. By contrast, the bulges of disk galaxies (Kormendy & Illingworth 1982) and less luminous elliptical galaxies have rotation velocities roughly in line with the predictions of simple models. Rather recently Bender *et al.* (1991) have shown that the low–surface–brightness ellipticals, like the most luminous ellipticals, rotate anomalously slowly.

In 1982 Efstathiou Ellis & Carter discovered that the central regions of NGC 5813 rotates rapidly, while the main body of this elliptical galaxy has a higher velocity dispersion and slow rotation. This discovery did not have much impact at the time, although Kormendy (1984) argued that the nucleus might be the remnant of a small compact galaxy that had spiralled to the centre of a larger one. Then in 1988 Franx & Illingworth found that the core of IC 1459 counter-rotates with respect to the galaxy's main body. Soon some similar new phenomenon was being reported every few months, culminating in the amazing discovery by Rubin et al. (1992) that the disk of the S0 galaxy NGC 4550 consists of two, roughly equally massive, co-spatial disks that rotate in opposite senses. It now seems that roughly a third of all elliptical galaxies have 'kinematically detached' cores. Recent work on metallicity gradients (Franx & Illingworth 1990, Davies et al. 1992) argues against Kormendy's proposal that these cores form by stellar-dynamical merging, since the nuclei turn out to be usually more metal rich than their surroundings, the reverse of what Kormendy's model suggests. One possibility is that kinematically detached cores are formed from residual gas that sinks rapidly to the system's centre when two early-type galaxies of comparable luminosity merge (Hernquist & Barnes 1991).

One of the most important discoveries of the Einstein x-ray satellite launched in 1978 was that giant elliptical galaxies are usually extended x-ray sources (Forman et al. 1979; Forman Jones & Tucker 1985). The observed spectra and the correlation between the detected x-ray and optical luminosities suggest that the x-rays are thermal radiation from gas at several million degrees Kelvin. Consequently, it now appears that elliptical galaxies have interstellar media almost as massive as those of spirals. The essential difference between the media is that those of ellipticals are at their system's virial temperature while those of spirals are much cooler and hence are supported against gravity by centrifugal force rather than by pressure gradients.

In fact, Larson (1974) had predicted that gas in ellipticals should be heated roughly to the virial temperature by supernovae – see below. But in Larson's models the gas flowed freely outwards and had much too small a central density to be detected by the Einstein observatory. It seems that Larson just missed a spectacular prediction because the central gas density depends rather sensitively on the circular speed and pressure at large radii, and Larson had no reliable way of choosing these boundary conditions. In particular, Larson was working before it was widely suspected that galaxies are enveloped in dark halos, and thus he under-estimated the circular speeds at large radii.

3 GALAXY PHOTOMETRY

Until the end of the 1960s, galaxy luminosity profiles were usually measured by aperture photometry. By 1950 Hubble and de Vaucouleurs had shown that the brightness profiles of elliptical galaxies fitted the laws that bear their names remarkably closely, and in 1959 de Vaucouleurs showed that the brightness profiles of disks follow the exponential law first proposed by Patterson (1940). The exponential nature of disks was established as standard lore by Freeman in 1970.

The installation of plate digitizers at several observatories in the mid 1970s led to a considerable increase in the number of galaxies with reliable photometric measurements, and for the first time made it possible to quantify the azimuthal brightness distributions of galaxies. Notable early studies include those of elliptical galaxies by King (1978) and Carter (1978), of spirals by Schweizer (1976), and of lenticular galaxies by Tsikoudi (1979).

Considerable interest focused on the brightness profiles of luminous elliptical galaxies (Carter 1978, Kormendy 1984). At large radii the profiles were generally found to decline more steeply with radius r than any power-law in r such as the asymptotic form $I \propto r^{-2}$ of the Hubble law. Thus the luminosities, $L = 2\pi \int I(r)r\,dr$, of most galaxies are well-defined rather than being formally divergent as the Hubble law predicts. However, a few galaxies such as the central galaxy of Abel 1413, were found to disappear into the noise from the night sky as power laws $I \propto r^{-\alpha}$ with $\alpha < 2$.

The nuclear profiles of elliptical galaxies gave rise to some controversy. King (1978) believed that he had resolved the essentially isothermal cores of most of the galaxies in his sample. However, Schweizer (1979) argued that if King's galaxies had profiles with singular central densities (as de Vaucouleurs' $r^{1/4}$ law predicts), they would *appear* to have isothermal cores of about the observed size when measured with King's resolution. Subsequent observations at higher resolution using linear detectors such a CCDs (Lauer 1985, 1989) tended to bear out Schweizer's claim that at least a large fraction of elliptical galaxies have unresolved central structures. On the other hand, such a structure often sits atop an approximate plateau in the brightness profile that resembles the core of an isothermal sphere.

Disk galaxies were generally assumed to comprise two or more essentially independent 'components', a disk, which, following Freeman's (1970) paper, was assumed to have an exponential brightness distribution, a bulge, which was frequently assumed to obey de Vaucouleurs' $r^{1/4}$ law, and perhaps additional components such as a bar, a lens, one or more rings. The essential assumption here is that a disk galaxy can be considered to be a small elliptical galaxy to which a disk has been added. Measurements of the velocity dispersions and of bulges tended to support this conjecture (Kormendy & Illingworth 1982). But the minor-axis profiles of galaxies such as NGC 4565 (Jensen & Thuan 1982) show that this picture is over-simplified; at least some bulges, especially those with marked peanut morphology, do not resemble small elliptical galaxies.

An important discovery was that the isophotes of elliptical galaxies are not nested, coaxial ellipses as had hitherto been assumed. In fact the major axes of successively larger isophotes are often not aligned because isophotes tend to be 'twisted' with respect to one another. This phenomenon would be a natural consequence of the galaxies not being axisymmetric objects, as had hitherto been assumed by almost everyone except Contopoulos (1954), but are triaxial bodies. This conjecture rapidly became standard lore because it fitted nicely with the anomalously low rotation velocities of giant ellipticals described above and with dynamical developments discussed below.

A similar twist of the isophotes of M31's bulge with respect to the disks's isophotes had been suspected by Baade and was confirmed by the Stratoscope II measurements of Light *et al.* (1974). This suggested that the bulges of disk galaxies are also triaxial (Stark 1977). Analysis of the velocity fields of the disks of spirals suggested that disks are also sometimes elliptical in shape near their centres (Bosma 1978).

Binney & de Vaucouleurs (1981) concluded that the distribution of the apparent ellipticities of elliptical galaxies in the *Second Reference Catalogue of Galaxies* could be satisfactorily interpreted in terms of either axisymmetric or triaxial galaxies. However, there were significantly fewer round disk galaxies in the catalogue than would be expected if all disk galaxies were axisymmetric. This result was considerably extended by Lambas *et al.* (1992), who showed that in the A.P.M. survey there are too few round galaxies of every morphological class for galaxies of any class to be wholly axisymmetric. Moreover, if galaxies are triaxial, one expects to detect rotation velocities along some apparent minor axes. By the mid-1980s a few such cases of minor-axis rotation were known and suggested that elliptical galaxies are maximally triaxial (Binney 1985). By 1990 this had become a firm conclusion (Franx *et al.* 1991).

Faber (1973) showed that the colours and luminosities L of elliptical galaxies are correlated in the sense that more luminous galaxies are redder. Moreover, central velocity dispersion σ is correlated with luminosity in that more luminous galaxies have larger velocity dispersions. It follows that colour is correlated with velocity dispersion and in 1987 Dressler *et al.* (1987b) showed that colours are also correlated with the difference between the actual velocity dispersion of a galaxy and that predicted from its luminosity and the mean $\sigma - L$ relation. In effect, elliptical galaxies lie on something approximating a plane in colour–σ–L space. This plane came to be called the 'fundamental plane'.

This discovery excited considerable interest because (i) it seemed likely to prove an important clue to understanding how elliptical galaxies formed, and (ii) an accurate relation between two distance-independent quantities such as velocity dispersion and colour, and luminosity, which can normally be inferred from apparent magnitude only if one knows the galaxy's distance, would provide a powerful distance indicator with which to determine the Hubble constant and the large-scale streaming velocities. Indeed, in the form of the so-called $D_n - \sigma$ relation, the fundamental plane has now been extensively used to map large-scale motions (Dressler et al. 1987a).

On the other hand, fifteen years after its discovery, it still is not clear precisely what the fundamental plane should be telling us about galaxy formation. Much more is now known about the variation of metallicity (which is the underlying cause of colour variations), and velocity dispersion *within* galaxies. This additional knowledge confuses the original issues and makes it necessary to think in terms of more sophisticated models than seemed appropriate fifteen years ago. But it remains likely that the final picture will be similar to that introduced by Larson (1974) twenty years ago. In this metallicity is determined by the onset of a galactic wind. The latter is controlled by the stellar density and by the depth of the galactic potential well, which are dynamically related to L and σ.

4 GALACTIC NUCLEI

Radio astronomy really took off after the second world war as technology developed for military purposes was recycled into astronomy. For about fifteen years from 1950 radio telescopes provided our deepest probe of the Universe. That this was so was not immediately obvious – the interpretation of radio surveys, such as the xC sequence of Cambridge surveys, was fraught with uncertainties generated by source confusion. But there was no doubt that the radio sky is markedly different from the optical sky, and gradually it emerged from the source counts that most faint radio sources are synchrotron sources at cosmological distances. As the positions of these objects improved, and more of them were optically identified, it emerged that the most intrinsically luminous sources are 'radio galaxies' – giant elliptical galaxies with anomalously high radio luminosities.

By the early 1970s the new aperture-synthesis telescopes were resolving radio galaxies into a compact nucleus and vast lobes that can be 100 kpc or more from the nucleus. Theoretical arguments combined with the morphology of the objects at different wavelengths support the conjecture that the nucleus causes the lobes to shine by blasting them with jets of relativistic plasma.

This model was particularly attractive because in 1963 Hazard et al., Schmidt and Greenstein & Matthews had discovered that certain blue stellar sources are actually at cosmological distances and thus fabulously luminous – the first such 'quasar', 3C 273, has a redshift $z = 0.158c$ and luminosity in excess of $10^{12}\, L_\odot$. Thus, by the mid 1960s it was clear from optical work that the Universe contains stellar sources with luminosities comparable to those of entire galaxies. This made it plausible that the prodigious powers associated with the most luminous radio galaxies could have emerged from mere galactic nuclei.

As linear detectors came into more widespread use in the early 1980s the 'fuzz' associated with the emission of the host galaxy was detected around more and more quasars (Wyckoff *et al.* 1981; Smith *et al.* 1986), thus confirming the conjecture that radio galaxies and quasars are hyperluminous galactic nuclei.

The next question was clearly 'what powers these nuclei?' Most of their luminosities derive ultimately from relativistic plasma, and the energy associated with such plasma in the most powerful sources is equivalent to the rest-mass energy of millions of solar masses. These facts suggest that the central power-house is a relativistic object or objects of some type. Are their luminosities produced by the combined efforts of hundreds or thousands of conventional stellar sources such as neutron stars, or is the luminosity of a quasar produced by a single super-massive object? An obvious discriminant between these two classes of theory is provided by the sources' variability; if a quasar is powered by many smaller sources, its luminosity cannot easily vary by a factor of two in a short period, while no such restriction applies in the case of a quasar powered by a single supermassive object. Since the luminosities of some quasars do vary by large factors in a few years, many quasars must be powered by supermassive objects rather than by clusters of less luminous sources.

By simply summing the energy output of all the observed quasars and assuming a plausible efficiency ($\lesssim 10\%$) for the conversion of rest-mass energy into radiation, one concludes (Soltan 1982) that roughly one in three galaxies more luminous than the characteristic galaxy luminosity $L_* = 1 \times 10^{10} h_{100}^{-2}\, L_\odot$, must contain a dead quasar with mass $M \approx 10^8\, M_\odot$. From the late 1970s on an important goal of galactic astronomy has been the detection of such dead quasars.

Caltech led the way with two important studies (Young *et al.* 1978, Sargent *et al.* 1978) of the nucleus of M87, the galaxy which sits at the centre of the Virgo cluster. M87's nucleus is enveloped in a great cloud of x-ray emitting gas at a temperature of $\sim 10^7$ K and shoots out a small blue jet of synchrotron-emitting plasma. So it is a prime candidate for the location of a dead or quiescent quasar. The Caltech group used a then new CCD detector to measure M87's luminosity and velocity-dispersion profiles and announced that they had detected a black hole with mass $\sim 5 \times 10^9\, M_\odot$.

This dramatic claim naturally stimulated a great deal of activity, and M87's nucleus has been intensively studied since, as have the nuclei of nearer, less luminous galaxies such as M31, M32 and the Sombrero galaxy, NGC 4594. It now seems unlikely (Dressler & Richstone, 1990) that M87 has a black hole as massive as that originally claimed.

Given that want of spatial resolution is a limiting factor in attempts to detect a black hole at the centre of M87, it is natural to look carefully at the nuclei of less distant galaxies – systems such as the Andromeda nebula, M31, its dwarf elliptical companion M32 and the Sombrero galaxy, NGC 4594, would be expected to sport smaller black holes than M87, but they may nonetheless be easier to detect because of their greater proximity (e.g. Merritt, 1988; Kormendy & Richstone 1992).

The core of M87 is fairly spherical and any rotational velocities are too small to be dynamically significant. This circumstance greatly simplifies modeling work by comparison with that necessary to interpret observations of M31, M32 and NGC 4594, all of which show non-negligible nuclear rotation. Even in the case of M87 our understanding of the relevant theory is by no means complete and much more remains to be done on more rapidly rotating galaxies.

5 POPULATIONS AND COMPONENTS

Baade introduced the concept of stellar populations in 1944 as a result of his work on M31 and its companions. Through the 1950s, the application of this concept to the Milky Way was extensively developed (O'Connell 1958). The key result was the strong correlation between the spectroscopic and kinematic properties of stars. This correlation is apparent in globular clusters, which can be divided into metal-poor halo clusters, which form a rather spherical, very slowly rotating body, and metal-rich 'disk' clusters, which are more strongly concentrated to the Centre and the Plane and have significantly larger rotation velocities. The correlation is also apparent in RR-Lyrae stars (Plaut 1965) and in the stars of the solar neighbourhood (Roman 1952, Wielen 1977).

In the 1960s the Galaxy's stars were often divided into the most metal-poor stars, which were assigned to 'halo Population II', distinctly metal-poor stars, assigned to 'intermediate Population II', and stars of more-or-less solar metallicity, which were assigned to 'extreme Population I' if young, and simply 'Population I' otherwise.

Eggen, Lynden-Bell and Sandage (1962) interpreted these correlations in terms of a very influential picture in which the Galaxy formed rapidly by the collapse of a spinning, axisymmetric cloud. As the cloud collapsed, it spun up and became more metal-rich as stars formed and supernovae exploded. The first stars to form were metal-poor and on highly eccentric orbits, while the last stars to form were metal-rich and on nearly circular orbits. Hence the observed correlations. A remark which made this kind of formation picture plausible was that of Mestel (1963), who pointed out that a disk with a flat rotation curve has roughly the same distribution dM/dL_z of mass with each specific angular momentum L_z as a rigidly rotating uniform sphere. Hence the Milky Way's disk could have formed by the collapse of an initially uniform and rigidly rotating cloud which remained axisymmetric at all times.

In the 1970s this picture was extensively developed by Larson (1974), who used moment equations to model the collapse and metal-enrichment of spinning, axisymmetric proto-galactic clouds. On the whole these models provided good fits to the photometry of disk galaxies, but Larson did find it necessary to endow his clouds with significant viscosity. This viscosity had the effect of moving angular outwards through the infalling matter. Since the discovery of the bar instability in rapidly rotating, self-gravitating systems (see below) it has seemed likely that such angular-momentum redistribution is achieved by the gravitational field of a bar rather than by turbulent viscosity of the type envisaged by Larson.

Spitzer & Schwarzschild (1951, 1953) offered a very different interpretation of the correlations between metallicity, age and peculiar velocities. They suggested that stars were always born on nearly circular orbits in the Plane and are scattered from these onto more highly eccentric and/or inclined orbits by giant gas clouds. Until the 1970s there was no direct evidence for the existence of such clouds, but Spitzer & Schwarzschild (1953) inferred that they must have masses near $10^6 \, M_\odot$. This is just the upper limit of the masses of the giant molecular clouds discovered twenty years later (Solomon *et al.* 1987).

In 1967 Barbanis & Woltjer suggested that spiral structure might be responsible for scattering stars from circular orbits in the Plane onto more eccentric and/or inclined orbits. Since steady spiral structure cannot accomplish this job, they suggested that many spiral arms might be ephemeral structures. It now seems likely that both molecular clouds and ephemeral spiral structure play rôles in 'heating' the Disk (Carlberg & Sellwood 1985, Sellwood & Kahn 1991, Jenkins 1992).

However, neither spiral structure nor molecular clouds appear equal to the task of producing the highest-velocity component of the old Disk (sometimes called the 'thick disk'), let alone the Galactic halo. Some workers continue to suppose that these components were already hot at birth as envisaged by Eggen *et al.* and Larson (Burkert *et al.* 1992). Others have conjectured that stars are scattered onto appropriately eccentric orbits by massive halo black holes (Lacey & Ostriker 1985) or by satellites accreted by the Milky Way (Quinn & Goodman 1986, Binney & Lacey 1988, Tóth & Ostriker 1992).

6 DYNAMICS

The foundations of stellar dynamics were laid by Jeans and Eddington in a remarkable series of papers published in *Monthly Notices of the Royal Astronomical Society* during the 1914–18 war. At that time too little was understood about the nature of the Milky Way or external nebulae to go very far in the application of the theory. After Oort had established the nature of the Milky Way's disk, it was possible to apply the theory to the solar neighbourhood, as Oort did in his classical analysis of the mass-density near the Sun. Bertil Lindblad took the first steps towards a theory of spiral structure. But by 1950 remarkably little had been accomplished in the way of interpreting observational data on a dynamical basis.

6.1 Hot stellar systems

In the 1960s and early 1970s elliptical galaxies were generally interpreted in terms of the lowered isothermal models introduced by Michie (1963) and associated with the name of King, who in 1966 used them to interpret observations of globular clusters. Prendergast and Tomer (1970) started the study of flattened axisymmetric systems, which was what elliptical galaxies were then thought to be. Wilson (1975) extended this work and investigated a wider class of lowered isothermal models than Woolley & Dickens (1961) and Michie (1963) had studied.

The papers of Bertola and Capaccioli (1975), Illingworth (1977), Williams & Schwarzschild (1979) and others alluded to above, made it essential to consider more general models than the rotationally flattened lowered isothermal models that had been the focus of attention. N-body models with several hundred particles were then becoming fairly easy to construct and Binney (1976) and Aarseth & Binney (1978) found that n-body models relaxed from initial conditions inspired by Zel'dovich's pancake theory (Sunyaev & Zel'dovich 1972), naturally formed slowly rotating, flattened ellipticals of the type demanded by the observations that were just then beginning to come in. At that stage it was not apparent exactly how these systems functioned, but it did seem likely that the flattening of slowly rotating system's was some-how connected with the 'third integral' discussed much earlier by Contopoulos (1960) and others (e.g. Ollongren 1962) in the context of the solar neighbourhood.

In the 1915 paper that introduced his eponymous theorem, Jeans concluded that the Milky Way could not be in a steady state because the velocity dispersions of stars towards the galactic centre and perpendicular to the plane were not equal, as his new result required if the distribution function depended on the obvious integrals of motion. In the 1930s Oort had simply assumed that the vertical motions of stars were decoupled from those in the plane, and in the 1950s Contopoulos used Hamiltonian perturbation theory to obtain a formal series expansion for the extra integral that enforces this decoupling. This integral came to be called the 'third integral' because it came third after energy and angular momentum about the system's symmetry axis.

In 1964 Hénon & Heiles used one of the electronic digital computers which were then just beginning to become wide-spread, to integrate numerically the equations of motion of a star in a very simple model of the effective potential experienced by a star in a flattened, axisymmetric galaxy. This study has been immensely influential throughout the burgeoning field of chaos because it vividly demonstrates cohabitation of order and chaos. Orbits close to certain closed orbits seem to respect a third integral, while others further removed from stable closed orbits do not. The Hénon and Heiles paper seemed to mesh nicely with the so-called KAM theorem, jointly and semi-independently developed by Kolmogorov, Arnold & Moser. In essence this demonstrates that in certain restrictive circumstances it is possible to construct convergent perturbation series for quasi-periodic motions in Hamiltonians that consist of an integrable zeroth-order term and a small perturbation. Moreover, the orbits that can be expressed in terms of such perturbation series cover a non-negligible volume of phase space.

Unfortunately, KAM have nothing to say about when a given Hamiltonian lies close to an integrable one. Very few realistic integrable Hamiltonians are known, and in practice Hamiltonian systems are nearly always treated as either perturbed harmonic oscillators or perturbed Kepler problems. The Hénon & Heiles potential is that of a degenerate two-dimensional harmonic oscillator perturbed by cubic terms. In view of the KAM theorem it is natural to conclude that the chaos discovered by Hénon & Heiles at high energies is an inevitable consequence of the potential being significantly anharmonic at these energies. If this view were right, one would expect chaos to dominate the orbital structures of *all* realistic galactic potentials, because these are rarely nearly harmonic or nearly Keplerian.

That this conclusion is false was demonstrated in 1979 by Schwarzschild, who, inspired by the n-body experiments of Aarseth & Binney (1978) and others, integrated large numbers of orbits in the potential of a realistic model of a triaxial elliptical galaxy. Schwarzschild found that his orbits were of three types, 'tube orbits', which have definite senses of circulation about the potential's longest or shortest axes, and 'box orbits', which are modified Lissajou figures. All these orbits appear to have two effective isolating integrals in addition to their only classical integral, energy.

Moreover, by linear programming Schwarzschild was able to show that the density distribution that generates the potential can be constructed by populating the orbits appropriately, thus demonstrating the dynamical feasibility of self-consistent triaxial systems.

The importance of Schwarzschild's work was underlined by two subsequent developments: (i) Wilkinson & James (1982) showed that the orbital structure of the triaxial endpoint of an n-body simulation was very similar to that described by Schwarzschild. (ii) De Zeeuw (1985) showed that the potentials introduced by Stäckel (1890) in his study of the separability of the Hamilton-Jacobi equation, have precisely the orbital structure described by Schwarzschild. Or, more precisely, the structure Schwarzschild recognized after de Zeeuw had suggested that Schwarzschild's long-axis tube orbits should really be divided into two sub-families, the inner- and outer-tube orbits.

From Stäckel's work it is clear that his eponymous potentials are very special, and if the dynamics of triaxial ellipticals were due to their potentials being slightly perturbed Stäckel potentials, the possible forms of galactic potentials would be strongly constrained. So it was important to determine whether potentials that are not approximately Stäckel ones also have many orbits that respect non-classical integrals.

Binney & Spergel (1982, 1984) showed that orbits that have no Stäckel counterparts are quasi-periodic like orbits in integrable potentials such as Stäckel ones. In particular they showed that 'minor' orbit families are quasi-periodic, as are many orbits in rotating potentials, but that the actions of these orbits do not fit together to form a scheme of global angle-action variables as do the actions associated with de Zeeuw's various families. Gerhard & Binney (1985) showed that the phase spaces of very centrally concentrated potentials, such as that of a triaxial elliptical galaxy with singular central density, are dominated by the 'boxlet' minor families. Recently, Schwarzschild (1992) has shown that a triaxial system with $\rho \propto r^{-2}$ density profile can be self-consistently constructed by orbits of these minor families in just the same way that his original elliptical of finite central density can be constructed from orbits of de Zeeuw's major families. Thus it appears that galactic potentials do not have to be perturbed Stäckel potentials in order to be self-consistently generated by quasi-periodic orbits

6.2 Disk dynamics

In the 1950s it was generally thought that spiral structure was a magnetohydrodynamic phenomenon. Bertil Lindblad was one of the few astronomers who believed that gravity might be solely responsible, and he still laboured under the delusion that spiral arms lead rather than trailed galactic rotation. However, towards the end of his long career he achieved two important insights. First, he was struck by the fact that the difference $\Omega - \kappa/2$ between the circular frequency Ω and half the epicyclic frequency κ generally varies very slowly with radius (Lindblad 1961). He conjectured that this difference frequency might be the frequency with which a spiral pattern rotated rigidly while individual stars streamed through it. Second, he understood how gas would be concentrated by a spiral gravitational field into thin, dense arms (Lindblad 1963).

The next major advance was the study by Lin & Shu (1964) and Kalnajs (1965) of the propagation of density waves through differentially rotating disks of stars. Their central result was the Lin–Shu–Kalnajs (LSK) dispersion relation connecting frequency to wavelength for tightly-wound disturbances. At first it was thought that this demonstrated that disks could support self-sustaining spiral disturbances indefinitely. But in 1969 Toomre pointed out that waves obeying the LSK dispersion relation have non-zero group velocity, with the result that in a time comparable to the time for a simple-minded spiral pattern to 'wind up', the wave energy would be carried to a Lindblad resonance and Landau damped away.

From Toomre's work it was clear that if the LSK waves were to explain long-lived spiral structure, they would have to be continuously regenerated. By 1981 two re-generation mechanisms had been proposed, both associated with the corotation res-onance: Toomre (1981) argued that the responsible agent was the 'swing amplifier' of Goldreich & Lynden-Bell (1965) and Julian & Toomre (1966), while Mark (1974, 1976) proposed a mechanism he dubbed the 'WASER'.

In an important and sadly unpublished thesis, Zang (1976) furnished persuasive ev-idence that the normal modes of stellar disks could be understood in terms of LSK waves bouncing around within the barrier formed by the corotation resonance (see Toomre 1981), and, moreover, that when a mode is unstable, its growth rate can be reliably estimated from the swing amplifier model of what happens at corotation; the WASER predicts growth rates that are too small.

In parallel with this largely analytic work on spiral structure, n-body experimentalists were raising important questions about the whole intellectual basis of spiral structure theory. In 1971 Hohl used 100,000 particles to simulate a disk in centrifugal equi-librium and found it to be violently unstable to the formation of a bar. Ostriker & Peebles (1973) showed that Hohl's bar instability was not simply a consequence of his disk being initially in a state of approximate rigid rotation, but rather a consequence of its being rather 'cold' in the sense of having a small ratio of random to ordered kinetic energy. Several subsequent studies confirmed this conclusion. As Ostriker &

Peebles pointed out, the random velocities of stars in the solar neighbourhood are so small that one would naïvely predict the Milky Way's disk to be bar unstable. Either it must be embedded in a remarkably massive hot component, or it must be appreciably hotter near its centre than it is at the Sun.

For more than a decade after Hohl's 1971 paper, n-body experimentalists were preoccupied with the bar instability. By the mid 1980s enough was understood about the stability of disks for work to begin in earnest on n-body simulations of the fine detail of disk dynamics which is spiral structure. This field was dominated by Sellwood, who by developing techniques for limiting the effects of particle noise, was eventually able to obtain n-body results in agreement with linear perturbation theory (e.g. Sellwood & Athanassoula 1986). However, as in Hohl's pioneering study, unexpected processes continue to turn up in n-body experiments and the analytic interpretation and astronomical significance of some of these it is still not clear (e.g. Sellwood & Kahn 1991).

Spiral structure is most obvious in the gaseous components of disks. In the standard density-wave theory of spiral structure, the gaseous component of responds strongly to the spiral potential perturbation because it is the coldest component of the disk. Consequently it shocks in the neighbourhood of the potential troughs. This leads to the formation of young, blue stars and to enhanced synchrotron emission from the shocked interstellar medium. The radio observations of Mathewson *et al.* (1972) and Visser (1980) confirmed these predictions of Fujimoto (1968) and Roberts (1969).

The irregularity of the spiral patterns in the extreme population I tracers in most galaxies suggests that there might be a purely stochastic element in spiral structure. In a delightful paper Mueller & Arnett (1972) demonstrated that some remarkably convincing spiral patterns could be generated by allowing star formation to propagate like a forest fire through a differentially rotating disk. Seiden & Gerola (1982) have extensively explored this idea of 'self-propagating star formation'. Undoubtedly real spiral structure owes much to the tendency of differential rotation to shear any luminous region into a trailing spiral; infrared photometry (Kennicutt & Edgar 1986) clearly demonstrates that spiral structure is not confined to extreme population I objects.

REFERENCES
Aarseth, S. J. & Binney, J. J. 1978. *MNRAS*, **185**, 227
Baade, W. 1944. *ApJ*, **100**, 137
Bahcall, J. N. & Soneira, R. M. 1980. *ApJS*, **44**, 73
Barbanis, B. & Woltjer, L. 1967. *ApJ*, **150**, 461

Bender, R., Paquet, A. & Nieto, J.-L. 1991. *A&A*, **246**, 349

Berry, M. V. 1978. In *Topics in nonlinear dynamics*, ed. S. Jorna (New York: American Institute of Physics)

Bertola, F. & Capaccioli, M. 1975. *ApJ*, **200**, 439

Binney, J. J. 1985. *MNRAS*, **212**, 767

Binney, J. J. 1976. *MNRAS*, **177**, 19

Binney, J. J. & de Vaucouleurs, G. 1981. *MNRAS*, **194**, 679

Binney, J. J. & Lacey, C. G. 1988. *MNRAS*, **230**, 597

Binney, J. J. & Spergel, D. N. 1982. *ApJ*, **252**, 308

Binney, J. J. & Spergel, D. N. 1984. *MNRAS*, **206**, 159

Blitz, L., Binney, J. J., Lo, K. Y., Bally, J. & Ho, P. T. P. 1992. *Nature*, in press

Blitz, L. & Spergel, D. N. 1991a. *ApJ*, **370**, 205

Blitz, L. & Spergel, D. N. 1991b. *ApJ*, **379**, 631

Bosma, A. 1978. Ph.D. thesis, University of Groningen

Burkert, A., Truran, J. W. & Hensler, G. 1992. *ApJ*, **391**, 651

Burbidge & Burbidge 1975. In *Stars and Stellar Systems*, vol. 9, ed. A. and M. Sandage, p. 81 (Chicago: University of Chicago Press)

Burke, B. F. 1957. *AJ*, **62**, 90–90

Burton, W. B. 1970. *A&A*, **10**, 76

Carlberg, R. G. & Sellwood, J. A. 1985. *ApJ*, **292**, 79

Carter, D. 1978. *MNRAS*, **182**, 797

Contopoulos, G. 1954. *Zeitschrift f. Ap.*, **35**, 67

Contopoulos, G. 1960. *Zeitschrift f. Ap.*, **49**, 273

Davies, R. L., Sadler, E. M. & Peletier, R. S. 1992. *MNRAS*, in press

de Vaucouleurs, G. 1959. *ApJ*, **130**, 728

de Vaucouleurs, G. & Pence, W. D. 1978. *AJ*, **83**, 1163

de Zeeuw, P. T. 1985. *MNRAS*, **216**, 273

Dressler, A., Faber, S. M., Burstein, D., Davies, R. L., Lynden-Bell, D., Terlevich, R. J. & Wegner, G. 1987. *ApJ*, **313**, L37

Dressler, A., Lynden-Bell, D., Burstein, D., Davies, R. L., Faber, S. M., Terlevich, R. J. & Wegner, G. 1987. *ApJ*, **313**, 42

Dressler, A. & Richstone, D. O. 1990. *ApJ*, **348**, 120

Efstathiou, G. P., Ellis, R. S. & Carter, D. 1982. *MNRAS*, **201**, 975

Eggen, O. J., Lynden-Bell, D. & Sandage, A. R. 1962. *ApJ*, **136**, 748

Ewen, H. I. & Purcell, E. M. 1951. *Nature*, **168**, 356

Faber, S. M. 1973. *ApJ*, **179**, 731

Forman, W., Schwarz, J. Jones, C. Liller, W. & Fabian, A. 1979. *ApJ*, **234**, L27

Forman, W., Jones, C. & Tucker, W. 1985. *ApJ*, **293**, 102

Franx, M. & Illingworth, G. 1988. *ApJ*, **327**, L55

Franx, M. & Illingworth, G. 1990. *ApJ*, **359**, L41

Franx, M., Illingworth, G. & de Zeeuw, P. T. 1991. *ApJ*, **383**, 112

Freeman, K. C. 1970. *ApJ*, **160**, 811

Fujimoto, M. 1968 *Proc. IAU Symp. 29*, 453

Gerhard, O. E. & Binney, J. J. 1985. *MNRAS*, **216**, 467

Gilmore, G. & Reid, I. N. 1983. *MNRAS*, **202**, 1025

Goldreich, P. & Lynden-Bell, D. 1965. *MNRAS*, **130**, 125

Greenstein, J. L. & Matthews, T. A. 1963. *Nature*, **197**, 1041

Hauser, M. G., *et al.*, NASA photograph G90-03046 (1990).

Hazard, C, Mackey, M. B. & Shimmins, A. J. 1963. *Nature*, **197**, 1037

Hénon, M. & Heiles, C. 1964. *AJ*, **69**, 73

Hernquist, L. & Barnes, J. E. 1991. *Nature*, **354**, 210

Hohl, F. 1971. *ApJ*, **168**, 343

Illingworth, G. 1977. *ApJ*, **218**, L43

Jenkins, A. 1992. *MNRAS*, **257**, 620

Jenkins, E. B. & Savage, B. D. 1974. *ApJ*, **187**, 243

Jensen, E. B. & Thuan, T. X. 1982. *ApJS*, **50**, 421

Julian, W. H. & Toomre, A. 1966. *ApJ*, **146**, 810

Kalnajs, A. 1965. *The stability of highly flattened galaxies*, PhD. thesis, Harvard University

Kennicutt, R. C. & Edgar, B. K. 1986. *ApJ*, **300**, 132

Kent, S. M. 1991. *ApJ*, **370**, 495

Kerr, F. J. 1957. *AJ*, **62**, 93–93

King, I. R. 1966. *AJ*, **71**, 64

King, I. R. 1978. *ApJ*, **222**, 1

Kormendy, J. 1984. In *Morphology & dynamics of galaxies*, 12th Advanced Course Swiss Society of Astronomy & Astrophysics, eds L. Martinet & M. Mayor, (Geneva: Geneva Observatory

Kormendy, J. 1984. *ApJ*, **287**, 577

Kormendy, J. & Richstone, D. O. 1992. *ApJ*, **393**, 559

Kormendy, J. & Illingworth, I. 1982. *ApJ*, **256**, 460

Kuijken, K. & Tremaine, S. D. 1991. In *Dynamics of Disk Galaxies*, ed. B. Sundelius, p. 71 (Göteborg: Chalmers University)

Lacey, C. G. & Ostriker, J. P. 1985. *ApJ*, **299**, 633

Lambas, D. G., Maddox, S. J. & Loveday, J. 1992. *MNRAS*, **248**, 404

Larson, R. 1974. *MNRAS*, **169**, 229

Lauer, T. 1985. *ApJ*, **292**, 104

Lauer, T. 1989. In *Dense stellar systems*, ed. D. Merritt (Cambridge University Press) p. 3

Light, E. S., Danielson, R. E. & Schwarzschild, M. 1974. *ApJ*, **194**, 257

Lin, C. C. & Shy, F. H. 1964. *ApJ*, **140**, 646

Lindblad, B. 1961. *Stockholm Obs. Ann.*, 21, No. 8

Lindblad, B. 1963. *Stockholm Obs. Ann.*, 22, No. 5

Liszt, H. S. & Burton, W. B. 1980. *ApJ*, **236**, 779

Mark, J. W.-K. 1974. *ApJ*, **193**, 539

Mark, J. W.-K. 1976. *ApJ*, **206**, 418

Mathewson, D. S., van der Kruit, P. C. & Brouw, W. N. 1972. *A&A*, **17**, 468

Mathewson, D. S., Schwarz, M. P. & Murray, J. D. 1977. *ApJ*, **217**, L5

Merritt, D. 1988 (ed.) *Dense stellar systems*, (Cambridge University Press)

Mestel, L. 1963. *MNRAS*, **126**, 553

Michie, R. W. 1963. *MNRAS*, **125**, 127

Minkowski, R. 1961. In *IAU Symposium No 15*, ed. G. C. McVittie, p. 112 (New York: Macmillan)

Mueller, M. W. & Arnett, W. D. 1976. *ApJ*, **210**, 670

O'Connell, D. J. K. (ed.) 1958. *Stellar populations*, (New York: Interscience)

Ollongren, A. 1965. In *Galactic structure*, eds A. Blaauw & Schmidt, M., p. 501 (Chicago University Press)

Ostriker, E. C. & Binney, J. J. 1989. *MNRAS*, **237**, 785

Ostriker, J. P. & Peebles, P. J. E. 1973. *ApJ*, **186**, 467

Patterson, F. S. 1940. *Harvard Bul.*, **914**, 9

Penzias, A. A., Jefferts, K. B. & Wilson, R. W. 1971. *ApJ*, **165**, 229

Plaut, L. 1965. In *Galactic structure*, eds A. Blaauw & Schmidt, M., p. 267 (Chicago University Press)

Prendergast, K. H. & Tomer, E. 1970. *AJ*, **75**, 674

Quinn, P. J. & Goodman, J. 1986. *ApJ*, **309**, 472

Rubin, V. C., Graham, J. A. & Kenney, J. D. P. 1992. *ApJ*, **394**, L9

Roberts, W. W. 1969. *ApJ*, **158**, 123

Roman, N. 1952. *ApJ*, **116**, 122

Sargent, W. L. W., Young, P. J., Boksenberg, A., Shortridge, K., Lynds, C. R. & Hartwick, F. D. A. 1978. *ApJ*, **221**, 731

Schmidt, M. 1963. *Nature*, **197**, 1040

Schwarzschild, M. 1954. *AJ*, **59**, 273

Schwarzschild, M. 1979. *ApJ*, **232**, 236

Schwarzschild, M. 1992. *ApJ*, in press

Schweizer, F. 1976. *ApJS*, **31**, 313

Schweizer, F. 1979. *ApJ*, **233**, 23

Seiden, P. E. & Gerola, H. 1982. *Fund. Cosmic Phys.*, **7**, 241

Sellwood, J. A. & Athanassoula, E. 1986. *MNRAS*, **221**, 195

Sellwood, J. A. & Kahn, F. 1991. *MNRAS*, **250**, 278

Sinha, R. P. 1979. Ph.D. thesis, University of Maryland

Smith, E. P., Heckman, T. M., Bothun, G. D., Romanishin, W. & Balick, B. 1986.

ApJ, **306**, 64

Solomon, P. M., Rivolo, A. R., Barrett, J. & Yahil, A. 1987. *ApJ*, **319**, 730

Soltan, A. 1982. *MNRAS*, **200**, 115

Spitzer, L. & Savedoff, M. P. 1950. *ApJ*, **111**, 593

Spitzer, L. & Schwarzschild, M. 1951. *ApJ*, **114**, 385

Spitzer, L. & Schwarzschild, M. 1953. *ApJ*, **118**, 106

Stäckel, P. 1890. *Math. Ann.*, **35**, 91

Stark, A. A. 1977. *ApJ*, **213**, 368

Sunyaev, R. A. & Zel'dovich, Ya. B. 1972. *A&A*, **20**, 189

Toomre, A. 1969. *ApJ*, **158**, 899

Toomre, A. 1981. In *The structure and evolution of normal galaxies*, eds S. M. Fall & D. Lynden-Bell, P. 111 (Cambridge University Press)

Tóth, G., Ostriker, J. P. 1992. *ApJ*, **389**, 5

Trumpler, R. J. 1930. *Lick Observatory Bulletin*, **14**, 154

Tsikoudi, V. 1979. *ApJS*, **43**, 365

van Albada, T. S., Bahcall, J. N., Begeman, K. & Sancisi, R. 1985. *ApJ*, **295**, 305

Visser, H. C. D. 1980. *A&A*, **88**, 159

Wannier, P. & Wrixon, G. T. 1972. *ApJ*, **173**, L119

Wielen, R. 1977. *A&A*, **60**, 263

Wilkinson, A. & James, R. A. 1982. *MNRAS*, **199**, 171

Williams, T. B. & Schwarzschild, M. 1979. *ApJ*, **227**, 156

Wilson, C. P. 1975. *AJ*, **80**, 175

Woolley, R. & Dickens, R. J. 1961 *Royal Greenwich Observatory Bulletin*, No. 42

Wyckoff, S., Wehinger, P. A. & Gehren, T. 1981. *ApJ*, **247**, 750

Young, P., Westphal. J. A., Kristian, J., Wilson, C. P. & Landauer, F. P. 1978. *ApJ*, **221**, 721

Zang, T. A. 1976 *The stability of a model galaxy*, PhD. thesis, Mass. Inst. Technology

Galaxy Distribution Functions

WILLIAM C. SASLAW

This is mainly a review of the properties of gravitational galaxy distribution functions. It discusses their theoretical derivation, comparison with N-body simulations, and – perhaps most importantly – their observed features. The observed distribution functions place strong constraints on any theory of galaxy clustering.

1 INTRODUCTION

The galaxy distribution function $f(N, v)$ is the probability of finding N galaxies in a given size volume of space (or in a projected area of the sky) with velocities between v and $v + dv$. It is the direct analog of the distribution function in the kinetic theory of gases. For perfect gases, the spatial distribution is provided by a Poisson distribution at low densities and a Gaussian distribution at high densities, along with a Maxwell-Boltzmann distribution for the velocities. It is only in the last few years that we have discovered the comparable distribution for galaxies interacting gravitationally in the expanding universe. There are still many aspects of this problem which need to be understood.

Distribution functions had their origin in the observations and speculations of William Herschel two hundred years ago. In his catalog of nebulae he noticed that their distribution was irregular over the sky. Although we now know that some of these nebulae were galaxies and others resulted from stars, HII regions and planetary nebulae, and that some of the irregularities are intrinsic while others are due to local obscuration by the interstellar matter in our Milky Way, Herschel tended to view them all as a single class of objects. As an explanation for their irregular distribution, Herschel (1785) suggested that " ... a few stars, though not superior in size to the rest, may chance to be rather nearer each other than the surrounding ones; for here also will be formed a prevailing attraction in the combined center of gravity of them all, which will occasion the neighboring stars to draw together; not indeed so as to form a regular globular figure, but however in such a manner as to be condensed towards the common center of gravity of the whole irregular cluster. And this construction admits of the utmost variety of shapes, according to the number and situation of the stars which first gave rise to the condensation of the rest." Herschel's speculations, based on incomplete, faulty but pioneering observations have turned out to be very much in advance of their time.

Herschel worked after the real Renaissance, but the renaissance in this subject actually coincides rather well with Dennis Sciama's life and career. We all know that Dennis is distinguished, not only for his own work, but also for stimulating the work of many other astronomers. But even I was surprised to learn how early his influence began. No sooner had Dennis been born, in 1926, than Hubble started the first modern program to determine the galaxy distribution function. He used the 60″ and 100″ telescopes at Mt Wilson to make a uniform survey of galaxies brighter than 20th magnitude. Five years later he had preliminary results based on 900 plates with 20,000 galaxies (Hubble, 1931). He measured the spatial distribution function $f(N)$, but only for large areas on the sky which contained many galaxies. They did not fit a Poisson or Gaussian distribution, but they did fit a lognormal distribution. We will return to Hubble's result near the end of this review. If Hubble had also looked at small areas with few galaxies, he would have found that they did not fit the lognormal, but he had the good sense to quit while he was ahead.

In his seminal book, *The Realm of the Nebulae*, Hubble (1936) summarized his results and pointed out the next major step:

> "It is clear that the groups and clusters are not superimposed on a random (statistically uniform) distribution of isolated nebulae, but that the relation is organic. Condensations in the general field may have produced the clusters, or the evaporation of clusters may have populated the general field. Equations describing the observed distribution can doubtless be formulated on either assumption, and, when solved, should contribute significantly to speculations on nebular evolution."

Solving these equations has turned out to be much more difficult than Hubble expected. Half a century later, we are beginning to understand them. First, Milne (1935) made an attempt to calculate $f(N,v)$ from purely kinematic arguments in his non- gravitational theory of the expanding universe. He tried to solve what we would now call the collisionless Boltzmann equation by examining families of trajectories of orbits. Even within the context of his approach, this led to undetermined functions. In a larger context, it was unable to take dynamical gravitational correlations into account. But reading through these chapters of Milne's book shows what a truly valiant analysis it was.

Shortly afterward, Zwicky (1937) suggested adapting statistical mechanics to explain the galaxy distribution. The basic principles of his approach were: "1. The system of

external nebulae throughout the known parts of the universe forms a statistically stationary system. 2. Every constellation of nebulae is to be endowed with a probability
weight $f(\epsilon)$ which is a function of the total energy ϵ of this constellation. Quanti-
tatively the probability P of the occurrence of a certain configuration of nebulae is
assumed to be of the type

$$P = A\left(\frac{V}{V_0}\right) f\left(\frac{\epsilon}{\epsilon_k}\right).$$

Here V is the volume occupied by the configuration or cluster considered. V_0 is the
volume to be allotted, on the average to any individual nebula in the known parts of
the universe, and ϵ is the total energy of the cluster in question, while ϵ_k will probably
be found to be proportional to the average kinetic energy of individual nebulae."
Zwicky then went on to discuss, in general terms, how the functions $A(V/V_0)$ and
$f(\epsilon/\epsilon_k)$ might be found semi-empirically from observations. Neither Zwicky, nor
anyone else to my knowledge, ever carried out this program. His first principle is not
quite accurate since the distribution evolves significantly as it clusters. However, a
simple generalization of this principle, as we shall see, turns out to be very useful.
The functions $A(V)$ and $f(\epsilon)$ are rather different than Zwicky expected, but the
separability of $P(V, \epsilon)$ is indeed a very good approximation.

In the 1950s, Neyman and Scott (1959) used distribution functions in a purely descriptive manner to try to explain the Lick galaxy counts (which had superceded
those of Hubble). They postulated simple *a priori* statistical models of clustering,
not based on any dynamics, and tried to match them to observations. Generally
these did not match very well, even when they increased the number of free parameters. Nonetheless, their work was very important in raising the level of the statistical
discussion and some of their basic ideas have been revived in a more general context.

Lemaître (1961) realized, also in the 1950s, the importance of combining gravitational dynamics with galaxy statistics and spent his last years trying to solve this
problem. (He was partly motivated by Shapley who kept asking him whether he
could reconcile galaxy clustering with the expanding universe). His approach was
both theoretical and numerical. Lemaître developed new numerical techniques to integrate the equations of motion of the galaxies, but died before he could apply them
extensively. At about the same time, van Albada (1960) succeeded in analytically
solving the collisionless Boltzmann equation, including gravity for discrete galaxies
in the linear regime. He did this by calculating position and velocity moments of
the collisionless Boltzmann equation, and truncating them at fourth order. About
ten years later I extended this approach to the BBGKY hierarchy of gravitational
clustering to take correlations into account (Saslaw, 1972) and these results were
subsequently extended further and made more rigorous (e.g. Inagaki 1976; Fall and

Saslaw 1976). They provided some understanding of the dynamical evolution of the two-particle correlation function in the linear regime.

Although the two-particle correlation function had been discussed earlier as a purely descriptive statistic in the 1950s and 60s by Limber, Layzer, Gamow, Rubin, Neyman, Scott, Kiang, myself and others, and Totsuji and Kihara (1969) had accurately determined its value for the galaxy distribution, its greater physical understanding awaited the theoretical analyses and N-body simulations of the 1970s and 80s. From about 1975 to the late 1980s, the two-particle correlation function dominated the discussion of galaxy statistics. It soon became clear that rigorous mathematical analyses could not get much beyond the linear regime because the kinetic evolution equations became intractable, but more powerful computers made large N-body simulations possible. These could be done well into the regime of non-linear clustering for a variety of models and compared directly with analyses of observations. A major industry has developed around these simulations and comparisons, and the two-particle correlation function remains one of the mainstays of our understanding of the galaxy distribution.

But it is not enough. Around the mid 1980's it became clear that the two-particle correlation function did not give an adequate description of the large scale filaments and voids which many astronomers were finding. Determining the two-particle correlation accurately on scales larger than 10-20 Mpc becomes very difficult. Even if it could be measured accurately, it mainly emphasizes the structure on small scales. To characterize large voids and clusters would require using many-particle correlation functions, and these have been impossible to calculate for gravitational clustering even in the linear regime.

Therefore, other statistical descriptions of galaxy clustering are being explored. These include: 1) topological genus as a function of density level (e.g. Gott *et al.*, 1989, 2) Fourier power spectra, 3) percolation (e.g. Dominik and Shandarin, 1992), 4) minimal spanning trees (e.g. Barrow, Bhavsar and Sonoda, 1985), 5) multipole analyses (Scharf, *et al.* 1992), 6) Voronoi polygons (van de Weygaert, 1991), 7) fractal descriptions (Valdarnini, 1992), 8) transect number counts (Broadhurst, *et al.* 1990), 9) moments of counts such as skewness and kurtosis (e.g. Lahav, Itoh, Inagaki and Suto 1992), and 10) distribution functions. Some of these statistics complement one another by emphasizing different aspects of the galaxy counts; many of them can be related to one another. All have their virtues and deficiencies.

Here I shall mainly discuss the galaxy distribution function. It seems to have the overriding virtue 1) that at least for the case of simple gravitational galaxy clustering it can be derived analytically in the non-linear regime from more fundamental dynamics. This solves the problem that Milne and Lemaître sought to understand, and provides some of the equations that Hubble foretold. None of the other statistical descriptions has yet developed to this point, although they can also be explored by numerical simulations. These numerical experiments are very powerful and useful, but we must remember that replicating the galaxy distribution numerically is not the same as understanding it. Understanding requires a broader and deeper network of physical concepts.

Other virtues of the distribution function are that: 2) it is relatively easy to measure accurately over a wide range of scales, 3) it usually contains much more information than the low order correlations because it is related to all the correlation functions including those of very high order, 4) if the distribution is statistically homogeneous, when fairly sampled, its three-dimensional distribution function can easily be related to its two-dimensional form (Itoh, Inagaki and Saslaw 1988; Saslaw 1989), 5) it describes the whole range of clustering from the richest clusters to the field galaxies, without prejudice caused by the definition of a "cluster", and 6) the gravitational velocity distribution function can be derived from the spatial distribution function.

The main disadvantage of distribution functions is that they lack information about the detailed location of galaxies within a volume or area. So, at least in their simple form, they are not very good for determining the shapes of clusters. To some extent, this information can be captured by spatial filtering, but we shall not discuss that here. Distribution functions are also subject to observational and physical selection effects, although in many cases these can be calculated explicitly to provide useful information.

My own interest in galaxy distribution functions arose soon after Dennis started supervising my PhD thesis in 1965, and so perhaps I could add a few biographical remarks to this very brief, incomplete, and all too sketchy introductory survey of the development of the subject. The other excuse for these remarks is that they may help illustrate the sense of this special occasion, and some of the organizers have suggested short reminiscences of research student life a quarter century ago. We have to remember that it was a time when quasars had just begun to illuminate astrophysicist's imaginations, when the microwave background was casting a new glow over cosmology, when X-ray and infrared sources were just a glimmer in astronomy, and before pulsars were discovered.

I was just finishing up as an undergraduate and wondering what to do next when Peter Strittmatter who was then a sort of roving Post Doc at Princeton and several other places suggested going to Cambridge. George Field, who was my undergraduate advisor thought it a good idea and both he and Peter were enthusiastic about Dennis Sciama's work, style and approach to astrophysics. Now it happened that I had met Dennis even before becoming an undergraduate, although of course he would not remember this. As a high school student, I'd spent a summer at Cornell on one of their programs to encourage potential scientists. Cosmology then had the same fascination for kids that it has now – which doesn't quite mean that we have not made much progress – and I'd read some of the classic popular accounts such as Gamow's *One, Two, Three, Infinity*, Hoyle's *Nature of the Universe*, and Sciama's *The Unity of the Universe*. So when a poster appeared announcing that Dennis, who was visiting Cornell, would give a public talk on the title of his book, I thought "why not go?"

Of course the lecture was clear, entertaining, thoughtful and at just the right level for the audience, all trademarks of Dennis's talks. But these were not the main aspects I remembered, impressive as they were. My main recollection is that three of us, who wanted to find out more, stood around and asked Dennis questions which he patiently answered – for an hour and a half! This was certainly beyond the call of duty, and persuaded me that cosmologists can also be very friendly people. So when the suggestion to go to Cambridge and work with Dennis came along, I was already half convinced.

Fortunately, at this time Dennis was again visiting the U.S. at the Goddard Institute in New York City, persuading me that cosmologists are also very peripatetic people. So at George Field's suggestion I phoned him and we agreed to meet for lunch at Lindy's, a restaurant which, although no longer famous for its food, remained famous for having been famous. Our discussion about Cambridge and cosmology went on so long that the waiter hastened us out, persuading me that cosmologists don't worry about events on New York restaurant timescales. By the end of the meal we had agreed that I would go to Cambridge, provided a fellowship came through to make it financially possible; luckily it did.

Those were delightfully creative years at Cambridge, as many of us well remember. Theories often came and went daily, and sometimes observations did too. At times there was also a certain local tension between the DAMTP theoreticians trying to understand the nature and distribution of radio galaxies and the Cavendish observers who had the most recent maps and catalogs. On Saturday mornings in the Cavendish, graduate students and senior members of the radio astronomy group usually gathered to discuss recent papers and preprints. Theoreticians from DAMTP were welcomed, although we noticed when we arrived as a group, after the session

had already started, that the ongoing discussion ceased, papers were quickly shuf-
fled, and a new topic was taken up which was never about instrumentation or recent
results. On one of these occasions while a less interesting paper was being discussed,
I was reading my mail which contained a questionnaire from a sociologist studying
astronomers. I reached the question: What do you consider to be the best training
for theoretical astrophysics? 1. Physics, 2. Mathematics, 3. Astronomy, 4. Other
.... I was sitting next to John Faulkner who was showing some interest in this
question, so I handed him the paper and under the entry for "other" he wrote down
– "espionage". Although relations among the senior theoreticians and the Cavendish
radio astronomers may have had their ups and downs, relations among the research
students of the two groups were generally very friendly and information would sometimes leak out through some of the Cavendish research students who thought all the
secrecy was a bit silly.

Among the students in Dennis's group, and the other groups in DAMTP, the flow
of information was free and easy so we all learned a great deal from each other.
Dennis encouraged this exchange, and even if it resulted in working on projects which
evolved quite differently from his original suggestions, it did not matter as long as
we tried to do something new and important. So stimulating was this astronomical
atmosphere, that almost all of us tried to reproduce it elsewhere after we left. Some
former students and others who had been actively connected with the group helped to
propagate it more generally by catalysing the foundation of new astronomical research
institutes in Toronto, Virginia and Pune. These were also inspired by the Institute
of Astronomy in Cambridge which Fred Hoyle founded, and where Martin Rees and
Donald Lynden-Bell have maintained the tradition of free-range astrophysical inquiry.
It is all summarized by John Faulkner in a limerick he provided for this occasion:

> A renowned astrophysical charmer,
> Who went by the name of D. Sciama,
> With complete lack of prudence,
> Had oodles of students,
> Fulfilling his guru-like karma.

One of the significant results of the mid-1960s was that the spatial distribution function of radio galaxies in the 4C and related catalogs over the sky appeared to be
Poisson (Holden, 1966; Webster, 1976). At the time this argued against the existence
of structure on scales of several hundred megaparsecs. More recently, (Coleman &
Saslaw, 1990) a new radio galaxy sample about a hundred times as dense also showed
a Poisson distribution on much smaller scales of 50-100 Mpc. However, on even s-
maller scales of several tens of megaparsecs, departures from Poisson statistics occur
in both the distribution of the rich clusters of galaxies and of the galaxies themselves.

It is to the analyses of these non-Poisson distribution functions during the last few years that we now turn. Interactions among three major approaches have greatly stimulated the development of this subject. First, physical theory helps to suggest a conceptual structure for thinking about the problem. Second, computer experiments test the theory, determine its limits of validity and point toward new extensions. Third, the observed galaxy distribution functions determine the relevance of the theory and test the applicability of specific numerical simulations. Here I review and discuss some recent aspects of each of these approaches.

2 PHYSICAL THEORY
Gravitation provides the focus for much of the physical theory of galaxy distribution functions. It is simply inevitable. Many other physical processes have been proposed that may influence the galaxy distribution. Interest in them waxes and wanes, but gravity lasts forever. Moreover, it is only by first understanding gravitational clustering that we will be able to discover any non-gravitational aspects of the galaxy distribution with much certainty.

There is not yet a rigorous physical theory that starts from a Hamiltonian and calculates the dynamical evolution of the gravitational galaxy distribution function. Nevertheless, by making several physically motivated assumptions one can find a simple gravitational distribution function. We will see that it agrees well both with N-body simulations and with the observed galaxy distribution.

The basic assumptions of the theory (Saslaw and Hamilton, 1984; Saslaw 1985a; Saslaw, Chitre, Itoh and Inagaki, 1990) are:

1. Galaxies interact gravitationally as particles with a pairwise potential. This neglects tidal interactions and merging, which turn out to be of only secondary importance for the observed distribution function. The pairwise potential implies that the gravitational correlation energy depends directly on only the two-particle correlation function, ξ. However, the velocity distribution function, and therefore the velocity dispersion, will depend on all the higher order correlation functions since the acceleration vector of a given galaxy is produced by the locations of many other galaxies as well as by their separations.

2. The galaxy distribution is statistically homogeneous when fairly sampled over large regions. This does not preclude the existence of large voids or filaments. It merely supposes that the distribution functions of fair samples averaged over sufficiently large regions do not depend on the location of the region (apart from small variations). Statistical homogeneity can apply to small volumes provided a sufficient number are

sampled over a large enough region that any one volume is uncorrelated with most of the other volumes. The distribution in any one volume may be correlated with that of its near neighbors, but not with its distant neighbors. Statistical homogeneity implies that there is a very large scale on which the distribution is not coherent (e.g. it has random phases) and on which the distribution function has a Poisson form.

3. The ratio of gravitational correlation energy, $-W$, to twice the kinetic energy, K of peculiar velocities, i.e.,

$$b = \frac{-W}{2K} = \frac{2\pi Gm\bar{n}}{\langle v^2 \rangle} \int_0^\infty \xi\,(n,\,T,\,r,\,t)\,r dr \tag{1}$$

is scale invariant. Thus $b(\lambda r) = b(r)$ where λ is a scalar multiple. In general, b will depend on time because ξ will have a time dependent amplitude and scale length, and $\langle v^2 \rangle$ will also evolve. In a statistically homogeneous system, ξ depends on the relative separation $r = |\mathbf{r}_1 - \mathbf{r}_2|$ of two galaxies rather than on their six absolute coordinates. The value of $\langle v^2 \rangle$ is clearly scale independent since it is averaged over very large regions of a statistically homogeneous system. It includes the clusters as well as the isolated galaxies. Moreover the correlation energy integral is taken over the entire system, so in an unbounded statistically homogeneous system (or one that is much larger than the scale length of ξ) it is also scale independent. The scale invariance of this physical value of b (see also Itoh, Inagaki and Saslaw, 1992) should not be confused with the scale dependence of the pattern value of b found by fitting the distribution function of equation (5) below to counts in cells of different sizes (Itoh, et al. 1988; Suto et al. 1990; Itoh, 1990; Saslaw & Crane, 1991; Bouchet, et al. 1992; Lahav & Saslaw, 1992). The latter is simply the geometric effect that observations on scales much less than the correlation length tend toward a Poisson distribution.

4. The distribution evolves through a sequence of equilibrium states: a quasi-equilibrium evolution. There is no rigorous equilibrium state for an unbounded system of gravitating objects – they always like to cluster further. In the expanding universe, however, the clustering is generally slow compared with the expansion timescale. The system first evolves through a transient stage during which it relaxes from its initial state into a close approximation of the equilibrium state. The timescale for this stage and its degree of relaxation depend on how far the initial state departs from a Poisson distribution (Saslaw, 1985b; Itoh et al. 1988; Suto et al. 1990). Subsequent evolution can occur through these equilibrium states to a good approximation for a wide range of conditions.

It may seem surprising that from these four rather general assumptions we can deduce a very specific gravitational distribution function. So I will briefly outline the steps that lead to this result.

The assumption of a pairwise potential implies that the internal energy, U in some volume, V, can be related to the pressure, P, total number N, of galaxies, and the temperature $T = m\langle v^2 \rangle/3$ by equations of state having the form

$$U = \frac{3}{2}NT(1 - 2b) \qquad (2)$$

and

$$P = \frac{NT}{V}(1 - b) \qquad (3)$$

where b is given in equation (1). These equations of state are exact and hold for non-linear as well as linear clustering when the potential involves just pairs of particles (e.g. Belescu, 1975). (For more general potentials, equations (2) and (3) represent the first terms in a virial or cluster expansion.) Using the general thermodynamic relation $TdS = dU + PdV$ we can calculate the entropy S. Then inverting Boltzmann's relation $S = \ln f$ (in energy units with Boltzmann's constant $k = 1$) gives the probability f for finding a system in an ensemble which has a departure ΔS from the most probable state. Here the grand canonical ensemble is the one to use, since the different systems, each of which may represent a volume element in the universe, can exchange particles (galaxies) and energy (velocity correlations) across their boundaries. Applying standard thermodynamic fluctuation theory then relates the fluctuations in entropy to those of particle number and gives the distribution $f(N)$.

To carry out this procedure, we need to determine the dependence of b on n and T. Assumptions 2 and 3 enable us to bypass the kinetic theory which is essentially unsolvable in the non-linear regime. Scale invariance of b, assumption 3, implies that b can only be a function of n and T in the combination $G^3 m^6 n T^{-3}$. This result also follows from the requirement that the entropy is a perfect differential, so that the system's state does not depend on its thermodynamic path. Assumption 2, that correlations do not extend over the entire scale of the system can be shown to imply that $b(n, T)$ has the specific form

$$b = \frac{b_o n T^{-3}}{1 + b_o n T^{-3}} \qquad (4)$$

where b_o does not affect the form of the distribution (but may depend on time). The value of b ranges from 0 which represents a perfect gas, to 1 which represents a maximally clustered state.

The new statistical gravitational distribution function which results from this analysis is

$$f(N) = \frac{\overline{N}(1 - b)}{N!} \left[\overline{N}(1 - b) + Nb \right]^{N-1} e^{-\overline{N}(1-b)-Nb} \qquad (5)$$

where $\overline{N} = \overline{n}V$ is the expected average number of galaxies in the volume. This distribution has a number of interesting properties. In the limit $b \to 0$, it becomes a Poisson distribution as befits a perfect gas. The limit $b = 1$ represents a maximally clustered state where all the galaxies are contained in clusters occupying a total volume of measure zero. In this limit, only the void probability $e^{-\overline{N}(1-b)}$ survives, all other $f(N) = 0$. If $b \to 1$ and $\overline{N} \to \infty$ in such a way that the grand canonical potential $\overline{N}(1 - b) \to$ constant then the clusters continue to occupy a finite volume. Note that although b superficially resembles the "virial ratio", it involves the average gravitational correlation energy of inhomogeneities throughout the unbounded system, rather than the total gravitational energy of a bounded system as in the virial theorem. This is a fundamental property which makes the thermodynamics of galaxy clustering quite different from the thermodynamics of, say, star clusters. On large scales such that $N \to \infty$, $\overline{N} \to \infty$, and $\epsilon \equiv (\overline{N} - N)/\overline{N} \to 0$, the tail of the distribution has the Poisson form, normalized by the factor $(1 - b)$, as required by the assumption of statistical homogeneity. In the further "fluid limit" that $\overline{N}\epsilon \gg 1$, the distribution $f(N)$ becomes Gaussian with variance $\sigma^2 = \overline{N}/(1 - b)^2$ (discussed in Saslaw and Sheth, 1993).

One of the most useful properties of this distribution is that the distribution of the sum of two independent volumes, each containing this distribution also has the same distribution (Saslaw, 1989). This means that to a very good approximation (neglecting the slight correlations between neighboring volumes) a three-dimensional distribution with the form of equation (5) will also have this form in two-dimensions: i.e. it is projectively invariant. Numerical simulations (Itoh, et al. 1988) show that this projective invariance holds up reasonably well in $\Omega_0 = 1$ models and is very accurate in lower Ω_0 models where the correlations are less and therefore the independence of neighboring volumes is greater. This result means that a great deal can be learned from the projected distribution of galaxies on the sky, without the need for redshifts. Of course the full three-dimensional distribution (which is not the same as the redshift distribution, especially locally and in clusters where galaxies have relatively large peculiar velocities) may reveal further properties of the clustering.

Notice that there are no free parameters in this theory. The value of b can be determined directly from the positions and velocities of the galaxies if these are known (as in N-body simulations) and $\overline{n}$ follows by dividing the total number of galaxies by the total volume (or area projection). The absence of free parameters is a consequence of considering a grand canonical ensemble of systems, incorporating the dependence of both ξ and b on n and T. It is also possible to discuss the distribution in an individual given system (e.g. Balian & Shaeffer 1989) by using a scaling hypothesis for the void distribution and then deriving $f(N)$ for $N > 0$ by holding the volume integrals of ξ as well as higher order correlation volume integrals fixed for a particular set of points. However, the equivalent information regarding the dependence of ξ on n and T – which is implicit in the grand canonical ensemble – then must be added in the form of an ad hoc scaling function to represent the void distribution in terms of $\bar{n}$ and a volume integral of $\xi(r)$. For an adequate fit, these scaling functions which do not have a physical basis usually require several parameters (e.g. Bouchet, Shaeffer & Davis 1991). There have also been discussions of other distribution functions such as the lognormal (Hubble 1936; Crane and Saslaw, 1986) and negative binomial (Carruthers 1991), which are usually introduced in a rather ad hoc manner.

By making one additional simple assumption we can derive the velocity distribution function directly from equation (5) (Saslaw et al. 1990). This assumption is that, on average, the fluctuations in potential energy over a given volume (caused by correlations among the galaxies) are proportional to the local kinetic energy fluctuations. The resulting velocity distribution (in natural units with $G = M = R = 1$) is

$$f(v)dv = \frac{2\alpha\beta(1-b)}{\Gamma(\alpha v^2 + 1)} \left[\alpha\beta(1-b) + \alpha b v^2\right]^{\alpha v^2 - 1} e^{-\alpha\beta(1-b) - \alpha b v^2} v \, dv. \qquad (6)$$

Here $\beta \equiv \langle v^2 \rangle$, and Γ is the usual gamma function, and $\alpha = \beta^{-1}\langle 1/r \rangle$ where $\langle 1/r \rangle$ is the value of r^{-1} averaged over a Poisson distribution having an average density $\bar{n}$. This velocity distribution peaks at lower velocity and is considerably more skew than a Maxwell-Boltzmann distribution with the same value of $\langle v^2 \rangle$. It represents all the galaxies, including those isolated in the general field as well as those in the richest clusters.

The distribution functions given by equations (5) and (6) finally solve the problem posed by Hubble for the case of simple gravitational clustering. We will examine how they compare with N-body computer experiments and with the observed galaxy distribution. Naturally I would not be describing all this if there were not good agreement. First, however, we will see that it is possible to go even further and see how these distributions evolve with time.

All the arguments, so far, have been thermodynamic. Time does not enter. So how can we determine the system's time evolution without recourse to kinetic theory? Here is where the fourth assumption, quasi-equilibrium evolution, enters. In the form mentioned earlier, it just assumes that once the system has relaxed to the distribution of equation (5) it will retain the form of that distribution with evolving values of $\bar{n}(t)$ and $b(t)$. By mass conservation, $\bar{n}(t)$ is well-known to evolve proportionally to $R^{-3}(t)$ where $R(t)$ is the radius of curvature of the universe. To find $b(t)$, or equivalently $b(R)$, we need to refine the quasi-equilibrium hypothesis.

The refinement is to assume that, to a good approximation, quasi-equilibrium evolution occurs adiabatically. Heat flow caused by galaxy correlations growing across the boundary of a comoving volume element during a Hubble expansion time is small compared to the correlation energy within the volume. This assumption works best in low Ω_0 cosmologies, where the timescale for gravitational clustering is longer than the Hubble expansion timescale. But even for $\Omega_0 = 1$ where these timescales are equal, numerical experiments show that this is a good approximation. Combining it with equations (2)-(4) gives the relation (Saslaw, 1992)

$$R = R_* \frac{b^{1/8}}{(1-b)^{7/8}} \tag{7}$$

where R_* is a constant given by the initial state. Using $R(t)$ for any particular cosmological model, then shows how the physical value of b in equation (1) evolves with time. The curve rises rapidly, then levels off around $b \approx 0.8$, increasing very slowly after this value.

This physical value of b should not be confused with the pattern value of b which is found just by fitting equation (5) to a given distribution of galaxies (or particles in N-body simulations). The physical b contains both position and velocity information, while the pattern b contains only position information. As mentioned earlier, the pattern value will be scale dependent, and $b_{pattern} \to 0$ as the cell size becomes so small that it almost always contains either one or zero galaxies – the Poisson limit. In $\Omega_0 = 1$ models the distribution remains relaxed so that $b_{pattern} \approx b_{physical}$ as the system evolves. But in $\Omega_0 < 1$ models, the initially relaxed distribution freezes out at $z \approx \Omega_0^{-1}$ and $\langle v^2 \rangle$ decreases so that $b_{pattern} < b_{physical}$ as the universe expands (see Itoh, et al. 1988 where we denoted $b_{pattern}$ by b_{fit} and $b_{physical}$ by $b_{ab\ initio}$).

Equations (5), (6), and (7) for $f(N)$, $f(v)$ and $b(R(t))$ are three of the main results, so far, from the thermodynamic theory of simple gravitational clustering. To check these results, and explore the range of conditions for which the theory is valid, we turn to N-body experiments.

3 NUMERICAL SIMULATIONS

Computer experiments have long been used to follow the evolution of the two-particle correlation function and are increasingly being used to examine distribution functions. Although early examples (Aarseth and Saslaw, 1982) measured the void distribution $f(0)$ even before it was understood theoretically, subsequent analyses have shown that the $f(N)$ for all values of N are important since they emphasize aspects of clustering on different scales. The main topics relating simulations to the theory of simple gravitational clustering which I will review here are the effects of different cosmological expansion rates on clustering, the effects of a spectrum of galaxy masses, and the effects of different initial conditions.

Many numerical methods have been developed to simulate galaxy clustering, and each of them is well-adapted to explore particular aspects of the problem. For systematically studying distribution functions, the most useful method seems to be direct integration of the N equations of motion of the galaxies (Aarseth, 1985; Aarseth and Inagaki, 1986). Although the number of galaxies this method can follow is not large – 10,000 are easy at present and 100,000 are possible – few approximations are made and complete position and velocity information for each galaxy is available. The main approximations are to introduce a short range smoothing scale for the potential and to examine galaxies in a representative spherical volume of the universe. The smoothing scale does not expand. It represents the physical size of the galaxy and facilitates the numerical integration, but it does not have a significant effect on the distribution functions after their initial relaxation. It does affect the timescale for this initial relaxation somewhat by altering the non-linear interactions of neighboring galaxies (Itoh, *et al.* 1988). Since relatively few galaxies cross the boundary (and are reflected when they do) and all the distribution function statistics exclude galaxies that are too close to the boundary, this does not affect the results significantly. Comparisons of 1000, 4000 and 10,000-body experiments under the same conditions show that while increasing the number of galaxies smoothes the statistics, it does not affect the resulting distribution functions significantly.

Another numerical method smoothes the gravitational potential over a three-dimensional grid containing cells whose size expands with the universe. This is well-adapted for following large-scale linear and quasilinear structure, especially if the universe contains cold dark matter not associated with the galaxies, but it minimizes the important non-linear relaxation on small scales. Distribution functions obtained with this method depend significantly on cell size and resolution (e.g. Fry, *et al.* 1989). Moreover, this approach requires additional assumptions to define a galaxy out of the smoothed distribution. Usually a galaxy is defined as a region of high density contrast, although there is no independent theory for choosing a particular value of the density excess. Other possibilities such as the total mass in a volume, or the proximity of two (or more) regions of high density or mass excess could also be chosen. Presumably, with very high resolution and a suitable definition of what a galaxy is, this method will give similar results to the direct integration approach, although this problem has not been studied. Which approach best represents our actual universe is, of course, another question. Perhaps, as often happens in this subject, both will be suitable, but on different scales. We have mainly been concerned with understanding the physics of simple gravitational galaxy clustering, rather than advocating one or another model of cosmology and therefore have mainly developed computer experiments using the direct integration methods.

Early results (Saslaw and Hamilton, 1984; Saslaw, 1985b) showed that N-body simulations satisfying assumptions (1)–(4) had distribution functions which were described well by equation (5). This encouraged a more systematic exploration (Itoh *et al.*, 1988, 1990, 1993; Inagaki *et al.* 1992; Saslaw *et al.* 1990, Suto *et al.* 1990; Itoh, 1990), which is still being extended. First I will summarize some of the results for the spatial distribution function, then for the velocity distribution function, then for the evolution.

Figure 1 (from Itoh *et al.* 1993) illustrates the spatial distribution function for a 10000-body simulation with all particles (galaxies) having the same mass. They start with a Poisson distribution and zero velocity relative to the Hubble flow. The radius of the simulation is normalized to $R = 1$ and the results are shown here at an expansion factor $a/a_o = 8$. This time is chosen because the distribution has achieved a two-particle correlation function whose slope and amplitude agree reasonably well with the observed values. There are two ways to portray the distribution function. By selecting a given value of N and plotting $f_N(V)$ one obtains the distribution of radii for randomly placed spheres containing N galaxies. Here the void distribution $f_0(V)$ and $f_5(V)$ are illustrated, plotted in the first two panels as a function of the radius, r of the volume V. For larger scales, it is more informative to locate spheres of given radius r randomly throughout the simulation (but at least a distance r from the boundary) and plot $f_V(N)$, the probability for finding N galaxies in these spheres. This is illustrated for spheres of radius 0.1, 0.2, 0.3 and 0.4 in the last four panels. Histograms represent the simulation data, and dashed lines are the theoretical distribution of eq. (5). Figure 4, below, shows a Poisson distribution for comparison.

Figure 1 shows that under conditions where the assumptions of the theory apply, the theoretical and experimental distribution functions agree very well. The values of b are least square fits to eq. (5), and they vary by less than 3%. These are the pattern values of b. The physical value of b calculated directly from the positions and velocities of the particles in this simulation is 0.67, also in good agreement. From a number of simulations with the same initial conditions, we find that variations in b among the different realizations are usually about ±0.05.

Some general results (Itoh, *et al.* 1988) for spatial distribution functions follow:

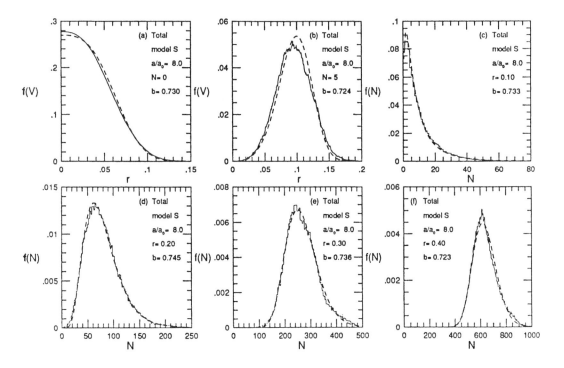

Figure 1. Spatial distribution functions (from Itoh, *et al.* 1993) for a 10000-body simulation. Distributions $f(V)$ are shown for the probability of finding $N = 0$ galaxies within a volume V of radius r (the void distribution function) and for $N = 5$. Distribution functions $f(N)$ are shown for the probability of finding N galaxies in a volume of radius r with $r = 0.1$, 0.2, 0.3 and 0.4. (The total radius of the simulation is normalized to unity.) Histograms are the N-body simulations; dashed lines are the best fits to the distribution function of equation (5).

1. Initial Poisson distributions which start off cold relative to the Hubble flow in $\Omega_0 = 1$ simulations relax to the distribution of equation (5) after the universe expands by a factor of about 1.5. This relaxation takes longer for systems whose initial random peculiar velocities are larger. Subsequent evolution occurs through a quasi-equilibrium sequence of states described by equation (5) with $b_{physical} \approx b_{pattern}$.

2. Decreasing the value of Ω_0, so that the gravitational clustering timescale becomes longer than the expansion time, does not prevent the distribution from relaxing to the form of equation (5). However it takes longer and is characterized by $b_{physical} > b_{pattern}$. This inequality becomes stronger as Ω_0 decreases. The reason is that there is less clustering at early times in the lower Ω_0 models and shortly after the distribution relaxes it becomes "frozen in". Most of the pattern expands essentially homologously, except within bound or deep potential wells were the distribution is relatively inde-

pendent of the expansion. The value of $b_{physical}$ increases faster in the lower Ω_0 models because there are more field galaxies which continue to lose energy to the adiabatic expansion. (See Figure 3 below in which $b_{ab\ initio} \equiv b_{physical}$ and $b_{fit} \equiv b_{pattern}$).

3. For $\Omega_0 = 1$ models, the three-dimensional fits to equation (5) give slightly higher values of $b_{pattern}$ than do two-dimensional fits to the distribution projected onto the sky. This is because galaxies are more likely to be correlated across the boundaries of neighboring volume elements. For lower Ω_0 models, the correlations are weaker and the spatial and projected values of $b_{pattern}$ agree very closely (See Figure 3 below).

4. Non-Poisson initial distributions take longer to relax to the form of equation (5), and some of them never reach it (Saslaw, 1985b; Suto, et al. 1990; Itoh, 1990; Bouchet, et al. , 1991). This is particularly true for simulations, such as some dark matter models, which have strong initial correlations, or anti-correlations, on large scales. These do not satisfy the statistical homogeneity requirement of assumption 2. They cannot relax rapidly enough on large scales relative to the expansion timescale. So the large N tail of the distribution, especially over large volumes, remains non-Poisson and does not satisfy equation (5), even though equation (5) may still be a good description of the relaxed bulk of the distribution.

5. Although the theory developed so far is for systems whose components all have the same mass, equation (5) still provides a good description of two-component systems provided the mass ratio of the two components is less than about 1 to 10 (Itoh, et al. 1990). The basic reason is that the relaxation leading to equation (5) is mainly collective at later times and does not depend on the masses of individual galaxies. For large mass ratios, however, the heavy galaxies tend to form the nuclei of satellite systems and if there are too many of these then the single component theory naturally cannot describe them.

6. Systems with a continuous mass spectrum, which is more representative of observed galaxy masses, cluster rather differently. They tend to form more extended large-scale patterns in the later stages of the evolution. On large scales these finite simulations become statistically inhomogeneous and equation (5) no longer applies, although it continues to represent the distribution well on smaller, more homogeneous scales.

We now turn to the velocity distribution function. Velocities of galaxies in computer simulations have also been compared with the velocity distribution predicted in equation (6) (Saslaw et al. , 1990; Inagaki, et al. 1992; Itoh, et al. 1993). Remarkably, it turns out that equation (6) describes the velocity distribution over an even greater

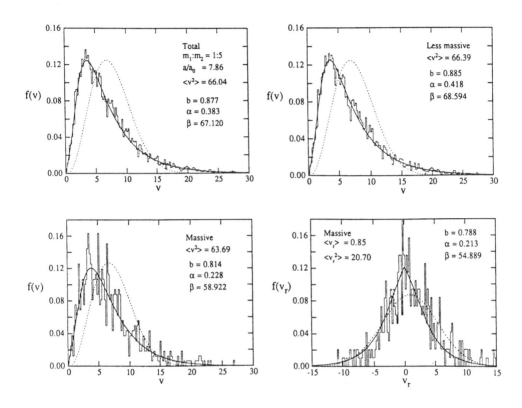

Figure 2. Velocity distribution functions (from Inagaki, *et al.* 1992) for a 10000-body simulation. Histograms are from the simulation, dotted lines are the Maxwell-Boltzmann distribution having the same $< v^2 >$ as the simulation, and solid lines are the best fits to the distribution function of equation (6).

range of conditions than those for which equation (5) describes the density distribution. Not only does it work very well for single mass systems, but also for systems with galaxy mass ratios much greater than 10:1. The velocity distribution depends sensitively on Ω_0, much more so than does the density distribution. Eventually it will be very important to compare equation (6) with observations of peculiar galaxy velocities, when they become sufficiently accurate. It stands as a clear prediction.

Figure 2 (from Figure 4 of Inagaki, *et al.* 1992) illustrates a typical example. This is a 4000-body experiment with $\Omega_0 = 1$ in which 500 galaxies each have relative mass 5 compared to the other 3500 galaxies each of mass unity. Initially both mass components have a Poisson distribution at rest relative to the Hubble expansion. They rapidly cluster and relax to the distribution of equations (5) and (6). The figure shows velocity distributions at an expansion factor of about 7.86 when the two-particle correlation function comes into reasonable agreement with the observations. The histograms represent the experimental 3-dimensional velocity distributions for all the galaxies, as well as for the two mass components separately. The lower right panel is the radial velocity distribution of the massive galaxies as seen by a representative observer at the center of the system (representing an arbitrary position in space). In each case, the dashed line is the Maxwell-Boltzmann distribution function having the same temperature (velocity dispersion) as the simulation, shown for comparison. It doesn't fit at all. The experimental distribution peaks at much lower velocities and is much more skew. The Maxwell-Boltzmann distribution cannot simultaneously represent both the field galaxies and those in rich clusters.

The solid line in Figure 2 is the best least squares fit of equation (6) to the experimental distribution, using b, α and β as free parameters (although $\alpha\beta$ becomes constant as the system relaxes). One measure of agreement is the ability of equation (6) to reproduce the entire distribution. Another measure is the comparison of the fitted value of β with the value of $< v^2 >$ which is computed directly from the peculiar velocities. For the total distribution and the less massive component, these two quantities agree within about 2%. For the more strongly clustered massive component they differ by about 8%. Comparing $< v^2 >$ with $3 < v_r^2 >$ for the massive component, measures the spatial isotropy of their velocities: these average values are 63.7 and 62.1 respectively – well within sample variations. Values of b measured from the velocity distributions are generally somewhat greater than $b_{physical}$ or $b_{pattern}$, especially for multi-mass systems. This property is not yet understood.

One of the main aspects of gravitational galaxy clustering from an initially unclustered state is that it starts growing rapidly, but after several expansion timescales the growth slows and clustering approaches an asymptotic state. This behaviour appears to occur for all values of Ω_0. It is shown in Figure 3 (from Saslaw 1992) which examines the evolution of single mass 4000-body simulations for three values of Ω_0 and an initial Poisson distribution moving with the Hubble expansion (cold). The crosses, with sampling error bars, plot the physical values of b (obtained ab initio from the positions and velocities of the particles), and the open circles and filled squares plot the pattern values of b (obtained by fitting equation 5 to the spatial distribution using b as a free parameter, since $\overline{N}$ is known) for the 3-dimensional and projected 2-dimensional experimental distributions. These are all plotted as a function of the expansion radius R of the universe relative to its initial value R_i.

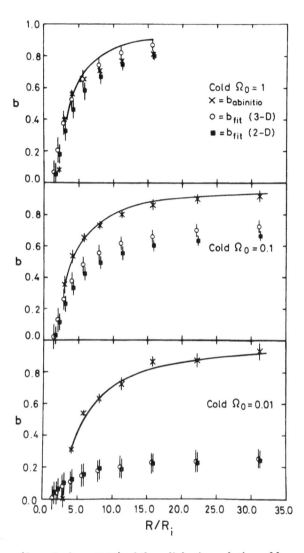

Figure 3. Comparison (from Saslaw, 1992) of the adiabatic evolution of $b_{physical} \equiv b_{ab\ initio}$ given by equation (7) with the values (crosses) from an N-body simulation described in the text.

The solid line in Figure 3 shows the evolution predicted by equation (7) based on the adiabatic clustering hypothesis described earlier. For $\Omega_0 = 1$ the evolution is about 10% subadiabatic at large expansion times (R_* is determined by the initial value of $b_{ab\ initio} \equiv b_{physical}$ when the system has relaxed). This is because for $\Omega_0 = 1$ many more galaxies are in clusters whose gravitational perturbations increase the velocities of the field galaxies. This counteracts the tendency for velocities of field galaxies to decrease adiabatically and it therefore increases the denominator in $b_{physical}$ above its adiabatic value, lowering the value of $b_{physical}$. For lower values of Ω_0, the adiabatic hypothesis is a better approximation and agreement with equation 7 is very good all the way into the highly non-linear regime.

The value of $b_{pattern}$ is much less than $b_{physical}$ for these $\Omega_0 < 1$ cases because the spatial clustering relaxes early in the evolution and then freezes out as the expansion becomes faster than the gravitational clustering timescale. This evolution of $b_{pattern}$ can be understood from linear perturbation theory for the growth of density fluctuations in the system (Zhan, 1989; Zhan and Dyer, 1989; Inagaki, 1991). The evolution of $b_{pattern}(R/R_i)$ and $b_{physical}(R/R_i)$ is reasonably well understood for systems with components of the same mass, but needs to be extended to systems with a mass spectrum. It would also be useful to tie the adiabatic and linear perturbation theories more closely to the fundamental dynamics.

4 OBSERVED GALAXY DISTRIBUTION FUNCTIONS

When the gravitational statistics of equation (5) were discovered, no one knew whether they would describe the observed distribution of galaxies in our Universe. It was one of the few genuinely robust predictions for large scale structure. Many astronomers doubted that it would be applicable since it did not involve any of the more speculative processes involving dark matter or the early universe which were then popular. So it was rather a surprise to find subsequently (Crane and Saslaw, 1986) that the projected distribution of galaxies in the Zwicky catalog agreed very well with equation (5). Its value of $b_{pattern} = 0.70\pm0.05$ even agreed with the expected value from the N-body simulations. This indicates that on scales up to at least about 10 megaparsecs (where the sampling noise becomes large since there are relatively few cells) the galaxy distribution is primarily gravitational. It has lost most of the memory of its initial state. This also implies that the initial state was close enough to a Poisson distribution that it had time to relax to the form of equation (5) on these scales.

The agreement holds throughout the entire range of distribution functions, from the void distribution $f_0(V)$ to the largest cells (containing ~ 100 galaxies) which have been measured. It incorporates information about the ~ 100-particle correlation functions. Other statistics, such as Hubble's (1934) lognormal distribution, do not seem to apply so broadly with so few free parameters. Several years after examining the Zwicky catalog, it occurred to me that it would be rather fun to go back to Hubble's original data, which was just for large cells, and see how well it fit equation (5). So Eric Lufkin, then a graduate student at Virginia, used Bok's (1934) linear plot of Hubble's data and found that equation (5) matched it with the results shown in Figure 4. The solid dots on the main part of the diagram show Hubble's original

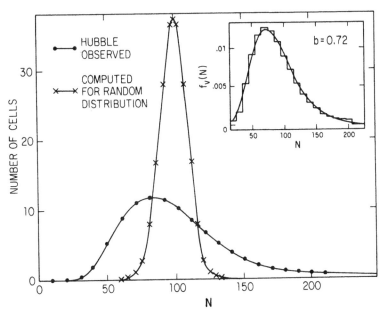

Figure 4. The main figure shows the observed distribution $f(N)$ which Hubble measured just for large N (as plotted by Bok, 1934), and a Gaussian distribution with the same average and dispersion. The inset shows Hubble's data renormalized and compared with the distribution function of equation (5).

counts in cells, connected by a curve simply put through the points. The more sharply peaked curve is a Gaussian distribution that Bok drew for comparison. (It was part of a Hubble-Shapley debate on the homogeneity of the universe with Zwicky and Bok respectively as seconds to the two principals.) The inset shows the least squares fit of equation (5) to Hubble's counts after they are normalized. It gives $b_{pattern} = 0.72$, just as in the earlier analysis of Zwicky's catalog.

Our early analyses of the Zwicky catalog (and later of the N-body simulations) showed a small but fairly systematic tendency for $b_{pattern}$ to increase with the scale of the cells. More extensive analyses over a wider range of scales showed that the effect is caused by sampling on scales less than the correlation lengthscale. This does not fairly sample the distribution, and in the extreme case where the cells are so small that they almost always contain either zero or one galaxy the distribution appears to be Poisson with $b_{pattern} = 0$. Figure 5 (from Saslaw and Crane 1991) shows this for sampled cells whose areas range from 0.25 to 25 square degrees in the Zwicky catalog. The different symbols are for different magnitude subsamples. Squares are for the entire catalog above galactic latitude 30° to avoid regions of obscuration within our galaxy, triangles are for the subset brighter than $m_{zw} \simeq 15$, and open ovals for those brighter than $m_{zw} \simeq 14.5$. The values of $b_{pattern}$ for these subsamples are renormalized to test the hypothesis that the luminosity of a galaxy is independent of

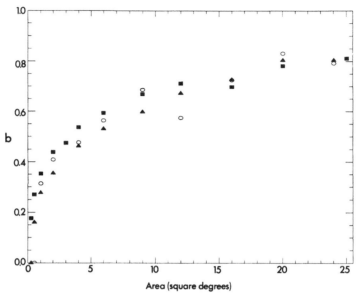

Figure 5. The scaling of $b_{pattern}$ for three magnitude-limited subsamples of the Zwicky catalog described in the text (from Saslaw and Crane, 1991).

the number of galaxies around it. To the extent this hypothesis is true, the three symbols should agree, which they do quite well, within the usual sampling uncertainty in $b_{pattern}$ of ± 0.05. This puts limits on the effects of environment on galaxy luminosity, though the detailed nature of these limits is not yet fully understood.

Returning to the dependence of $b_{pattern}$ on scale, we see that the effect of sampling on too small a scale for the entire catalog begins to level off above about 10 square degrees where the distribution becomes more fairly sampled. The best estimate of $b_{pattern}$ would be the peak of this curve. Here it still seems to rise slightly over the largest scales for which these statistics are meaningful. Hence $b_{pattern} = 0.8$ would be a better estimate for our Universe than our earlier value of 0.7 which was based on scales of about 12 square degrees in the original analysis. The dependence of $b_{pattern}$ on scale can be understood by relating $b_{pattern}$ to the volume integral of the two-particle corelation function, and this relation agrees reasonably well with the observations (Saslaw and Crane, 1991; Lahav and Saslaw, 1992). This scale dependence, which is essentially a sampling effect, should not be confused with the scale invariance of $b_{physical}$.

The question of fair sampling is especially important for redshift determinations of the three-dimensional $f(N)$. The redshift catalogs which have been analyzed so far, generally have relatively few members compared to the large two-dimensional projected samples. Moreover, these redshift catalogs, such as the CfA slice or the Pisces-Perseus

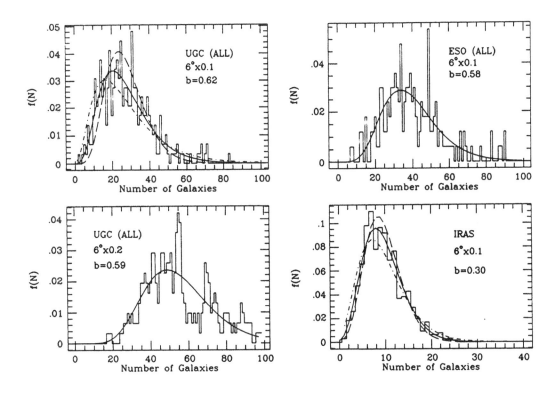

Figure 6. a) Observed distribution functions (histograms) for all galaxies in the UGC, ESO and IRAS catalogs (from Lahav and Saslaw, 1992). The UGC cell-counts are shown for two different size cells. Solid lines are the best fits of the distribution function of equation (5). The dashed and dot-dash lines show equation (5) for $b = 0.70$ and 0.50 in the UGC case and for $b = 0.40$ and 0.20 in the IRAS case for comparison.

region (e.g. Fry, et al. 1989) are preselected to be unusual regions, often with un- typically large clusters. This, of course, biases the statistics. In addition, the poorly determined peculiar velocities of galaxies in redshift catalogs, make the transforma- tion between redshift space and physical space uncertain. Analyses of numerical simulations (Fry et al. 1989) show that this uncertainty can lead to significant dis- tortions of the distribution function. Nevertheless, when larger, more complete and representative redshift catalogs are obtained in the next few years, they will add much valuable information about the distribution functions.

Recently a rather simple and general approach to the problem of selection and biasing in the observed galaxy distribution functions has been developed and applied to the distribution of different morphological types in the UGC and ESO catalogs, as well as to infrared galaxies in the IRAS catalog (Lahav and Saslaw, 1992). Figure 6 (from

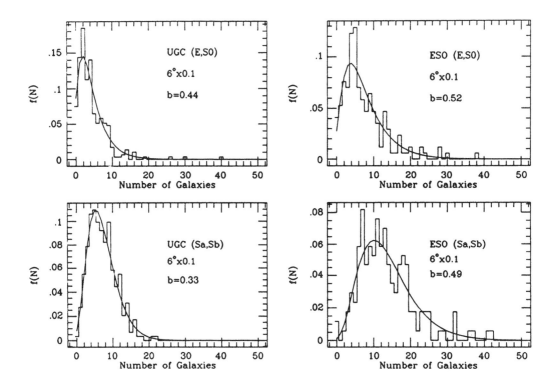

b) Observed distribution functions (histograms) for E + SO and Sa + Sb subsamples of the UGC and ESO catalogs, compared with the best fits of the distribution function of equation (5).

the last reference) shows the observed distribution functions in these catalogs for cells of $\Delta\alpha = 6°$ and $\Delta \sin \delta = 0.1$, about 36 square degrees. They are also shown for subsets of different morphological types in the ESO and UGC catalogs. The solid line curve is the best least square fit of equation (5) to the distributions, and the resulting value of $b_{pattern}$ is also given. For all the UGC sample, the dashed and dot-dash lines illustrate equation (5) with values of $b_{pattern} = 0.7$ and 0.5, respectively, for comparison. For the IRAS sample these comparison lines have $b_{pattern} = 0.4$ and 0.2.

These results show that the fits to equation (5) are good over the whole range from isolated field galaxies to the richest cluster galaxies. They also describe the elliptical and the spiral subsamples. Values of $b_{pattern}$ for all the galaxies in these catalogs are less than for the Zwicky catalog because these catalogs are subsamples of galaxies based on the additional criteria of large angular size for the UGC and ESO catalogs, and on bright infrared emission for the IRAS catalog. There are differences in detailed selection procedures between the UGC and ESO catalogs, and this may be reflected in their different values of $b_{pattern}$, although this is not fully understood (see the discussion in Lahav and Saslaw, 1992).

All these observed distributions are essentially gravitational. The value of $b_{pattern}$ for the spiral subsample is consistent with them being selected randomly from the whole catalog. The ellipticals, however, behave differently. Their value of $b_{pattern}$ is about twice what it would be if they were selected randomly. This result helps quantify the tendency of ellipticals to cluster more strongly than spirals, a tendency much discussed ever since it was discovered by Hubble and Zwicky in the 1930's. Therefore theories which attempt to explain morphological differences by general physical processes such as gas stripping or galaxy mergers are strongly constrained by having to fit these observed distribution functions. Such theories will have difficulties. Although there is good evidence in specific cases that these processes occur, it is far from clear that they can provide a general explanation for the different galaxy types. Simple models along the lines previously proposed fail to reproduce the observed distributions for the ellipticals (Lahav and Saslaw, 1992). More complicated versions of those models may work, though some aspects may have to be introduced ad hoc. However, it is also possible that simple models of gravitational segregation may account for the difference of $b_{pattern}$ between ellipticals and spirals. This could succeed as an explanation if the ellipticals in the sample are, on average, several times more massive than the spirals.

5 NEW PATHS AND UNSOLVED PROBLEMS

By developing the gravitational physics of galaxy distribution functions it has become possible to understand and relate many of their observed properties, and to reproduce major features of N-body computer simulations. In which directions might the subject advance? Here I have only had space to outline some of our understanding of galaxy distribution functions. But there are many areas that need to be clarified and explored. I'll suggest a "baker's dozen" of theoretical and observational problems:

T1. Can the gravitational distribution function be derived from more rigorous microscopic principles? Is there a statistical "counting argument" for the number of microcells in macrocells of phase space which gives $f(N)$? Does it follow from a Hamiltonian approximation which allows the gravitational partition function to be summed explicitly, even if not exactly?

T2. Can the theory be extended to systems having components of different masses? This could be tested by existing N-body simulations.

T3. How can distribution functions be modified to provide useful information about the shapes of clusters?

T4. Can one develop a kinetic theory, or other approach, to describe the departures of $b_{physical}(t)$ from adiabatic evolution, especially in the $\Omega_0 = 1$ universe? This departure seems sufficiently small that a perturbation approach may work.

T5. How is the gravitational evolution of distribution functions modified by substantial departures from Poisson initial conditions?

T6. What amounts and distributions of dark matter, interacting gravitationally with the galaxies, would destroy the substantial agreement between the observations and simple gravitational clustering?

T7. How can other descriptive statistics best be combined with distribution functions to provide insight into their underlying physics?

01. How does the distribution function behave on very large scales? The largest angular scale on which it has been measured so far is 100 square degrees (Lahav and Saslaw, 1992) where there seems to be a small decrease of $b_{pattern}$, but the sampling statistics are not definitive.

02. Is the observed three-dimensional distribution function consistent with the distribution function found for projected distributions, or does it contain significant additional non-gravitational information? Only recently have enough redshifts been measured that one can begin to observe the representative three-dimensional distribution function. One still needs to be cautious since peculiar velocities confuse the redshift-distance relation, especially for nearby galaxies.

03. Can observed large scale "filamentary" or "wall" structures be sufficiently well defined (in terms of total mass overdensity, number overdensity, continuity, etc.) to decide whether they are causal structures that preserve a memory of initial conditions, or just structures which emerge by chance from gravitational clustering? This is a combined observational-theoretical problem in which the theory should also describe chance patterns having an interesting appearance but of no causal significance.

04. Does the peculiar velocity distribution of galaxies agree with the prediction of gravitational clustering? This requires good secondary distance indicators, which are being developed. Is it possible to measure a reasonably accurate value for $b_{physical}$?

05. Do galaxies of all types follow the gravitational distribution function, or are there some which have been strongly influenced by other formation, evolutionary, or environmental processes?

06. At high redshifts, do good samples also satisfy the gravitational distribution, but with smaller $b_{pattern}$, and does $b_{pattern}$ evolve as expected from gravitational perturbation theory and N-body simulations?

Clearly there is a great deal to learn, though much has been accomplished in a subject which is both one of the oldest and one of the newest in extragalactic astronomy and cosmology. I would especially like to thank and acknowledge many collaborators who have helped explore the landscape beyond the correlation functions which dominated discussions of large scale structure in the 1970s and 1980s: Sverre Aarseth, H. M. Antia, Suketu Bhavsar, Kumar Chitre, Paul Coleman, Phil Crane, Andrew Hamilton, Shogo Inagaki, Makoto Itoh, Ofer Lahav, Eric Lufkin, Yoel Rephaeli, Ravi Sheth, Trinh Thuan, and David Valls-Gabaud. It has been great fun. Discussions with many astrophysicists have helped over the last few years to clarify ideas in this area, and on this occasion it is a particular pleasure to thank Dennis Sciama for his continuing interest, enthusiasm and perceptive advice.

The National Radio Astronomy Observatory is operated by Associated Universities, Inc., under cooperative agreement with the National Science Foundation.

REFERENCES

Aarseth, S.J. and Saslaw, W.C. 1982. *Astrophys. J.* , **258**, L7.

Aarseth, S.J. 1985, in *Multiple Time Scales*. ed. J.U. Brackbill and B.I. Cohen (New York: Academic Press), p.377.

Aarseth, S.J. and Inagaki, S. 1986, in *The Uses of Supercomputers in Stellar Dynamics*, ed. P. Hut and S. McMillan (Berlin: Springer-Verlag), p.203.

Albada, G.B. van, 1960. *Bulletin Astron. Inst. Neth.*, **15**, 165.

Balian, R. and Shaeffer, R. 1989. *Astron. Astrophys.*, **226**, 373.

Barrow, J.D., Bhavsar, S.P. and Sonoda, D.H. 1985. *Mon. Not. R. astr. Soc.* , **216**, 17.

Belescu, R. 1975. *Equilibrium and Nonequilibrium Statistical Mechanics* (New York: Wiley).

Bok. B.J. 1934. *Harvard Coll. Obs. Bulletin*, **895**, 1.

Bouchet, F.R., Shaeffer, R. and Davis, M. 1991. *Astrophys. J.* **383**, 19.

Broadhurst, T.J., Ellis, R.S., Koo, D.C. and Szalay, A.S. 1990. *Nature* , **343**, 726.

Carruthers, P. 1991 *Astrophys. J.* **380**, 24.

Coleman, P.H. and Saslaw, W.C. 1990. *Astrophys. J.* **353**, 354.

Crane, P. and Saslaw, W.C. 1986. *Astrophys. J.* **301**, 1.

Dominik, K.G. and Shandarin, S. 1992, to be published.

Fall, S.M. and Saslaw, W.C. 1976. *Astrophys. J.* **204**, 631.

Fry, J.N., Giovanelli, R., Haynes, M., Melott, A., and Scherrer, R.J. 1989. *Astrophys. J.* **240**, 11.

Gott, J.R. *et al.* 1989. *Astrophys. J.* **340**, 625.

Holden, D. J. 1966. *Mon. Not. R. astr. Soc.* **133**, 225.

Hubble, E. 1931. *Publ. Astron. Soc. Pacific* **43**, 282.

Hubble, E. 1936. *The Realm of the Nebulae*, (New Haven: Yale Univ. Press).

Inagaki, S. 1976. *Publ. Astron. Soc. Japan* **28**, 77.

Inagaki, S. 1991. *Publ. Astron. Soc. Japan* **43**, 661.

Inagaki, S., Itoh, M. and Saslaw, W.C. 1992. *Astrophys. J.* **386**, 9.

Itoh, M., Inagaki, S. and Saslaw, W.C. 1988. *Astrophys. J.* **331**, 45.

Itoh, M., Inagaki, S. and Saslaw, W.C. 1990. *Astrophys. J.* **356**, 315.

Itoh, M. 1990. *Publ. Astron. Soc. Japan* **42**, 481.

Itoh, M., Inagaki, S. and Saslaw, W.C. 1993. *Astrophys. J.* , in press.

Lahav, O., Itoh, M., Inagaki, S. and Suto, Y. 1992. *Astrophys. J.* in press.

Lahav, O. and Saslaw, W.C. 1992. *Astrophys. J.* in press.

Lemaître, G. 1961. *Astron. J.* **66**, 603.

Milne, E.A. 1935. *Relativity, Gravitation and World Structure*, (Oxford: Clarendon Press).

Neyman, J. and Scott, E.L. 1959 in *Encyclopedia of Physics* ed. by S. Flügge (Berlin: Springer-Verlag) **53**, 416.

Saslaw, W.C. 1972. *Astrophys. J.* **177**, 17.

Saslaw, W.C. and Hamilton, A.J.S. 1984. *Astrophys. J.* **276**, 13.

Saslaw, W.C. 1985a. *Gravitational Physics of Stellar and Galactic Systems* (Cambridge: Cambridge Univ. Press).

Saslaw, W.C. 1985b. *Astrophys. J.* **297**, 49.

Saslaw, W.C. 1989. *Astrophys. J.* **341**, 588.

Saslaw, W.C., Chitre, S.M., Itoh, M. and Inagaki, S. 1990. *Astrophys. J.* **365**, 419.

Saslaw, W.C. and Crane, P. 1991. *Astrophys. J.* **386**, 315.

Saslaw, W.C. 1992. *Astrophys. J.* **391**, 423.

Saslaw, W.C. and Sheth, R.K. 1993 in preparation.

Scharf, C., Hoffman, Y., Lahav, O. and Lynden-Bell, D. 1992. *Mon. Not. R. astr. Soc.* in press.

Suto, Y., Itoh, M. and Inagaki, S. 1990. *Astrophys. J.* **350**, 492.

Totsuji, H. and Kihara, T. 1969. *Publ. Astron. Soc. Japan*, **21**, 221.

Valdarnini, R., Borgani, S. and Provenzale, A. 1992 in press.

Webster, A. 1976. *Mon. Not. R. astr. Soc.* **175**, 71.

van de Weygaert, R. 1991. *Mon. Not. R. astr. Soc.* **249**, 159.

Zhan, Y. 1989. *Astrophys. J.* **340**, 28.

Zhan, Y. and Dyer, C. 1989. *Astrophys. J.* **343**, 107.

Zwicky, F. 1937. *Publ. Astron. Soc. Pacific* **86**, 217.

Nonlinear Galaxy Clustering

BERNARD J.T. JONES

... we seek a theory which describes all that
actually happens, and nothing that does not,
a theory in which everything that is not
forbidden is compulsory.

The Unity of the Universe
D. W. Sciama
(Faber & Faber, 1959)

The clustering of galaxies on scales $< 5h^{-1}\mathrm{Mpc}^1$ shows some remarkable scaling
properties which somehow arise out of nonlinear gravitational self-organisation. This
scaling is characteristic of structures that are referred to as *multifractals*. There are
several ways of looking at these structures each providing their own special insights
into the nature of the clustering. Multifractal scaling can be shown to be closely
associated with the fact that galaxy counts-in-cells are approximately Lognormally
distributed and with hierarchical fragmentation processes. Moreover, the statistical
moments of the galaxy distribution scale in a way that is reminiscent of the renor-
malization group. This may throw light on the nature of the underlying dynamics of
the nonlinear gravitational clustering process.

1 INTRODUCTION

When Dennis Sciama published his book *"The Unity of the Universe"* in 1959, the
great debate was which theory of the Universe was the correct one: the "Big Bang" or
the "Steady State"? Dennis had been a member of a group of Steady-State enthusiasts
at Cambridge in the early 1950's. However, when in 1965 the relict radiation was
discovered, he rapidly embraced the new Hot Big Bang Theory and had his students
working on the cosmic singularity, radio source evolution, galaxy formation, and other
relevant topics. This adaptation, or indeed opportunism, is a hallmark of Dennis'
scientific style from which his numerous students have benefitted. The short section
in Dennis' book on the Formation of Galaxies in the evolving model was the stimulus
for my doing my PhD Thesis on that subject, starting in 1969.

---

[1] h denotes the Hubble constant in units of 100 km s^{-1}Mpc^{-1}

Since the discovery of the cosmic background radiation galaxy formation has become a central problem of cosmology. However, the later star forming phases of the galaxy formation process still remain much of a mystery. Most progress has been made in understanding the large scale structure: structures on scales in excess of $10h^{-1}$ Mpc. The reason for this progress is twofold. Firstly, the physics of galaxy clustering is relatively straightforward, the only force involved being gravity. This means that we can at least simulate the clustering process numerically. Secondly, there is now a large amount of data available on large scale clustering, both in terms of the projected distribution of galaxies on the sky and in terms of their collective motions relative to the background Hubble flow.

When looking at the CfA slice of the universe (de Lapparent et al., 1986), a sample of galaxies having redshift data out to distances $\sim 100h^{-1}$ Mpc., we perceive the distribution of galaxies as being dominated by filamentary structures surrounding voids that extend over scales in excess of $20h^{-1}$ Mpc. Nevertheless, the distribution of galaxies on scales $> 10h^{-1}$ Mpc. is linear in the sense that this filamentary structure is severely attenuated if the sample volume is smoothed with a large scale filter. The filamentary structure is associated with high frequency (short wavelength $\sim 5h^{-1}$ Mpc.) structures which are organised relative to one another by the larger scale flows. The filamentary structure is indeed nonlinear. There is a potential difficulty in that we map this structure using luminous galaxies as beacons, and we are unsure how the galaxy formation process might bias our understanding.

What we see today on scales $> 10h^{-1}$ Mpc. is a direct reflection of the initial conditions in the Universe, though the pattern of clustering (voids and filaments) is the result of a complex interplay between linear and nonlinear effects. Nevertheless, we have a considerable number of analytic tools to help us understand what is going on here and we can always resort to N-body simulations.

By contrast, the clustering regime between $1h^{-1}$ Mpc. and $10h^{-1}$ Mpc. is characterised by substantial density excursions, and although numerical simulations show what is happening, it would be fair to say that we have little understanding of why things evolve in the way they do: the problem is "nonlinear" and arguably beyond any simple understanding.

It is this last problem of nonlinear structural organization that I wish to discuss here. It is in fact a problem in which many domains of physics meet and so I think it is particularly interesting. I shall emphasize things that are to be learned from attempts to understand critical phenomena. However, I will not attempt to be either rigorous or complete in my discussion. Rather, I try to paint an image of these fascinating structures from the point of view of cosmology, giving enough detail for the reader to approach more detailed mathematical papers.

2 THE OBSERVED CLUSTERING OF GALAXIES

The inhomogeneity of the distribution of galaxies on the sky has been appreciated since the first deep surveys of Hubble, who showed that galaxy counts in cells were roughly Lognormally distributed. The recent Oxford-APM Schmidt plate survey of 1,000,000 galaxies (Maddox et al., 1990) is a vastly superior data set allowing much deeper analysis. The three dimensional picture has only emerged recently with the advent of large redshift surveys and redshift independent distance estimates for large numbers of galaxies.

The 1950's saw several fine attempts to quantify the clustering of galaxies, notable among which were the models of Neyman and Scott, and the correlation function approach of Cooper-Rubin. It was the correlation function approach that eventually came to dominate cosmology during the 1970's and 1980's with an effort led mainly by Peebles and his collaborators. Peebles' book on Large Scale Structure in the Universe (1980) provides a thorough review, though it is now over a decade old and so lacks some of the more recent advances like density and velocity field reconstruction techniques.

The two point correlation function $\xi(r)$ of a sample of galaxies is defined as the excess probability, relative to a Poisson distribution having the same mean density of points, of finding galaxies in small volumes δV_1 and δV_2 a distance r_{12} apart:

$$\delta P_{12} = \frac{n \delta V_1}{N} \frac{n \delta V_2}{N}[1 + \xi(r_{12})], \qquad r_{12} = |\mathbf{r}_1 - \mathbf{r}_2|. \tag{1}$$

Here, n is the mean density of galaxies in the sample, and N is the total number of galaxies in the sample volume. If $\xi(r_{12}) \equiv 0$, the occupancies of the volume elements δV_1 and δV_2 are independent and the distribution of galaxies is a Poisson process.

As has been long recognised, the two point correlation function for the distribution of galaxies in three dimensions is well fit by a simple power law:

$$\xi(r) = \left(\frac{r}{r_0}\right)^{-\gamma}, \quad r < 10h^{-1}\text{Mpc}. \tag{2}$$

with the fitted values
$$\gamma = 1.77 \qquad r_0 \simeq 5h^{-1}Mpc.$$

On larger scales there is a steepening away from this power law, and there is some suggestion that the correlation function becomes negative on scales larger than $20h^{-1}$ Mpc.

Unfortunately, common usage refers to r_0 as a "correlation length" or an "amplitude", whereas it is in fact merely a normalizing constant telling us the length scale on which the correlation function $\xi(r)$ drops to the particular value unity. The range of r over which the scaling law (2) applies might have nothing to do with r_0 and the perceived pattern of clustering may be determined by scales far in excess of r_0. There is some difference between the above definition of the sample correlation function and what is referred to as the *structure function* in solid state physics. In the astrophysical context, we are talking about the fluctuating component of the density field relative to the mean cosmic density, not the density itself. This is a natural definition in the cosmological context where the clustering is thought to be driven by the force of gravity acting on fluctuations in density about the mean value.

The fractal measures we shall be using later on refer to the distribution of the total number of galaxies in cells, not the number relative to the mean cell occupancy. However, since the analysis is invariably applied on scales where the density excursions about the mean are large ($\xi \gg 1$), there is little practical difference.

3 THE GALAXY CLUSTERING HIERARCHY

The logical step after getting the two-point correlation function was to move on to the three-, four- and higher-point correlation functions. The three-point function is defined analogously with the two point function, but instead of involving the excess probability (relative to a Poisson distribution of objects) of finding a *pair* of galaxies with a given separation, it involves the excess probability of finding a *triplet* of galaxies in a particular triangular configuration having sides r_{12}, r_{23} and r_{31}. The excess is measured over and above a Poisson distribution of objects and over and above what would be expected simply on the basis of there being a nonzero two-point correlation function. The calculation of these higher order correlation functions involves increasingly more computational effort, so smart tricks have been developed to estimate them.

The three-point function for galaxies was found to be a power law, but what was perhaps surprising was that for a variety of differently shaped triangles it could be expressed as a simple sum of products of two point functions:

$$\xi^{(3)}(r_{12}, r_{23}, r_{31}) = Q[\xi(r_{12})\xi(r_{23}) + \xi(r_{23})\xi(r_{31}) + \xi(r_{31})\xi(r_{12})] + \overline{Q}\xi(r_{12})\xi(r_{23})\xi(r_{31}),$$

with

$$Q = 0.66 - 1.29, \qquad \overline{Q} = 0. \tag{3}$$

The quoted range of Q values covers the range reported in the literature. The absence of the triple-term $\xi(r_{12})\xi(r_{23})\xi(r_{31})$, expressed by $\overline{Q} = 0$, is significant. If this term were nonzero it would dominate the three-point function at small separations. As a consequence of $\overline{Q} = 0$, we have $\xi^{(3)} \propto r^{-2\gamma}$, where r here refers to some lengthscale associated with a set of similar triangles.

Several points should be noted about this result. Firstly, there is an identical result for the clustering of galaxy clusters. Secondly, the result has only been established in the nonlinear clustering regime. Thirdly, there is a similar result for the four-point function $\xi^{(4)}$, which can also be expressed as a sum of products of two point functions: $\xi^{(4)} \propto r^{-3\gamma}$.

An important generalization of this scaling behaviour to all orders of correlation functions is to postulate that the qth order function $\xi^{(q)}$ based on an q-agon of points $\{r_i\}$ scales as

$$\xi^{(q)}(r_1, \ldots, r_q) = \lambda^{(q-1)\gamma} \xi^{(q)}(\lambda r_1, \ldots, \lambda r_q) \qquad (4)$$

Such a scaling behaviour does not necessarily demand that the higher order correlation functions can be expressed as sums of products of two-point functions, but it is consistent with expectations based on the BBGKY hierarchy description of clustering. One feature of this hierarchy is that it is described by a single scaling index γ. We shall show that in models of the galaxy distribution having multifractal scaling, the constant scaling relationship (4) between successive orders is achieved, but only for large q. The consequences of the simple scaling law (4) have been extensively discussed in a series of papers by Balian and Schaeffer (1989).

4 SCALING ANALYSIS

The question arises as to whether it is possible to look for the scaling properties of the galaxy distribution directly in terms of the distribution of galaxy counts in cells, rather than through the intuitively less accessible hierarchy of correlation functions. This leads us to modelling the galaxy distribution by fractals. For an introductory review of this subject, see Martínez (1990). I shall take two approaches to this question, one is to look directly at the statistical moments of the cell counts (the partition function) and the other is to look at the local scaling properties of the distribution. We shall see how under quite general circumstances these two approaches can be linked.

The Partition Function

Divide a sample volume containing N galaxies into N_r cells of scale r. Let the ith cell contain $n_i(r)$ galaxies (we keep the r-dependence to emphasize how counts depend on the partition). Consider the "partition function" for this sample:

$$Z(q,r) = \sum_{i=1}^{N_r} \left[\frac{n_i(r)}{N} \right]^q \qquad (5)$$

Note that in the estimator of $Z(q,r)$ the ordering of the cells is not important and so the information on the relationship between neighbouring cells appears through the r-dependence of Z.

$Z(q,r)$ is related to the qth. moment of the distribution of the points as viewed in cells of size r. If the sample distribution is drawn from a probability density $p(n;r)$ for finding n galaxies in a randomly placed cell of size r, the qth. moment of the cell occupancy is

$$m_q = \sum_{n=0}^{\infty} p(n;r)n^q, \qquad \overline{Z}(q,r) = \frac{N_r}{N^q}m^q. \qquad (6)$$

Z in equation (5) is a sample estimate of $\overline{Z}$ in equation (6).

The situation of interest is when, for all values of q, the function $Z(q,r)$ scales as a power law in r: [2]

$$Z(q,r) \propto r^{\tau(q)} \qquad (7)$$

It is usual to put

$$\tau(q) = (q-1)D(q) \qquad (8)$$

The factor $(q-1)$ appears because $Z(q=1,r)$ must be scale independent: it is just the sum of the cell occupancies, which is clearly independent of how the domain is divided into cells. The function $D(q)$ defined this way is a measure of the "generalized dimension of order q" of the distribution.

Note the resemblance between (7) with (8) and the correlation function scaling hierarchy (4) with the constant γ replaced by the q-dependent $D(q)$. We shall return to this later. It is also of interest to note that this scaling of Z implies that the moments of the underlying distribution have a particular scaling behaviour (equation (6)).

The $D(q)$
It can be shown that $D(q)$ is a monotonic decreasing positive function. Subject to that constraint, the function $\tau(q) = (q-1)D(q)$ in equation (7) could in principle be any arbitrary function of q. However, it will be most convenient to restrict attention to that class of functions that also allows for interpretation of the distribution in terms of local scaling indices. The class of distributions that can so be described are then said to display *multifractal scaling*. This imposes conditions on $\tau(q)$, among which is the restriction that $\tau''(q) < 0$ everywhere. It is evident from equation (6) that the $D(q)$ merely specify the way in which the statistical moments of the distribution scale with cell size.

---

[2] Technically, this scaling is only required in the limit $r \to 0$. However, in practise we ask only that this scaling behaviour hold over some substantial range of r.

There are technical problems in calculating $D(q)$, especially in the case of negative q-values and one has to resort to tricks. These tricks are discussed fully elsewhere (van de Weygaert, et al., 1992), and in particular in the astrophysics context by Martínez et al. (1990). The details need not concern us here except to point out that getting $D(q)$ in the range $-1 < q < 4$ presents no real difficulties. Going outside that range needs caution.

$D(0)$ and $D(2)$

$D(0)$ merely characterizes how the number of occupied cells scales as a function of cell size and so tells us only about the geometry of the distribution of points. This is evident since when $q = 0$ we have

$$Z(q,r) = \sum_{i=1}^{N} \left[\frac{n_i(r)}{N} \right]^0 = \sum_{occupied\ cells} 1 \propto r^{-D(0)}.$$

(empty cells contribute nothing to the sum) and so the number, or fraction, of occupied cells scales as $r^{-D(0)}$. Consequently, we have the simple yet important formula for $D(0)$ [3]:

$$D(0) = \lim_{r \to 0} \frac{\log N_{occupied}}{\log r} \tag{9}$$

where $N_{occupied}$ is the number of occupied cells on scale r. $D(0)$ goes under various names: $D(0)$ is the "Box- Counting" or "Capacity Dimension" and is used as an estimator of the "Hausdorff Dimension".

$D(2)$ measures the scaling of the variance of the counts in cells, and is therefore related directly to the slope γ of the two point correlation function:

$$D(2) = 3 - \gamma. \tag{10}$$

$D(2)$ goes under the name of the "correlation dimension".

Large $|q|$-values

It can be seen that if we go to very large $q > 0$ then in a finite sized sample the formula (5) picks out the dominant cluster in the distribution and we have

$$Z(q,r) \simeq \left[\frac{n_*(r)}{N} \right]^q \propto r^{(q-1)D(q)}$$

where $n_*(r)$ is the mass profile of the densest cluster. If the density profile is $\rho(r) \propto r^{\alpha_* - 3}$ then the mass profile is

$$n_*(r) \propto r^{\alpha_*}$$

---

[3] The limit as $r \to 0$ has been reinstated here to conform with the usual definition of the Capacity of the set, see the previous footnote.

and we have

$$D(q) \rightarrow \alpha_* \qquad q > q_* \gg 1.$$

A similar conclusion can be drawn for large negative q-values which describe the density distribution in the dominant voids. This means that estimating the $D(q)$ for large $|q|$ is strongly sample dependent. The estimates depend sensitively on the shape of the dominant voids or clusters in the particular realization of the galaxy distribution process (ie. the galaxy catalogue).

This asymptotic constancy of $D(q)$ for large q is a generic property of $D(q)$ and is not an indication that the distribution looks like a homogeneous fractal.

Homogeneous Fractals
If all the $D(q)$ are equal, then we have a situation where the distribution is char-acterised by one scaling index. We then have a homogeneous fractal (occasionally referred to as a "mono-fractal"). The standard models of clustering hierarchies are such fractals (Peebles, 1980). Martínez and Jones (1990) demonstrated that in the distribution of galaxies in the CfA catalogue, $D(0) \neq D(2)$, and so although the ac-tual distribution of galaxies displays scaling properties, they are more complex than the scaling properties of a homogeneous fractal. We shall discuss this below.

Large $|q|$-values again - bifractals and all that
An important consequence of the large $|q|$ behaviour $D(q) \rightarrow \alpha_*$ discussed above is that the high qth order moments all scale as $\lambda^{(q-1)\alpha_*}$. Thus the scaling of the high order correlation functions in multifractal models is as shown in equation (4) with $\gamma = \alpha_*$.

The constancy of $D(q)$ for large $q > 0$ does not by itself sufficient for homogeneous fractal scaling, which requires that $D(q)$ be constant for all q. Nevertheless, we might contemplate a $D(q)$ which takes the constant value D_- for $q < 0$ and the constant value D_+ for $q > 0$. Such a distribution has been referred to as a *bifractal*, intending to convey the impression that the distribution looks like two interwoven fractals: a fractal distribution of voids ($q < 0$) and a fractal distribution of clusters ($q > 0$). The argument is at best misleading since such a $D(q)$ does not describe a multifractal. There is, for multifractals, a mathematical constraint that the transition from D_- to D_+ be smooth enough to allow for the existence of a set of scaling indices. (This is the condition alluded to earlier that $\tau''(q) < 0$ everywhere). We shall return to this later.

5 SCALING IN GALAXY CATALOGUES AND N-BODY EXPERIMENTS

The CfA catalogue of galaxy redshifts (Huchra, 1988) provides a small sample of the local volume of the Universe which has been the subject of many statistical analyses. Scaling analysis of the kind just described was first done by Jones et al. (1988), with subsequent in-depth analysis by Martínez et al. (1990) and by Martínez and Jones (1990). The first conclusion is that the distribution of galaxies in that catalogue does display scaling properties (as expected from the two point correlation function analysis), but the scaling is not as simple as in a homogeneous fractal distribution of the kind first suggested by Mandelbrot: the function $D(q)$ does have a strong q-dependence. The key result is that

$$D(0) = 2.1 \pm 0.1 \qquad (11)$$
$$D(2) = 1.3 \pm 0.1$$

The important fact is that $D(0) \neq D(2)$. The value of $D(2)$ is in accord with what is expected from the slope $\gamma = 1.8$ of the two-point correlation function: $D(2) = 3 - \gamma$.

Valdarnini et al. (1992) have studied the scaling properties of a variety of N-body models that include biased galaxy formation. The models display multifractal properties on scales where the clustering is nonlinear. The asymptotic values of $D(q)$ are $D(q) \simeq 1$ for large $q > 0$ and $D(q) \simeq 3$ for large $q < 0$. It certainly appears possible to reproduce the observed $D(q)$ curve.

6 SIMPLE AND MULTIFRACTAL SCALING

The scaling distribution can be described explicitly in terms of "local scaling indices" α that describe the variation of mass about a point: $n(r) \propto r^\alpha$, with α varying from place to place in the sample. That description is an important dual description of clustering with multifractal scaling properties. Here I shall follow the discussion of Kadanoff et al. (1989) who begin by distinguishing two kinds of scaling.

Consider the probability $p(X, L)$ of measuring a value X for some property of the system when the system has been binned into cells of scale L. In "Simple" or "Finite size" scaling, we have

$$p(X, L) = L^{-\beta} g\left(\frac{X}{L^\nu}\right) \qquad (12)$$

for some function $g(x)$ and constants β and ν. In this case, the quantity X is distributed on a fractal set with a single scaling index.

There is a more complicated kind of scaling, "Multifractal Scaling" in which we have

$$\frac{\log p(X, L)}{\log \frac{L}{L_0}} = -f\left(\frac{\log \frac{X}{X_0}}{\log \frac{L}{L_0}}\right) \tag{13}$$

Here, X_0 and L_0 can be thought of as the physical units in which the quantities X and L are to be measured. It is convenient to define the "local scaling index"

$$\alpha = \frac{\log \frac{X}{X_0}}{\log \frac{L}{L_0}}. \tag{14}$$

With this we have $X \propto L^\alpha$, though α may vary as a function of position. This is why α is referred to as a *local* scaling index. Then the probability of finding value X in a cell of size L is just

$$p(X, L) = p(X_0, L_0)\left(\frac{L}{L_0}\right)^{-f(\alpha)} \tag{15}$$

We have power law scaling with the cell size, but the scaling index is itself an arbitrary function of the quantity X and the cell scale L. These two forms of scaling (12) and (15) agree only when $g(x)$ is a power law and $f(x)$ is linear.

Although α may vary as a function of position, its very definition says that the set of points with $\alpha = $ const. is a fractal of dimension $f(\alpha)$ (compare equation (14) with equation (9)). Since the set is not characterised by a unique value of α, the distribution can be thought of as a set of interwoven fractals having different dimensions. We call this structure a "multifractal".

A set of points distributed in power law clusters does not necessarily constitute a multifractal distribution. In order to be a multifractal distribution the scaling indices have to be constant on homogeneous fractal sets. Thus not all distributions of points are multifractals, even if they are distributed in power law clusters.

The scaling transformations (13) have the important property of forming a group. Thus rescaling the distribution first by a factor L_1 and then by a factor L_2 is equivalent to a single rescaling of the original distribution by a factor $L_1 L_2$. The qth. moment of the quantity X scales under this transformation as

$$\sum p(X, L)X^q \propto L^{(q-1)D(q)} \tag{16}$$

This is essentially a restatement of the assumed scaling of the partition function (7). However, its importance lies in recognizing that the dynamical processes that organize large scale structure must preserve this invariance.

7 THE $f(\alpha)$ SPECTRUM

Relating $f(\alpha)$ and $D(q)$

The function $f(\alpha)$ introduced in the last section provides another very important way of characterising the scaling properties of a distribution. It is in fact closely related to the generalized dimension $D(q) = (q-1)^{-1}\tau(q)$ derived from the partition function via the equations

$$\tau(q) = \alpha q - f(\alpha) \tag{17}$$

$$\alpha(q) = \frac{d\tau}{dq}$$

These are recognised as a Legendre transform $(q, \tau(q)) \to (\alpha, f(\alpha))$. This means that we can think of the distribution of points either in terms of the scaling of the statistical moments of the distribution, or in terms of the fractal distribution of local scaling indices, α.

The existence of this Legendre transform requires several conditions, among which is the negativity of the derivative $f''(\alpha)$. This is related to the requirement that the mapping $\alpha \leftrightarrow q$ be one to one and invertible. Thus if $D(q)$ took only two different constant values for $q < 0$ and $q > 0$, then the rapid jump at $q = 0$ between these values would result in a multivalued $q(\alpha)$. The bifractal distribution cannot be expressed in terms of scaling indices and is not a multifractal distribution. This does not in itself invalidate the bifractal as a model for the galaxy distribution, but it does show that being a multifractal distribution is quite special.

Deriving the $(q, \tau) \leftrightarrow (\alpha, f)$ transformation

We can estimate the sum (5) for a given value of q by looking at the cells of the partition that make the greatest contribution to the sum. Which terms dominate the sum will in general depend on the scale r. We may suppose that the counts in those cells have the scaling behaviour

$$\frac{n_i}{N} \propto r^{\alpha q}. \tag{18}$$

The question then is how many such cells contribute to the sum? If these cells make up a fractal of dimension f_q, the number of such cells scales as r^{-f_q} and the sum in (5) is approximated by

$$Z(q, r) \simeq \sum_{dominant\ cells} \left(\frac{n_i(r)}{N}\right)^q \propto r^{q\alpha_q - f_q} \tag{19}$$

and so

$$\tau(q) = (q-1)D_q = q\alpha_q - f_q. \tag{20}$$

This brings in the quantities f_q and α_q, but we need another equation relating α and q directly so that q can be eliminated in favour of α everywhere.

The trick is to write the partition function as an integral rather than a sum, using α as an independent variable. Up to logarithmic terms

$$Z(q,r) \simeq \int r^{q\alpha' - f(\alpha')} d\alpha' \simeq \left[\frac{2\pi}{f''(\alpha(q))}\right]^{1/2} r^{q\alpha(q) - f(\alpha(q))}. \tag{21}$$

This approximation of the integral using Laplace's method works provided $f(\alpha)$ satisfies certain conditions, among which is the negativity of $f''(\alpha)$ alluded to earlier. The main contribution to the integral comes from the place where $q\alpha' - f(\alpha')$ is a maximum, that is where

$$\left.\frac{df(\alpha)}{d\alpha}\right|_{\alpha=\alpha_q} = q. \tag{22}$$

Differentiating (20) and using (22) to cancel some terms, we get

$$\frac{d}{dq}(q\alpha_q - f_q) = \alpha_q + q\frac{d\alpha_q}{dq} - \frac{df_q}{d\alpha_q}\frac{d\alpha_q}{dq} = \alpha_q = \frac{d\tau}{dq}. \tag{23}$$

which is the second of equations (17). Equations (20) and (22) comprise the Legendre transform relationship between the pair of variables (q,τ) and $(\alpha, f(\alpha))$. If we calculate the function D_q, we can use (22) to get $\alpha(q)$ and then (20) to get $f(\alpha)$.

The function $f(\alpha)$ is called the "$f(\alpha)$ spectrum" and it is by construction the same function as appears in the discussion of scaling in section 6.

Properties of $f(\alpha)$
Given that $D(q)$ is monotonic decreasing, equations (17) can be used to show that $f(\alpha)$ is a convex function with a maximum value equal to $D(0)$, the Hausdorff dimension. The fact that $D(q)$ is a monotonic decreasing function also shows that $f(\alpha)$ spans a finite range of values of α:

$$\alpha_{min} = \lim_{q\to\infty} D_q, \qquad \alpha_{max} = \lim_{q\to-\infty} D_q. \tag{24}$$

The maximum value of α describes the scaling properties of the distribution in the underdense regions, while the minimum values of α describe the distribution in the densest regions.

The curvature of the $f(\alpha)$ curve in the vicinity of the maximum can be expressed in terms of the scaling index $\tau(q) = (q-1)D(q)$ of the partition function:

$$\frac{d^2f}{d\alpha^2} = \left(\frac{d^2\tau}{dq^2}\right)^{-1} = \frac{1}{\tau''(q)}. \tag{25}$$

If $\tau''(q)$ is zero, the function $f(\alpha)$ reduces to a single point and $D(q)$ is constant, we have a fractal described by a single dimension - a homogeneous fractal. The value of $\tau''(q)$ is thus a measure of the richness of the scaling structures within a fractal set.

We can use the curvature of the $f(\alpha)$ curve in the vicinity of its maximum to approximate it by a parabola. This brings in the relationship between multifractals and multiplicative statistical distributions like the Lognormal (Jones et al., 1992).

8 THE STATISTICS OF COUNTS-IN-CELLS

What does the fact of multifractal scaling imply for the statistical distribution of the galaxy cell counts? In the past most descriptions of clustering were given in terms of such cell count statistics. The existence of multifractal scaling has implications for how the cell count distribution depends on the cell size. I will tell how the fact of being a multifractal implies that the galaxy counts in cells distribution approximates a Lognormal distribution (Jones et al., 1992). Thus the galaxy distribution is characterised by a single distribution function (as opposed to a family of $N-$particle distribution functions) with rather specific and verifiable scaling properties. The $N-$point correlation functions are a consequence of these scaling relations.

The starting point is the $f(\alpha)$ spectrum which describes the local variation of cell occupancy with cell size. For the CfA Catalog of Galaxies we observe that, at least in the vicinity of its maximum, $f(\alpha)$ is reasonably approximated by a parabola:

$$f(\alpha) = D_0 - \frac{(\alpha - \alpha_0)^2}{4(\alpha_0 - D_0)} \tag{26}$$

with

$$\alpha_0 \simeq 2.6, \qquad D_0 = 2.1 \pm 0.1$$

It is a simple exercise to show that the $D(q)$ curve for this $f(\alpha)$ is simply a straight line:

$$D(q) = (D_0 - \alpha_0)q + D_0. \tag{27}$$

This is a good fit to the data (Martínez and Jones, 1990) on the interval $-1 < q < 2$ and has correlation dimension

$$D_2 = 1.1 \pm 0.3$$

which is consistent with the observed slope of the galaxy 2-point correlation function $\gamma = 3 - D_2$.

Of course, the simple straight line approximation (27) to $D(q)$ should not be taken too far: it goes negative for $q > 4$ and is therefore unacceptable as an approximation for large q. Bearing this reservation in mind, we can proceed to calculate the underlying probability density governing the distribution of galaxies. Jones et al. (1992) find that the probability of finding η_r galaxies in a cell of size r is

$$P(\eta_r) \propto \frac{N}{\eta_r} \frac{1}{\log \dfrac{r}{r_s}} \exp\left(-\frac{[\log(\eta_r/N) - \alpha_0 \log(r/r_s)]^2}{4(\alpha_0 - D_0)\log(r_s/r)}\right) \qquad (28)$$

with mean μ and variance σ^2 scaling as

$$\mu = \log N - \alpha_0 \log(r_s/r), \qquad \sigma^2 = 2(\alpha_0 - D_0)\log(r_s/r). \qquad (29)$$

The distribution (29) is the Lognormal distribution, which fits both the observed galaxy distribution and the N-body simulations reasonably well. Its major failing is when describing the distribution of the rare dense objects (Coles and Jones, 1991). These scaling laws are in principle verifiable, but this has not been done yet.

9 MODEL FOR MULTIFRACTAL DISTRIBUTIONS

What kind of point distributions lead to Multifractal distributions of the kind that are being claimed here? We can give two simple examples, one which throws some light on limitations of the standard "power law cluster" model and the other showing a hierarchical clustering model that does the trick.

Power Law Clusters

In the original power-law cluster model (Peebles, 1980, section 61) the galaxies are randomly placed in randomly distributed spherical clumps, all having the same radius R and having the same density profile $\rho \propto r^{-\epsilon}$. The model gives the correct 2-point function with $\gamma = 2\epsilon - 3$. The model is claimed to fail because it gives the wrong slopes for the 3- and 4-point functions.

Instead, we can place the galaxies in spherical clusters with a density profile $\rho \propto r^{-\alpha}$, where α is scale dependent, but nevertheless constant on some homogeneous fractal set. We are free to choose these fractal sets as we please, and we can write their dimension as $f(\alpha)$. This is no more than a restatement of equation (15), but we still need a prescription for constructing a density distribution.

Consider a patch of universe of scale R. If we take the value of α to be constant on sets such that $r/R = \text{const} < 1$ we can calculate a density distribution function for the counts of galaxies in cells of scale r:

$$P(n_r) \propto \frac{1}{\eta_r} \frac{1}{\log \dfrac{r}{r_s}} \left(\frac{r}{r_s}\right)^{D_0 - f(\alpha(r/R))} \qquad (30)$$

(see, for example, Jones et al. 1992). For one patch, this matter distribution would look like a cluster surrounded by smaller clusters, and so on, but the distribution would not be space filling since the Hausdorff dimension would be less than 3. The structure would look somewhat "dendritic". Many such patches can be randomly superposed to give a global picture of the matter distribution.

Models having different scaling behaviours on different scales have been discussed by Castagnoli and Provenzale (1991). The difficulty with such models is linking them to a physical mechanism. The multiplicative models discussed next may have more appeal in this respect.

Multiplicative Rescaling Hierarchy

A model which suggests a gravitational collapse hierarchy was presented by Jones et al (1992). In that model, a cubic volume is divided into 8 subcubes, and the matter in the parent volume is redistributed among the subcubes so that 4 of the eight are empty, and the other four receive fractions f_1, f_2, f_3, f_4 of the matter in the parent[4]. The assignment of which subcubes are empty and which subcubes get which fraction f_i is random. The matter in each of the nonempty subcubes is then redistributed in exactly the same way. Each subcube is divided into eight and the matter in the subcube redistributed so that half its subcubes are empty and the other half get fractions f_i of the available material. This is a multiplicative random process.

Emptying half the volume at each stage forces a Hausdorff dimension $D_0 = 2.0$. Choosing $f_1 = f_2 = 0.07, f_3 = 0.32, f_4 = 0.54$ gives an $f(\alpha)$ curve that fits the data reasonably well over the range $-1.5 < q < 2.5$ and acceptably over a larger range of q. Two of the cells get most of the material, but it is important that the other two cells be not empty, otherwise the Hausdorff dimension of the final distribution would be wrong. This corresponds to our intuitive view of a gravitational collapse hierarchy in which most but not all of the material fragments into two parts as it collapses, and then each fragment in turn goes through the identical process. Since gravitation is scale independent it seems not unreasonable to have the same matter redistribution at each level of the hierarchy.

10 CONCLUDING REMARKS

The fact that the distribution of galaxies appears to display multifractal scaling is an important observation. Taken together with the fact that the same kind of scaling is seen in N-body experiments, the question arises as to what does this scaling tell us about the dynamical processes that drive the distribution? It seems clear that inventing *ad hoc* distribution functions that fit the data is not very edifying unless the distribution function is motivated by a dynamical argument.

---

[4] The number of subcubes, 8, is only a matter of geometric convenience. Any number would do provided half the volume is swept clean at each stage.

The BBGKY approximation has provided a dynamical motivation for the simple hierarchical models, but that is only mildly nonlinear. The nonlinear alternative appears to be a hierarchical fragmentation model along the lines originally proposed by Hoyle, but modified so as to produce multifractal scaling. This model exploits the self similarity that inevitably arises because the gravitational force is unshielded and of infinite range.

ACKNOWLEDGEMENTS
My thanks go first of all to Dennis Sciama, who taught me how to be a physicist and through whom I learned that there was more to scientific understanding than displaying a collection of equations or computer generated points. I also thank my own students, Dennis' "grandstudents", for spending their valuable time discussing the Universe, and in particular those who have collaborated with me on the multifractal nature of cosmic clustering.

References
Balian, R. and Schaeffer, R., 1989, *Astron. & Astrophys.*, **220**, 1 and **226**, 373.

Castagnoli, C. and Provenzale, A., 1991, *Astron. & Astrophys.*, **246**, 634.

Coles, P. and Jones, B.J.T., 1991, *Mon. Not. R. astr. Soc.*, **248**, 1.

de Lapparent, V., Geller, M.J. and Huchra, J.P., 1986, *Astrophys. J.*, **302**, L2.

Halsey, T.C., Jensen, M.H., Kadanoff, L.P., Procaccia, I. and Shraiman, B.I., 1986, *Phys. Rev. A*, **33**, 1141.

Huchra, J., 1988, *The CfA Redshift Catalog* available on E-mail from the author.

Jones, B.J.T., Coles, P. and Martínez, V.J., 1992, *Mon. Not. R. astr. Soc.*, to appear.

Jones, B.J.T., Martínez, V.J., Saar, E. and Einasto, J, 1988, *Astrophys. J. (Letters)*, **332**, L1.

Kadanoff, L.P., Nagel, R., Wu, L. and Zhou, S., *Phys. Rev. A*, **39**, 6524.

Maddox, S.J., Efstathiou, G., Sutherland, W.J. and Loveday, J., 1990, *Mon. Not. R. astr. Soc.*, **242**, 43P and **243**, 692.

Martínez, V.J., *Vistas in Astronomy*, **33**, 337.

Martínez, V.J. and Jones, B.J.T., 1990, *Mon. Not. R. astr. Soc.*, **242**, 517.

Martínez, V.J., Jones, B.J.T., Dominguez-Tenreiro, R. and van de Weygaert, R., 1990, *Astrophys. J.*, **357**, 50.

Peebles, P.J.E., 1980, *Large scale Structure of the Universe*, Princeton.

Valdarnini, R., Borgani, S. and Provenzale, A., 1992, *Astrophys. J.*, **394**, 422.

van de Weygaert, R., Jones, B.J.T. and Martínez, V.J., 1992, *Phys. Lett. A*, to appear.

Quasars: Progress and Prospects

MARTIN J. REES

Quasars offer important clues to the process of galaxy formation and the epoch when it occurred. Although they almost certainly involve relativistic processes close to a collapsed object, quasars have unfortunately not yet given us any real tests of strong-field gravity.

1 INTRODUCTION

In December 1963, the first Texas Conference on Relativistic Astrophysics was held in Dallas. Quasars had just been discovered, and were already being interpreted as gravitationally-collapsed massive objects. In his after-dinner speech, Thomas Gold said that relativists were "not only magnificent cultural ornaments, but might actually be useful to science .... What a shame it would be if we had to dismiss [them all] again". We haven't had to do so – on the contrary, 'relativistic astrophysics' is a subject with ever-widening scope. It burgeoned with the detection of the microwave background in 1965, of neutron stars in 1967, and of the first stellar-mass black hole candidates in 1971. Dennis Sciama's research group was at the centre of all the key debates throughout that exciting period. I was myself fortunate to begin research in 1964, when these developments were just gaining momentum. It was my great good fortune to have been assigned as one of Dennis' students, and he has been a valued mentor and advisor ever since.

It has to be admitted that our understanding of quasars has advanced rather slowly. On the theoretical front, we have sometime had the illusion of rapid progress, but there has really been a rather slow advance, with large amplitude sawtooth fluctuations superimposed on it as fashions have come and gone. On the observation front, progress has at least been 'monotonic', but only in the last few years have observers provided quasar samples selected by well-defined criteria, which reveal the shape of the luminosity function and show how it has evolved with cosmic time. And there are now much more extensive data on the spectra, structure, and time-variation of individual quasars.

The most remarkable feature of the quasar population is the sharp decline, by a factor $10^2 - 10^3$ in comoving density, between $z = 2$ and the present epoch ($z = 0$). This has been known for 20 years, and was prefigured even earlier by the radio source counts, to whose interpretation Dennis himself contributed. The extension to still higher redshifts has been more recent – five years ago, no quasars were known with redshifts exceeding 4, whereas there are now about 40 in this category. There is some evidence that the comoving density declines at $z > 3.5$, but the steepness (and even the reality) of this trend is still controversial.

I shall touch on several interlinked questions:
(a) How much of a problem for theories of galaxy formation are the highest-z quasars, now being found in increasing numbers?
(b) How quickly and efficiently can a black hole form and grow in the core of a young galaxy?
(c) What can we learn about quasars from the dead remnants that may lie in the centres of nearby galaxies?

2 FORMATION OF THE EARLIEST QUASARS

To put the high-redshift objects into a cosmological context, it is useful to recall the relation between time and redshift in the standard models. For an Einstein-de Sitter model

$$t(z) = 13.5h_{50}^{-1}(1 + z)^{-3/2} \text{ yrs.} \tag{1}$$

The highest-z known quasars formed when the Universe was only 10^9 years old, when t was as little as $0.07t(0)$.

The bigger redshifts now being discovered enable us to probe earlier epochs than could have been done in the pioneering days of quasar research. The 'record' redshift has, year by year, gone up. But at the same time, the realisation that galaxies (with their dark halos) are more extended and diffuse than was previously suspected has lowered theoretical estimates of the redshift of galaxy formation. In the 1970s, it was widely believed that 'galaxy formation' happened at earlier epochs than were directly observable. But current ideas on extended halos, and on the origin of galactic rotation, suggest – almost irrespective of the cosmogonic model – that galaxy formation must still have been going on at eras that we can directly observe (corresponding to redshifts $z > 4$.

At the epoch corresponding to $z = 5$, the cosmic expansion timescale (eqn (1)) is long compared with the dynamical timescale within the 'luminous' part of a typical galaxy. But it is *not* long compared to the timescales for *extended halos*: the free-fall time from a radius r in a galactic halo like our own is $\sim 10^9$ ($r/100$ kpc) years, and the Universe must be at least twice as old as this before material at radius r can virialise.

The angular momentum of galaxies is believed to have been acquired by tidal inter-actions with their neighbours at the epoch of turnaround. This process can only, however, impart a transverse velocity that is 5–10 per cent of what is needed for rotational support. If the discs acquired their angular momentum from tidal torques, this implies (e.g. Fall and Efstathiou 1980) that the material now residing at ~ 10 kpc in a disc (e.g. the Solar Neighbourhood in our own Galaxy) must have fallen in from ~ 100 kpc and then cooled down. At redshifts of 4 or 5, galaxies would not have acquired extended virialised halos – indeed, they would not yet have developed discs.

The fact that quasars formed so early in cosmic history is an important constraint on models for galaxy formation, particularly on 'top down' models in which large-scale structures develop before individual galaxies; for instance, the simple adiabatic ('pancake') model dominated by neutrinos cannot, when the amplitude is normalised to fit clustering data or the microwave background anisotropies revealed by COBE, account for collapsing systems at such high redshifts.

I shall focus my discussion on the incorporation of quasars into the standard Cold Dark Matter model for galaxy formation. This is just one of a number of alternative models currently 'in play' – it has suffered several alleged 'deaths', but had as many resurrections. 'CDM' makes definite predictions about the fluctuation spectrum; it probably cannot, without some modification or supplementation, account for the ob-served large scale structure (see Bernard Jones's contribution), but it offers gratifying agreement with the properties of galactic halos, groups and clusters.

A feature of the CDM model is that structures build up hierarchically – from small scales to large – but that bound systems develop rather late. Mergers are important even at redshifts of 1 or 2, and large galactic halos would generally have assembled rather recently. In the 'standard' version of the model the fluctuations are Gaussian; only the most exceptionally high amplitude fluctuations would have led to galactic-mass systems that had already virialised as early as $z > 4$.

Even the purely gravitational (non-dissipative) features of galaxy formation pose a severe challenge to numerical simulations. This is especially the case in the CDM model where, because of the flat fluctuation spectrum on subgalactic mass scales, a

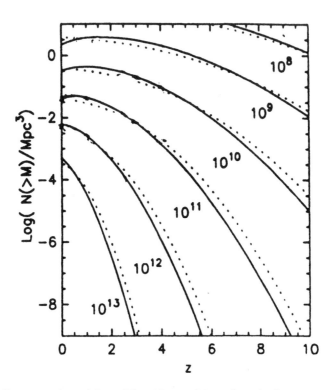

Figure 1. This diagram, adapted from Efstathiou and Rees (1988), shows, as a function of epoch, the comoving number-density of virialised halos of different masses that is expected on the basis of a CDM cosmogony. The build-up of structure is hierarchical, so the larger-mass systems form more recently, and their density 'thins out' towards higher redshifts because only exceptionally high peaks of an initially Gaussian distribution would already have collapsed at high redshifts.

very large dynamic range is required. For this reason, much can still be learnt from analytical studies. There have been many of these, going back to the classic paper by Press and Schechter (1974). Later authors, especially Bardeen *et al.* (1986) and Peacock and Heavens (1990), have discussed the statistics of Gaussian fluctuations, and the quantitative details of the hierarchical build-up of collapsing and virialised systems.

The earliest quasars would be expected to develop in the first sufficiently massive and deep potential wells that virialise. These would arise from high amplitude peaks in the initial dark matter distribution.

Figure 1 shows the comoving density of virialised halos of various masses, as a function of z. At high z, the curves drop precipitously, reflecting the extreme rareness of very high-amplitude peaks in a Gaussian distribution. The drop occurs at a smaller redshift for the higher-mass systems, because the amplitude of the CDM spectrum decreases towards bigger scales. Even though the hosts of quasars may now be $10^{12} M_\odot$ halos, the relevant curve in Fig 1 may be the one corresponding to a lower mass because quasar activity could in principle occur in a sufficiently deep and compact potential well of only $10^{10} M_\odot$. A massive galaxy can build up via 'nucleation' around a high-amplitude peak; this may amount to only about 1 per cent of the final galaxy, but the baryons that participate in the quasar activity are predominantly those which were originally in this central 1 per cent.

Consider a system which collapses by a redshift 5. The inner $10^{10} M_\odot$ would attain virial equilibrium with a characteristic (half-mass) radius of 5 kpc. What happens to the fraction Ω_b/Ω_{total} (0.05–0.1 for "standardised parameters") of this mass that is in the form of baryons? The system is in the part of the mass-radius diagram where cooling is very efficient. The gas (if it did not turn immediately into stars) would condense towards the centre until angular momentum became important. The angular momentum is typically $\lambda \simeq 0.05$ of that needed for rotational support at the turnaround radius. This means that the gas can fall to $\sim 200(\lambda/0.05)^2$ parsecs even without getting rid of any angular momentum. Its self-gravitation becomes important after falling in by a factor that would be $\sim \Omega_b/\Omega_{total}$ if the dark matter had an r^{-2} density distribution (the Fall-Efstathiou case), but would be closer to $(\Omega_b/\Omega_{total})^{1/3}$ if the dark matter had a non-singular core of radius $\gtrsim 5$ kpc.

The free-fall timescale from 5 kpc is less than 10^8 years. Star formation could occur during this infall – indeed, this is the way the bulge of a large galaxy might form. Figure 2 shows schematically the processes involved. Production of heavy elements via high-mass stars can occur during this time. The 'branching ratio' between star formation and direct infall cannot be predicted.

The angular momentum of the material that is likely to form the quasar would prevent it from falling closer than 200 pc to the centre. The dynamical timescale at this radius is 3.10^6 years. The next question is therefore how quickly this material – unprocessed gas, together with material that has been processed through stars – can shed its angular momentum and accumulate closer to the centre. Specifically, it is important to know whether this gas can lose its angular momentum in less than 100 orbital periods (i.e. $\lesssim 3.10^8$ years). If it cannot, then it will not trigger a quasar as early as the epoch $z = 5$.

One certainly cannot give a rigorous answer to this question. However, it may be worth listing three types of viscosity that are expected to operate.

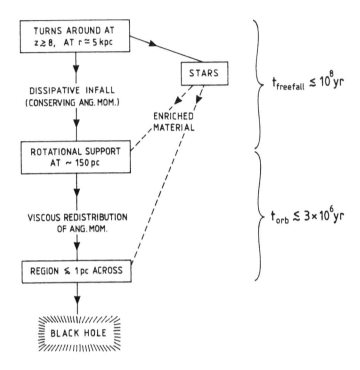

Figure 2. A schematic illustration of what happens in the central $10^{10} M_\odot$ ($10^9 M_\odot$ of baryons) of a bound system that collapses at $z \simeq 5$. The outcome will be a stellar bulge surrounding a massive black hole. The fraction of the mass going into the hole depends on the speed with which angular momentum can be lost, on the efficiency of star formation, and on feedback effects from high-mass stars. Note that heavy elements can be synthesised within the period over which the black hole forms.

(a) Gravitational instabilities

Provided that star formation doesn't irreversibly consume too much of the gas during the infall from ~ 5 kpc (see Figure 2), it would be self-gravitating by a large margin by the time it collapsed into a rotationally-supported configuration. The non-axisymmetric instabilities to which it would then be vulnerable would redistribute angular momentum on a timescale of only a few t_{orb} (see for, instance, Hernquist (1990) Navarro and White (1992)).

(b) Gaseous viscosity

The usual types of viscosity that operate in accretion discs, characterised by the α–parameter, would operate in this context too. The stars themselves would create extra modes of viscosity. Supernovae would stir up the remaining gas so that it would have high bulk random motions (cf. Charlton and Salpeter 1990). The estimate of the time is $t_{orb}(h/r)^{-2}\alpha^{-1}$, where h is the disc thickness.

(c) Poynting-Robertson drag
There is one effect which sets an (albeit rather low) minimum to the effective viscosity. This is radiation drag (or the Poynting-Robertson effect). The light from the stellar component will exert a drag on the gas, which causes infall on a timescale of order $(c/v_{orb})\,t_{orb}$.

The only option which would preclude the formation of a central mass concentration would be the efficient conversion of all the gas into stars that are all of such low mass that they neither expel nor recycle any of their material within 5.10^8 years. This possibility aside, there is no impediment to the accumulation of $10^8 - 10^9\,M_\odot$ of baryons (already enriched with heavy elements) in a central region less than a few parsecs in size within a few times 10^8 years of the initial collapse – indeed such an occurrence seems almost inescapable.

3 HOW RAPIDLY CAN A BLACK HOLE GAIN MASS?
The foregoing discussion shows that, even in the CDM cosmogony, compact self-gravitating aggregates containing as much as $10^9\,M_\odot$ of baryons can form by $z \simeq 5$, when the age of the universe (from equation (1)) is only 10^9 years. The formation of such systems could be even earlier in other cosmogonies – indeed, even if the dark matter is CDM, there is the possibility of a non-gaussian distribution of fluctuation amplitudes, triggered by textures, strings, fluctuations in the baryon-to-photon ratio, etc.

But one has to ask whether it is sufficient that the baryons have concentrated themselves within a few parsecs? A quasar cannot switch on until this material has condensed into a relativistic (probably fully collapsed) object. Maybe there is a much longer timelag before a black hole or relativistic object can form?

Because a black hole does not have a hard surface, the efficiency is less well-defined, being a function of the accretion rate $\dot{M}$ as well as of other parameters (angular momentum, radiative efficiency, etc.) In the case of spherically-symmetric accretion, one can show explicitly how, if $\dot{M}$ is high, radiation emitted near the hole will be trapped and itself swallowed (see, for instance, Begelman 1978). The 'trapping radius' moves out as $\dot{M}$ increases; the efficiency automatically drops as $\dot{M}$ rises, so that the hole can accept an arbitrarily large amount of mass without the emergent luminosity exceeding L_{Ed}.

[An extreme instance is the collapse of a supermassive star when it becomes subject to the well-known post-Newtonian instability. It collapses almost in free-fall. The radiation within it, which supplies almost all the internal pressure, is carried down the hole, because its outward diffusion relative to the flow is negligibly slow compared to the speed with which the bulk flow advects the radiation inwards.]

It is not so clear what happens when there is angular momentum. There are explicit solutions for thick discs, with narrow funnels along the axis, in which the binding energy of the orbit from which material is swallowed (and therefore the efficiency) tends to zero, again allowing $\dot{M}$ to become very large without the luminosity much exceeding L_{Ed}.

It is, admittedly, possible that the excess mass inflow can be stemmed and the efficiency – the amount of radiation escaping per unit mass swallowed – cannot become very low when angular momentum is important. However, the cited examples should at least suffice to show that there is no *obvious* limit on the rate at which a black hole can grow: there are certainly specific models where there is *no* limit other than the rate at which material falls towards it.

There would clearly be a problem in explaining the high-z quasars if the radiative efficiency were always as high as ~ 0.1. The 'e-folding' time would then be $\sim 5 \times 10^7$ years and it would take longer than the (then) Hubble time for a hole that started off with only a stellar mass to grow by the requisite 20-odd powers of e. As Turner (1990) has emphasised, an optimum model requires an efficiency that is low in the early stages of the hole's growth; but high during the quasar phase itself – otherwise the active lifetime will be $\ll 4.10^7$ years. The number of quasar generations would then need to be correspondingly larger, and the mean remnant mass per galaxy would become too high to be consistent with the recent evidence on black holes in galactic nuclei (even if these are interpreted as upper limits).

These requirements – low efficiency while the hole grows, higher efficiency when it has achieved a mass $\gtrsim 10^8 \, M_\odot$ – are naturally met in the type of picture outlined here. Once the inner part of a galaxy has virialised, the infall rate into the centre is likely to be constant. Therefore, a natural assumption is that $\dot{M}$ should be roughly constant, at least until infall and star formation are nearly complete. The three parts of Figure 3 illustrate this. When the hole mass is still small, the accretion rate is highly supercritical. But the hole can accept the inflowing material because the efficiency adjusts to a sufficiently low value that L is below the appropriate L_{Ed}. When M has risen to $10^8 \, M_\odot$, the efficiency would rise, and the quasar phase would begin. The value of L/L_{Ed} would thereafter decline, because M continues to grow but the efficiency cannot rise above ~ 0.1; at some stage, moreover, $\dot{M}$ will eventually

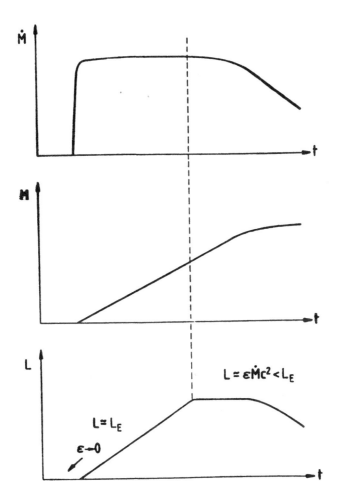

Figure 3. Schematic diagrams indicating the radiative efficiency of accretion onto a black hole at different stages in its growth. The infall rate of material towards the hole is assumed to have the form indicated in the top panel, remaining roughly constant until it eventually declines either because the galaxy completes its growth or because the matter gets incorporated into stars. (Typical values may be $\sim 10\,M_\odot\ \mathrm{yr}^{-1}$). There is no reason why the black hole cannot accept all the material falling towards it, so that its mass grows at a rate M, as shown in the second panel. In the early stages of its growth, the radiative efficiency is low because most of the radiation is advected inwards, enabling the hole to accept the high influx of matter without exceeding L_{Ed}. At a certain stage (at the time indicated by the vertical dotted line) the mass has grown so large that the inflow rate $\dot{M}$ cannot supply a luminosity L_{Ed}, even if the efficiency is no longer reduced by trapping of radiation. The luminosity thereafter stays constant until $\dot{M}$ decreases, though the character of the emergent radiation may change, since this could be a function of L_{Ed}, which decreases, even for constant $\dot{M}$, as the mass M continues to grow.

decrease because the infall stops or because the gas gets used up by star formation. (Though it may be reactivated by, for instance, a merger).

The amount of material accumulating in a central object would naturally depend on the density and on the depth of the potential well. Tentative corroboration comes from Dressler and Richstone's (1988) claim that the masses of the putative quiescent black holes in nearby galaxies, remnants of an early phase of quasar activity, are proportional to the bulge mass. The processes depicted in Figure 2 would result, almost concurrently, in the formation of both the bulge and the central hole. The timelag between virialisation of the first $10^{11}\,M_{\odot}$ systems and the ignition of the first quasars is short enough that even quasars with redshifts as large as (say) 8 would be no great embarrassment for CDM.

Simulations show that large halos at the present epoch have a variety of histories – some would have nucleated around a single peak; others may result from a relatively recent merger of systems that coagulated around separate high peaks in the initial density distribution. But in general they form from the inside outwards: the inner $10^{10}\,M_{\odot}$ may virialise at $z > 8$ even though the halo is not fully assembled until $z < 2$. The baryons most likely to aggregate into a central compact object are precisely those associated with this inner material: it is by no means improbable that a large fraction of *these* baryons ($10^{9}\,M_{\odot}$) would participate in the AGN. Thus, the quasar could switch on before the halo was assembled. The host systems would by now have accumulated $\gtrsim 10^{12}\,M_{\odot}$ halos. The most promising nearby sites for dead quasars are therefore the centres of very big galaxies (see section 5), but this does not mean that the onset of quasar activity had to await the assembly of the entire halo.

4 EVOLUTION OF THE QUASAR POPULATION

The luminosity function of AGNs has a break or bend which occurs at a higher L at high z. There are compelling reasons for believing that the lifetime of an individual source (at least for those with high L) is $\ll 10^{9}$ years. The luminosity function must therefore involve an integral over the life-cycle of each source, as well as being a function of other parameters.

For black hole processes, the key parameter is L/L_{Ed}. (This ratio obviously controls the importance of radiation pressure relative to gravity. It also controls another key parameter – the ratio of the cooling and inflow timescale – which may determine the efficiency, the ratio of thermal and non-thermal radiation, etc.). It is therefore natural to identify the break in the luminosity function with a specific value of L/L_{Ed}.

But if this is so, we must conclude that *higher-z* quasars involve *bigger* black holes. This would be impossible if the z-dependence resulted from the evolution of a single generation of objects, which lived more than 10^9 years. But it *would* be possible if there were many generations of short-lived AGNs, whose luminosity peaked soon after a black hole formed, and if conditions were more propitious for forming really big holes at high z. This proposal, though perhaps counter-intuitive, is actually not implausible.

At first sight, one might think that a hierarchical cosmogony that accounts for powerful quasars at high redshifts would predict even bigger black holes, and hence even more powerful AGNs, at recent epochs, contrary to what is observed. But optimal black hole formation requires not just a deep potential well, but also a high density. These joint requirements may be better fulfilled by a $10^{11} M_\odot$ perturbation that virialised early than by a bigger mass that turned around more recently.

The fraction of the mass going into a black hole will depend on the depth of the potential well, and also on the gas density. The latter is expected to be a parameter because, for a potential well of given depth, the retention of gas will be more efficient at high densities, since dissipative processes are then more efficient; also the angular momentum of a system of given mass would be less if it collapsed at higher density. As an illustrative example, let us suppose (Haehnelt 1992) that the hole mass is related to that of the virialised halo around it by a formula of the form;

$$\frac{M_{Hole}}{M_{Halo}} \propto (1+z)^a \exp\left[-\left(\frac{V_*}{V_c}\right)^B\right] \qquad (2)$$

In this formula, z is the collapse redshift of a halo with velocity dispersion V_c. The first term incorporates the density-dependence, the characteristic density scaling as $(1+z)^3$; the second describes how the retention is more efficient in deeper potential wells. The standard CDM model predicts how many halos of mass M collapse at redshift z (where $V_c \propto M^{1/3}(1+z)^{1/2}$). Combining this with relation (2), one derives the number of black holes of mass M that form at each cosmic epoch. Figure 4 shows, for a particular choice of parameters, how this can reproduce a rise and fall in the black hole formation rate. Moreover, the rate peaks at a somewhat earlier epoch for the more massive holes.

We would need to specify a lot more detail in order to infer a luminosity function. However, if each newly-formed black hole were to radiate at around the Eddington limit for a few times 10^7 years, and then fade, the number of quasars, as well as their luminosity function, is roughly reproduced. It is not realistic to push this model too far, because the luminosity function is influenced by several further factors – certainly by beaming effects, and perhaps also (at the bright end) by gravitational lensing.

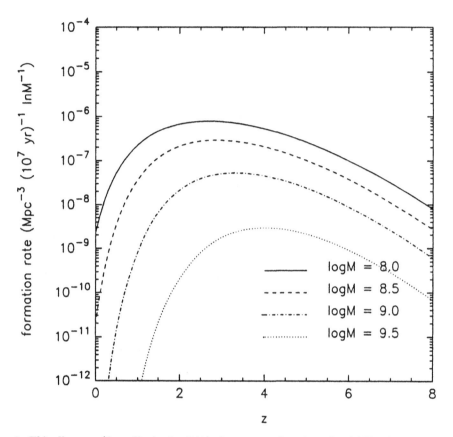

Figure 4. This diagram (from Haehnelt 1992) shows, as a function of redshift, the rates at which black holes of various masses might form, as a function of z. The calculation is done for a specific CDM model, with a biasing factor $b = 2.5$ The fraction of the virialised mass that goes into the hole is assumed to be governed by equation (2). Note that the production rate of the most massive holes peaks at $z \simeq 3$, even though halos build up to larger masses at lower redshifts. The actual masses depend, of course, on a particular choice of the constant of proportionality in (2).

The brightest quasars will be associated with 3σ peaks collapsing at redshifts $z = 3-4$ to form halos with $V_c \simeq 400$ km s^{-1}. At higher redshifts, no sufficiently large masses (with deep potential wells) would have formed; systems turning around later would have lower densities, and this would militate against black hole accumulation if there is indeed a dependence on density as steep as $\sim \rho^2$ (i.e. $a \simeq 6$ in equation (2)). Specific models of this kind are explored and discussed more fully by Haehnelt and Rees (1992).

5 DEAD AGNS IN NEARBY GALAXIES?

One very important development in the last few years has been the strengthening evidence for massive black holes in nearby galaxies. Indeed, most nearby galaxies may have dark central mass concentrations; moreover the mass seems to scale with that of the stellar bulge. Theoretical implications have been reviewed elsewhere (Rees 1990) and the latest observational situation is described by Kormendy (1992). The evidence corroborates the view that there were many generations of quasars, rather than just one. (The latter option would predict hugely massive remnants, each of $\gtrsim 10^{10}$ $M_\odot$, in only 1-2 per cent of galaxies.)

Such holes could reveal their presence by the flares that occur when a star passes sufficiently close that it gets disrupted. The flare is energised by accretion onto the hole of some of the debris from the disrupted star; the remainder of the debris is ejected. The accretion process is very complicated – it is unsteady, the orbits are highly elliptical, and relativistic effects are important. (Moreover, because the orbital angular momentum of the disrupted star will generally be misaligned with the hole's spin, Lense-Thirring precession would destroy axisymmetry even if the debris had acquired circular orbits as a result of viscous effects.)

Because of these difficulties, the timescale for the flare is uncertain; so also is its spectrum, and the consequent bolometric correction. Further work on this problem is in progress, and more is certainly needed. All that can be said at the moment is that, for a characteristic timescale of a few months, the bolometric luminosity could attain a level of order the Eddington luminosity corresponding to the hole's mass – higher, therefore, than a supernova. Moreover, although the time-dependence is uncertain, there is no reason why it should mimic the light curve of a supernova. Any such event should therefore be recognisable, and should be detected by supernova searches, especially those using non-photographic techniques which do not discriminate against events in the high-surface-brightness central parts of galaxies.

The event rate in a galaxy harbouring a 10^8 $M_\odot$ central hole depends, of course, on the properties of the stellar system surrounding it: typical estimates are that there should be one disruption every $10^3 - 10^4$ years. The Berkeley supernova search being carried out by Pennypacker, Perlmutter and their collaborators, with an exposure of several thousand "galaxy years" already, has not yet found any event of this kind. The significance of these limits cannot be fully quantified until the bolometric correction is better known. Indeed, it may turn out that these events have an X-ray signature. Any evidence for transient X-ray activity in the nuclei of normal galaxies would also be interesting.

6 STARS IN RELATIVISTIC ORBITS & PERIODICITIES IN AGNS

In the 1970s, several authors considered whether the fuel for AGNs could be supplied by tidal disruption of stars, and concluded that the disruption rate could not be high enough unless the star cluster were so dense and massive that even more gas would be released by star-star collisions than by interactions with the central black hole itself (see especially Frank 1978). Stellar disruptions may nevertheless be important in quiescent galactic nuclei or low-level AGNs, as was mentioned in §5 – in some nearby galaxies, we know enough about the star density in the central few parsecs to be able to predict the rate of captures with some confidence.

Even if stars are not important for fuelling powerful AGNs, they may have other observable effects, especially if they move in orbits very close to the central object and can somehow modulate the power output. They offer, in particular, the only way in which an AGN could display a regular period. There is tantalising evidence that the X-rays from the Seyfert galaxy NGC 6814 have a component which has maintained a steady periodicity close to 12130 seconds over the last decade (Done *et al.* 1992 and earlier references cited therein). This new evidence motivates us to explore whether it is possible for a star to get into a close (perhaps even relativistic) orbit around a massive black hole?

At first sight, one might think that the tidal capture process could do the job. For a $10^6 - 10^8$ M$_\odot$ black hole, the tidal radius r_T, within which a solar-type star would be disrupted, is only $(10 - 100)$ times the Schwarzschild radius r_s. A star captured into an orbit with pericentre at (say) $3r_T$ could not be disrupted on first passage, but would be captured into a highly eccentric orbit with pericentre $\sim 3r_T$, and would undergo further tidal dissipation at each subsequent pericentre passage (Rees 1988, 1990). But to get into a tight circular orbit with radius $\sim r_T$, and orbital binding energy $m_* c^2 (r_s/r_T)$, it must lose many times its self-binding energy $\sim Gm_*^2/r_*$, and the energy lost via tidal distortion must be dissipated internally by the star. It would take many thermal (Kelvin) times for the star to radiate so much energy, and since tidal dissipation releases energy much more rapidly than this, the star would be puffed up and destroyed before getting onto a tightly bound circular orbit.

The star's lost orbital energy could, however, be dissipated *externally* if the star were exposed to a gradual drag force – if for example, it impacted repeatedly on an accretion disk (Syer, Clarke & Rees 1991). The energy then goes mainly into the "plug" of disk material hit by the star, rather than into the star itself. It may, however, take a long time for a captured star's eccentric orbit to be "ground down" by the disk. The capture of an unbound star requires either that a single impact should dissipate a specific energy $\frac{1}{2}v_\infty^2$, or else that there should be a succession of impacts which cumulatively dissipate this energy before the stellar orbit diffuses away from an orbit that is sufficiently radial to intersect the disk. Alternatively, a star could be injected into a bound orbit if it was originally a member of a binary star system which approached close to the hole: a three body slingshot effect then expels one member of the binary, leaving the other star bound to the hole (Hills 1988).

Once bound to the hole, and in the absence of further two-body interactions, the stars are dragged into orbits in the plane of the disk, where their interaction with the accretion flow can be quite subtle. In the case of a very massive hole ($\gtrsim 10^9 \, M_\odot$) a solar-type could be on an almost relativistic orbit without being tidally disrupted. Such a star would lose orbital energy via gravitational radiation at a very slow rate (on timescale $\gtrsim 10^6$ yrs) because the mass ratio is so high.

If such a star were not in the hole's equatorial plane, it could modulate the emission. The radiation from direct impacts on the disk would be small compared with the overall luminosity. However, the impacts might create a 'spray' of material which could expand to cover a substantial solid angle around the X-ray source, while still being opaque enough to scatter the continuum radiation at particular orbital phases. Alternatively, the non-thermal emission arising from the Blandford-Znajek process, which requires strong electric fields, could be modified by the passage of a star: even modest rates of mass loss, resulting (say) from irradiation of the stellar surface, could release enough plasma to "short out" the field in a relativistic jet.

It is important that periodic behaviour can, indeed, be explained in principle in the context of conventional black hole models. If such periodicities really exist, the exciting prospect opens up of testing "strong field" gravity for the first time, by checking whether there is evidence for the precession, etc. expected if a star is orbiting in a Kerr metric.

7 CONCLUDING COMMENTS

To the 'root-mean-square' participant in this meeting, it probably comes as a disappointment that quasars have still not yielded any firm tests of strong-field gravity. Nor, despite their great distances, do they offer any firm evidence on the geometry of the Universe – the z-dependence of their properties is dominated by complex astrophysics that is still inadequately understood.

However, it is remarkable that we can now use quasars to study more than 90 per cent of cosmic history. We can directly observe the early stages of galaxy formation. And quasars serve as probes of the intervening gas via their rich absorption spectra; moreover, they can probe the total mass distribution via gravitational lensing. There is certain to be great progress in this field during the present decade; and Dennis Sciama and his academic progeny will doubtless be contributing a good share of this progress.

REFERENCES

Bardeen, J.M., Bond, J.R., Kaiser, N. and Szalay, A.S. 1986. *Astrophys. J.* **304**, 15.

Begelman, M.C. 1978. *Mon. Not. R. astr. Soc.* **184**, 53.

Charlton, J.C. and Salpeter, E.E. 1989. *Astrophys. J.* **346**, 101.

Done, C., Madejski, G.M., Mushotzky, R.F., Turner, T.J., Koyama, K. and Kunieda, H. 1992. *Astrophys. J.* (in press).

Dressler, A. and Richstone, D.O. 1988. *Astrophys. J.* **324**, 701.

Efstathiou, G.P. and Rees, M.J. 1988. *Mon. Not. R. astr. Soc.* **230**, 5P.

Fall, S.M. and Efstathiou, G.P. 1980. *Mon. Not. R. astr. Soc.* **193**, 189.

Frank, J. 1978. *Mon. Not. R. astr. Soc.* **184**, 87.

Haehnelt, M. 1992 in *First Light in the Universe: Stars or Quasars* ed. B. Rocca-Volmerange.

Haehnelt, M. and Rees, M.J. 1992 in preparation.

Heckman, T.M. 1991 in *Massive Stars and Starbursts* ed. N. Walborn and C. Leitherer, Space Telescope Science Institute (in press).

Hernquist, L. 1990. Proc. 14th Texas Conference (Ann. N.Y. Acad. Sci.).

Hills, J.G. 1988. *Nature* **331**, 687.

Kormendy, J. 1992 in *Testing the AGN paradigm* ed. S. Holt *et al.* (AIP, N.Y.).

Navarro, J. and White, S.D.M. 1992 in preparation.

Peacock, J.A. and Heavens, A. 1990. *Mon. Not. R. astr. Soc.* **243**, 133.

Press, W.H. and Schechter, P. 1974. *Astrophys. J.* **187**, 425.

Rees, M.J. 1988. *Nature* **333**, 523.

Rees, M.J. 1990. *Science* **247**, 817.

Ryden, B. and Gunn, J.E. 1987. *Astrophys. J.* **318**, 15.

Syer, D., Clarke, C.J. and Rees, M.J. 1991. *Mon. Not. R. astr. Soc.* **250**, 505.

Terlevich, R.J. 1991 in *Relationships Between AGNs and Starburst Galaxies* ed. A. Filippenko (in press).

Turner, E. 1990. *Astr. J.* **101**, 5.

Decaying Neutrinos in Astronomy and Cosmology[†]

DENNIS W. SCIAMA

1 INTRODUCTION

In this talk I will discuss the hypothesis (Sciama 1990a) that most of the dark matter in the Milky Way consists of tau neutrinos whose decay into photons is mainly responsible for the widespread ionisation of hydrogen in the interstellar medium (outside HII regions). I introduced this hypothesis because there are several difficulties with the conventional explanation of the observed ionisation. This explanation involves photons emitted by O and B stars, supernovae etc. The two most important difficulties involve the large opacity of the interstellar medium to ionising photons and the large scale–height of the free electron density. The opacity arises mainly from the widespread distribution of atomic hydrogen in the interstellar medium, which makes it difficult for the ionising photons emitted by widely separated sources to reach the regions where the ionisation is observed. The scale–height of the electron density (as derived from pulsar dispersion measure data by Reynolds (1991)) is about 1 kpc, whereas the scale height of the conventional sources is only about one tenth of this.

Both of these problems would be immediately solved by my neutrino hypothesis since the neutrinos would be smoothly distributed throughout the interstellar medium and their scale–height would be expected to exceed 1 kpc. Similar considerations apply to the edge–on spiral galaxy NGC 891 (Sciama and Salucci, 1990) whose widespread ionisation is revealed by its H_α emission, which extends out to distances from its central plane of several kiloparsecs. This extended emission is clearly visible in the H_α picture of NGC 891, taken by Rand, Kulkarni and Hester (1990), which is shown on the front cover of this book alongside a broad band picture of the same galaxy due to Keppel et al (1991). According to my hypothesis, this H_α glow is mainly excited by photons emitted by decaying dark matter neutrinos in NGC 891.

[†] This article is dedicated to my wife Lidia whose love and support have sustained me throughout my career.

2 THE KINEMATICS OF ν_τ DECAY

The decay

$$\nu_\tau \to \gamma + \nu_{\mu,e}$$

would be expected to occur if $m_{\nu_\tau} \neq 0$ and $m_{\nu_{\mu,e}} < m_{\nu_\tau}$. Its possible astronomical importance was first pointed out by Cowsik (1977) (see also de Rujula and Glashow (1980)). The photon energy E_γ in the rest frame of the decaying neutrino is determined by energy and momentum conservation:

$$E_\gamma = \frac{m_{\nu_\tau}^2 - m_{\nu_{\mu,e}}^2}{2m_{\nu_\tau}}$$

Thus the emitted photon is monochromatic. This will be of great importance in what follows.

It is likely, on particle physics grounds, that $m_{\nu_{\mu,e}} \ll m_{\nu_\tau}$. Since these conditions would simplify the discussion we shall assume them from now on. We would then have that

$$E_\gamma = \frac{1}{2}m_{\nu_\tau}.$$

For the decay photon to be able to ionise hydrogen we need

$$E_\gamma \geq 13.6 \ eV,$$

and so

$$m_{\nu_\tau} \geq 27.2 \ eV.$$

Now it is a standard result for neutrinos surviving from the hot big bang that

$$m_{\nu_\tau} = 93 \ h^2 \ \Omega_{\nu_\tau} \ eV,$$

where the Hubble parameter h is given by

$$H_0 = 100 \ h \ km \ sec^{-1} Mpc^{-1}$$

and, from observation,

$$\frac{1}{2} \leq h \leq 1$$

while $\Omega_{\nu_\tau} = \rho_{\nu_\tau}/\rho_{crit}$. Thus, for example, if

$$\Omega_{\nu_\tau} \sim 1,$$

then $E_\gamma \geq 13.6eV$ if $h \geq 0.55$, that is, over nearly the whole of the observationally permitted range for h. Hence our requirement that the decay photons can ionise hydrogen would fit in very well with the assumption that tau neutrinos have essentially the critical density.

We must now consider what decay lifetime τ would be required to account for the observed electron density in intercloud regions of the interstellar medium. This question is discussed in the original paper (Sciama 1990a). We found there that

$$\tau \sim 2 \times 10^{23} \; secs.$$

This estimate is probably accurate to about 50%.

Eventually, of course, the value of τ will be determined entirely from particle physics considerations. The correct theory for τ is not yet known, but it is encouraging that a possible theory, namely supersymmetry with broken R parity, can give our required value of τ for $m_{\nu_\tau} \sim 30 \; eV$ (Roulet and Tommasini 1991, Gabbiani, Masiero and Sciama 1991). Moreover if our hypothesis comes to be established first, it would serve as a guide to selecting the correct theory underlying particle physics.

3 CONSEQUENCES OF THE HYPOTHESIS
The hypothesis has the merits of being versatile, testable and well-defined. A partial list of phenomena with which it would interact or even dominate is the following:
- Particle physics calculation of τ (Roulet and Tommasini 1991, Gabbiani, Masiero and Sciama 1991)
- See-saw model for neutrino masses (Sciama 1990c)
- Phase space constraint on m_{ν_τ} (Sciama 1990b)
- Shock propagation in supernovae (Fuller et al 1992)
- Flattening of the Milky Way halo (Sciama 1990a)
- Oort dark matter near the sun (Sciama 1990a)
- H ionisation in the local interstellar medium (Sciama 1991a)
- H ionisation in other galaxies (Sciama and Salucci 1990)
- Sharp HI edges of galaxies (Corbelli and Salpeter 1993))
- Mapping of $n_e(r,z)$ on $n_\nu(r,z)$ (Salucci and Sciama 1990, Sciama and Salucci 1990)
- Ionisation of the intergalactic medium (Sciama 1993b)
- Ionisation and temperature of Lyman α clouds (Sciama 1991b)
- Ionisation at large z: Implications for $\Delta T/T$ of the microwave background and galaxy formation (Scott, Rees and Sciama 1991)
- Extragalactic U-V background at ~ 1000 Å (Sciama 1991c)
- Ionisation of nitrogen in the Milky Way and other galaxies (Sciama 1992a)
- E_γ, m_{ν_τ} and H_0 constrained to 1-2% accuracy (Sciama 1990d, Sciama 1991a, 1992a)
- Intergalactic ionising flux and age of universe constrained to 10% accuracy (Sciama 1990d)
- Identity of dark matter in clusters of galaxies (Sciama, Persic and Salucci 1992a,b)

- The most decisive observational test would be a Future Satellite Experiment (Stalio, Bowyer, Sciama and Gimenez 1992)

I do not have time to deal with all these questions here and I shall confine myself to the last five. A full discussion will be given in my forthcoming book "Modern Cosmology and the Dark Matter Problem", to be published by the Cambridge University Press.

4 THE IONISATION OF NITROGEN

There are several observational reasons for believing that the decay photons must be able to ionise nitrogen as well as hydrogen (Sciama (1991a, 1992a)). This would require that

$$E_\gamma \geq 14.5 \ eV.$$

This apparently slight increase in the lower limit on E_γ in fact leads to a strong test of our hypothesis, and to precise constraints on E_γ and H_0. The argument involves consideration of the extragalactic flux F of hydrogen-ionising photons.

5 THE EXTRAGALACTIC FLUX OF HYDROGEN-IONISING PHOTONS

We must impose the constraint that the flux of decay photons from the cosmological distribution of tau neutrinos surviving from the hot big bang does not exceed the observational upper limit F_{obs} on F established by Reynolds et al (1986) from their H_α observations of a neutral intergalactic cloud in Leo. They found that

$$F_{obs} \leq 6 \times 10^5 \ photons \ cm^{-2} \ sec^{-1}.$$

To calculate the integrated ionising flux from cosmological tau neutrinos one must allow for the red shift of the emitted photons, which would eventually reduce their energy to below 13.6 eV. To do this, we put

$$E_\gamma = 13.6 + \epsilon \ eV.$$

The condition that the decay photons can ionise nitrogen then becomes

$$\epsilon \geq 0.9 \ eV.$$

The contribution of the decay photons to F is then given by

$$\frac{n_{\nu_\tau}}{\tau} \frac{c}{H_0} \frac{\epsilon}{13.6}$$

if $\epsilon/13.6 \ll 1$, where n_{ν_τ} is the present cosmological density of tau neutrinos $\sim$ 115 cm^{-3}.

Remarkably, for $\tau = 2 \times 10^{23} secs$ and $H_0 = 57 \ km \ sec^{-1} \ Mpc^{-1}$ (our final value), the observational constraint then gives

$$\epsilon \leq 0.9 \ eV.$$

A solution is thus just possible but only if

$$\epsilon = 0.9 \ eV$$

and

$$F = 6 \times 10^5 cm^{-2} sec^{-1}.$$

More realistically we should consider τ uncertain to about 50%. We therefore adopt

$$\epsilon = 1.1 \pm 0.2 \ eV.$$

Then

$$E_\gamma = 14.7 \pm 0.2 \ eV.$$

Thus, because of the arithmetical structure of our argument, although τ is uncertain to 50%, E_γ is uncertain only to 1.5%.

We therefore derive a similarly accurate estimate of m_{ν_τ} (subject to $m_{\nu_{\mu,e}} \ll m_{\nu_\tau}$), namely,

$$m_{\nu_\tau} = 29.4 \pm 0.4 \ eV.$$

This mass value agrees well with the value derived from the phase space constraint applied to the Milky Way (Sciama 1990b), lies in the range for which the MSW effect occurring in supernova envelopes might ensure that the outgoing shock wave does not stall (Fuller et al 1992), would fit in with a simple form of the seesaw model for ν masses (Sciama 1990c), and might be checked experimentally by both laboratory oscillation experiments at Fermilab and future observations of neutrinos from supernovae.

6 THE HUBBLE CONSTANT AND THE AGE OF THE UNIVERSE

We now show that we can derive precise values for H_0 and for the age of the universe. This argument was given in detail in Sciama (1990 d). Since n_{ν_τ} is known, our precise value for m_{ν_τ} leads to a precise value for $\Omega_{\nu_\tau} h^2$. We can add in the relatively small baryon contribution to Ωh^2 to argue as follows:

If $\Omega = 0.2$, (the lowest value allowed by analysis of large scale peculiar velocities), the value of h resulting from our derived value of Ωh^2 would lead to an age of the universe of 8×10^9 years. This is unacceptably low (we are here assuming that the cosmological constant is zero). Similarly, $\Omega = 0.6$ would lead to an age of 10×10^9 years, which is still too low. However, if $\Omega = 1$, the age would be 12×10^9 years. The observed age is controversial, but one school of thought would allow an age as low as this.

At this point therefore we assume that

$$\Omega = 1$$

exactly. (If Ω were significantly greater than 1, h would be less than 0.5, which is unacceptable). Note that our choice of $\Omega = 1$ is not based on a priori theoretical arguments, although it is, of course, pleasing that we have arrived at a theoretically desirable value. With this choice for Ω, and our previous value for Ωh^2, we obtain for H_0 the value

$$H_0 = 57 \pm 0.5 \ km \ sec^{-1} \ Mpc^{-1}.$$

Again the arithmetical structure of the argument implies that a 50% uncertainty in τ leads to a 1% uncertainty in another parameter, in this case the Hubble constant.

7 DARK MATTER IN CLUSTERS OF GALAXIES

This topic is particularly important for us because a recent negative search for a decay line at $\sim 15eV$ from the dark matter in the cluster of galaxies A 665 (Davidsen et al 1991) has been widely regarded as disproving our hypothesis (e.g. Trimble 1992). However, recent observations have suggested that the dark matter in clusters may be mainly baryonic (Sciama, Persic and Salucci, 1992a, b). For example, Eyles et al (1991), from accurate measurements of the radial dependence of the density and temperature of the X-ray emitting gas in the Perseus cluster, concluded that the dark matter in this cluster is more centrally condensed than the visible matter. They attributed this greater concentration to the action of dissipation, and suggested that the dark matter in the Perseus cluster is baryonic. A similar conclusion follows from the x-ray data on the clusters A 85 and A 2199 (Gerbal *et al* 1992), Coma (Briel *et al* 1992) and A 665 itself (Hughes and Tanaka 1992).

Further evidence of this kind can be derived from analysis of the giant arcs which have been found in several clusters of galaxies, and which result from the gravitational lensing of background galaxies by the cluster. I am particularly pleased to refer to this argument at a meeting so closely concerned with new developments in general relativity. As pointed out by Bergmann, Lynds and and Petrosian (1990), the gravitational lens theory requires the mean column density of the cluster within the position of the arc to be equal to a certain critical column density determined mainly by the red shift of the cluster, and weakly by the red shift of the lensed galaxy. If the core radius of the dark matter exceeded the radius of the arc, it would be an unlikely coincidence that the essentially constant column density within the arc would just equal the critical column density (especially since several clusters with arcs have been discovered). However, if the core radius of the dark matter is significantly less than the radius of the arc, then the column density would decrease appreciably with radius. In this case it only requires the column density at the centre of the cluster to exceed the critical column density for there to be a place where the mean column density drops down to the critical value. The arc would then appear at this place.

Bergmann, Lynds and Petrosian concluded from this argument that the core radius of the dark matter in these clusters is indeed less than the radii of the arcs. This would again imply that the dark matter is more centrally condensed than the visible matter in these clusters. A similar conclusion has been reached by Guhathakurta (1991).

I would stress that the same argument involving dissipation does not apply to our Galaxy (Sciama, Persic and Salucci, 1992a, b). The dark matter in this case must be less centrally condensed than the visible matter, since we need it in the halo in order to explain the observed extended flat rotation curve. It is therefore reasonable to continue maintaining our hypothesis, which requires the dark matter in our Galaxy to consist mainly of neutrinos.

There is a general point of importance involved here. According to current estimates we have

$$\Omega_{vis} < \Omega_b < \Omega$$

for all allowed values of h. If this is correct, then there must exist both baryonic and non-baryonic dark matter. The location of these two types of dark matter has still to be determined, but there is no obvious reason why their ratio should be the same in different types of object. In particular it could differ in rich clusters and individual normal spiral galaxies, as we require. Of course if this is true, the reason for it must be found in cosmogonical considerations.

8 FUTURE SATELLITE EXPERIMENT

The most decisive experimental test of my hypothesis would be to search for the predicted decay line from the neutrinos within one optical depth from the sun, that is, within about a parsec. There might also be a significant contribution to the signal from the neutrinos in the low density bubble which stretches out beyond the local hydrogen cloud to about 100 pcs from the sun. The predicted flux is discussed in Sciama (1991), and is about $10^3 cm^{-2} sec^{-1}$. Its width should correspond to a neutrino velocity dispersion $\sim 200 - 300 \ km \ sec^{-1}$, that is, $\delta\lambda/\lambda \sim 10^{-3}$.

I hope to search for this line in collaboration with R. Stalio, S. Bowyer and C. Morales. We hope to have space allocated on the Eureka 2 platform by ESA, and we plan to use the International Diffuse EUV Spectrometer (IDES). The sensitivity of our detectors should be adequate to measure the predicted line after 6 months of observation. The flight is scheduled for 1996. It would be a nice 70th birthday present from nature if in that year we succeed in observing the line with its predicted energy and flux.

9 THE C^o/CO RATIO PROBLEM – A NEW PRECISION TEST OF THE DECAYING NEUTRINO THEORY

After this article was submitted to the editors I discovered (Sciama 1993a) an important new test of the decaying neutrino theory, which they have graciously permitted me to include here. It is based on Tarafdar's (1991) suggestion that decay photons could explain a well–known anomaly in the abundance ratio of neutral atomic carbon to carbon monoxide in the interiors of dense molecular clouds in the interstellar medium of our Galaxy. This is known as the C^o/CO ratio problem, and has recently been reviewed by Sorrel (1992).

The anomaly consists in the fact that the observed abundance ratio is 10^5 times larger than expected on the basis of simple models. It is usually assumed that additional ultra–violet radiation must be present inside the clouds, which is dissociating the CO. Despite many suggestions, discussed by Tarafdar and Sorrell, the origin of this additional u–v radiation remains obscure.

Tarafdar has calculated that the decay photons from neutrinos in the clouds could have sufficient flux to solve this problems, but only if they are unable to dissociate the H_2 which is by far the main constituent of these clouds. Otherwise the clouds would be much too opaque to the decay photons.

Since the photodissociation threshold for H_2 is 14.7 eV we must then have

$$E_\gamma < 14.7 \ eV.$$

From the condition that the decay photons can ionise nitrogen we also have

$$E_\gamma > 14.5 \ eV.$$

(When Tarafdar published his paper the observational evidence concerning nitrogen ionisation had not been obtained, so he did not write down these inequalities). Taken together they imply that

$$E_\gamma = 14.6 \pm 0.1 \ eV,$$

in remarkable agreement with our previous result

$$E_\gamma = 14.7 \pm 0.2 \ eV,$$

which was based on the upper limit on E_γ coming from the argument involving the intergalactic ionising flux. It would then also follow that (if $m_{\nu_{\mu,e}} \ll m_{\nu_\tau}$)

$$m_{\nu_\tau} = 29.2 \pm 0.2 \ eV.$$

More important than this increase in precision is (a) the non–trivial fact that the upper limit does (just) exceed the lower limit and (b) the fact that now the upper limit as well as the lower limit is a rigid one, being based on an accurately measured threshold energy. In fact we can go to the next decomal place for each limit and write

$$14.53 < E_\gamma < 14.68 \ eV$$

or

$$E_\gamma = 14.605 \pm 0.075 \ eV$$

and

$$m_{\nu_\tau} = 29.21 \pm 0.15 \ eV,$$

corresponding to a precision of one part in 200. Of course it must be remembered that this new result does depend on the assumption that the C°/CO ratio problem is solved by means of decay photons.

I am grateful to countless colleagues for helpful discussions and to MURST for financial support.

REFERENCES

A.G. Bergmann, V. Petrosian, and R. Lynds 1990, *ApJ*, **350**, 23.

U.G. Briel, J.P. Henry and H. Bohringer 1992, *A&A*, **259**, L31.

E. Corbelli and E.E. salpeter 1993, to be published.

R. Cowsik 1977, *Phys. Rev. Lett.*, **39**, 784.

A.F. Davidsen et al. 1991, *Nature*, **351**, 128.

A. de Rujula and S.L. Glashow 1980, *Phys. Rev. Lett.*, **45**, 942.

C.J. Eyles, M.P. Watt, D. Bertram, M.J. Church, T.J. Ponman, G.K. Skinner and A. P. Willmore 1991, *ApJ*, **376**, 23.

G.M. Fuller, R. Mayle, B.S. Meyer and J.R. Wilson 1992, *ApJ*, **389**, 517.

F. Gabbiani, A. Masiero and D.W. Sciama 1991, *Phys. Lett.*, **259 B**, 323.

D. Gerbal, F. Durret, G. Lima–Neto, and M. Lachieze–Rey 1992, *A&A*, **253**, 77.

P. Guhathakurta, in *Clusters and Superclusters of Galaxies*, NATO Advanced Study Institute, Institute of Astronomy, Cambridge, July 1991

J.P. Hughes and Y. Tanaka 1992, *ApJ*, **398**, 62.

J.W. Keppel, R.-J. Dettmar, J.S. Gallagher, M.S. Roberts 1991, *ApJ*, **374**, 507.

R.J. Rand, S.R. Kulkarni and J.J. Hester 1990, *ApJ*, **352**, L1.

R.J. Reynolds 1991 in *IAU Symposium No. 144, The Interstellar Disk–Halo Connection* ed. H. Bloemen, Kleuver, Dordrecht, p. 67.

R.J. Reynolds, K. Magee, F.L. Roesler, F. Scherb, and J. Harlander 1986, *ApJ*, **309**, L9.

E. Roulet and D. Tommasini 1991, *Phys. Lett.*, **256 B**, 218.

P. Salucci and D.W. Sciama 1990, *MNRAS*, **244**, 9p.

D.W. Sciama 1982, *MNRAS*, **198**, 1p.

D.W. Sciama 1990a , *ApJ*, **364**, 549.

D.W. Sciama 1990b, *MNRAS*, **246**, 91.

D.W. Sciama 1990c, *Nature*, **348**, 617.

D.W. Sciama 1990d, *Phys. Rev. Lett.*, **65**, 2839.

D.W. Sciama 1991a, *A&A*, **245**, 243.

D.W. Sciama 1991b, *ApJ*, **367**, L39.

D.W. Sciama 1991c, in *The Early Observable Universe from Diffuse Backgrounds* ed. B. Rocca - Volmerange, J.M. Deharveng and J. Tran Than Van, Edition Frontiers (Gif–sur–Yvette Cedex) p. 127.

D.W. Sciama 1992a, *Int. Journ. of Mod. Phys. D*, **1**, 161.

D.W. Sciama 1992b, to be published.

D.W. Sciama 1993a, to be published.

D.W. Sciama 1993b, *Modern Cosmology and the Dark Matter Problem*, Cambridge University Press

D.W. Sciama, M. Persic and P. Salucci 1992a, *Nature*, **358**, 718.

D.W. Sciama, M. Persic and P. Salucci 1992a, *PASP*, , Dec issue.

D.W. Sciama and P. Salucci 1990, *MNRAS*, **247**, 506.

D. Scott, M.J. Rees and D.W. Sciama 1991, *A&A*, **250**, 295.

W.H. Sorrell 1992, *Comments Astrophys..*, **16**, 123.

S.P. Tarafdar 1991, *MNRAS*, **252**, 55P.

V. Trimble 1992, *PASP*, **104**, 1.

Cosmological Principles

JOHN D. BARROW

1 INTRODUCTION

I first met Dennis Sciama in 1974 whilst I was still an undergraduate. At our first meeting he told me about the challenge of explaining the large scale regularity of the Universe, along with other of its unusual features, like the existence of galaxies and its proximity to a state of "zero binding energy" that we now tend to call "flatness", without making special assumptions about initial conditions. Many of these issues remain a continuing focus of attention in cosmology. Here, my intention is to review a number of cosmological 'principles' and their interaction with a variety of cosmological developments that have taken place over the period during which Dennis has worked on cosmology. The talk on which this article is based formed a small part of these Proceedings which celebrate the huge contribution that Dennis has made and continues to make to general relativity, cosmology and astrophysics. Besides Dennis' personal contributions and those of his students, that of so many of his former students (and their students) exhibits the non-linear amplification in their effectiveness that was always created by the collaborations and contacts between them that have been catalysed by their shared associations with Dennis.

2 THE PERFECT COSMOLOGICAL PRINCIPLE

In 1948 Bondi, Gold and Hoyle (Bondi and Gold, 1948; Hoyle 1948) proposed a powerful cosmological symmetry principle which they called the 'Perfect Cosmological Principle'. It required the Universe to present the same aspect to observers in all places and at all times. Mathematically, this requires the Universe to be a homogeneous space–time and its metric geometry to be described by the famous de Sitter metric ($c = 1$)

$$ds^2 = dt^2 - \exp\{2H_o t\}\{dx^2 + dy^2 + dz^2\}. \tag{1}$$

This 'Principle' underpinned the 'steady–state' cosmology and the subsequent fate of this idea in the face of radio source counts and the discovery of the microwave background radiation is well known, (Sciama, 1971). But in retrospect there are some interesting aspects of the steady–state universe that have re–emerged in modern times. The problem of explaining the isotropy and average homogeneity of the

Universe was not widely recognised before 1967. Instead, attention focused upon the problem of explaining the existence of the small deviations from uniformity that subsequently amplified into galaxies. But Hoyle and Narlikar (1963) recognised that the isotropy of the Universe could be explained in the steady–state model but not in the Big Bang model. In modern parlance they proved the first "cosmic no hair theorem" for the de Sitter metric, showing that if one linearized the Einstein equations about the de Sitter metric (1) of the steady–state universe, then an observer would see exponentially rapid approach to isotropy and homogeneity, and as $t \to \infty$, the metric would approach

$$ds^2 \to dt^2 - \{1 + f(x)\} \exp\{2H_o t\}\{dx^2 + dy^2 + dz^2\}, \qquad (2)$$

where $f(x)$ is a function of the space coordinates alone. Thus within the event horizon of any geodesically falling observer the space–time approaches de Sitter exponentially rapidly. Herman Weyl also mentions this property of the de Sitter solution in his famous book *Philosophy of Mathematics and Natural Science* (1949). We see that Hoyle and Narlikar proved a cosmic no hair theorem for inflationary universes which possess a de Sitter–like phase of expansion. They recognised that the steady–state theory, unlike the Big Bang model could offer an explanation for the large scale isotropy and homogeneity of the universe because any perturbations to the steady–state model rapidly decayed. Of course the steady–state universe differs from the standard inflationary universe in that the latter assumes that there exists only a finite period of de Sitter expansion. The steady–state model has de Sitter expansion at all times. The problem with this is that it is far too effective at explaining the homogeneity of the Universe. The accelerated expansion does not allow any gravitational inhomogeneities to grow and survive. As a result the steady–state model needed to find unconventional ways of making galaxies. In fact, this was a problem that Dennis himself worked on at one time, (Sciama, 1965), proposing that local gravitational instabilities could be initiated in the gravitational wakes created by moving objects - an idea not dissimilar to that suggested recently to create galaxies from cosmic strings, (Brandenberger, 1991).

The steady–state theory has some further interesting connections with contemporary issues. We notice that in (1) the Hubble rate, H_o, in a steady–state universe is not associated with the "age" of the universe or any of the objects in it as it would be in a Big Bang universe where the age is always of order H_o^{-1}. As first pointed out by Rees (1972), the fact that the value of H_o^{-1} is just slightly bigger than the ages of the stars and the solar system is a natural state of affairs in the Big Bang model and offers suggestive support for it. By contrast, in the steady–state model it is a complete coincidence, (Barrow and Tipler, 1986).

At first the addition of the 'Creation ' or C–field to the theory of general relativity by Hoyle (1948) was regarded as creating a new theory of gravitation but McCrea (1951) showed that Hoyle 's C–field could be accommodated within general relativity simply by interpreting it as a contribution to the energy momentum tensor with a relation $p = -\rho$ between the pressure and the energy density. The fluid conservation equations then demand that ρ be constant and it becomes mathematically equivalent to the addition of a cosmological constant with $8\pi G\rho = \Lambda$. This creates a de Sitter space-time (1) with $3H_o^2 = \Lambda$. In fact the interpretation of the cosmological constant as a 'vacuum ' energy with a perfect fluid equation of state $p = -\rho$ was first given by Lemaître (1934).

An explanation for the presence or absence of a very small cosmological constant is still a key problem in cosmology. If the cosmological constant is not to dominate the dynamics of the universe then we require $3H^2 \geq \Lambda$ today, that is, roughly

$$\Lambda \leq 3t_o^{-2} \approx 10^{-55} \text{ cm}^{-2}. \tag{3}$$

In units of the Planck length $\ell_p \sim 10^{-33}$ cm this limit is (Hawking, 1983; Barrow and Tipler, 1986)

$$\Lambda/\ell_p^2 \leq 10^{-121}. \tag{4}$$

Today the smallness of this number is regarded as a major mystery. The "unnaturalness" of it could be captured by a statement to the effect that a small parameter is "unnatural" if setting it to zero *decreases* the symmetry of the problem. In this case this is the situation. Setting $\Lambda = 0$ moves us from the maximal symmetry of the de Sitter space–time to that of the Friedmann universe. By contrast a small microwave background anisotropy $\sim 10^{-5}$ is not "unnatural" in this sense. Setting it zero increases the symmetry of the cosmological model. It is worth remarking that in the 1930s the largeness of its reciprocal was regarded as a problem by Eddington and Dirac, (Barrow and Tipler 1986, Barrow 1990). The latter argued that one should require k and Λ to be zero in the Friedmann universes to avoid the creation of such large dimensionless numbers.

Before leaving the cosmological constant it is worth remarking upon its appearance in Newtonian theory. In most derivations of general relativity its appearance is an inevitable consequence of setting the most general linear combination of rank-2 tensors proportional to the energy-momentum tensor. Sometimes it is then discarded by claiming that it does not appear in the Newtonian limit. However the cosmological constant is as inevitable in Newtonian gravity as it is in general relativity so long as one formulates it appropriately. If one asks for the most general form of the gravitational potential Φ which has the "spherical" property which renders the external force due to a sphere of mass M identical to that of a point mass M located at its centre then the required form is, (Barrow and Tipler, 1986),

$$\Phi = Ar^{-1} + Br^2, \quad A, \ B \ \ constants. \tag{5}$$

If we identify $B \equiv \Lambda/6$ then the second term on the right gives the cosmological constant in Newtonian theory.

3 THE COSMOLOGICAL PRINCIPLE

Despite the ubiquity of the Cosmological Principle in modern cosmology, there is often some confusion as to what the evidence for it really is. Most books cite the microwave background isotropy and the approximate homogeneity of galaxies and radio sources. But the spate of observational surveys which revealed unexpected patterns (walls, filaments, voids, large scale flows..) in the luminous matter distribution have been interpreted by some as casting doubt upon the picture of large scale isotropy and homogeneity which the Cosmological Principle enshrines.

In practice, the Cosmological Principle has a single purpose: to justify the use of the Friedmann *metric* as a description of the large–scale universe (Barrow, 1989). As such it is a statement about the metric of space–time or, at the Newtonian level, about the gravitational potential Φ. Suppose we consider some deviation $\delta\rho$ from the mean density ρ over a length scale L. This creates a perturbation in the potential of $\delta\Phi$. Using Poisson's equation

$$\nabla^2\Phi = 4\pi G\rho \tag{6}$$

this gives the approximate relation

$$\delta\Phi \sim 4\pi G \delta\rho L^2 \tag{7}$$

Now, in the background universe $\Phi \sim 1$ and $6\pi G\rho t^2 \sim 1$ in a flat dust universe, so

$$\frac{\delta\Phi}{\Phi} \sim \frac{\delta\rho}{\rho}\left(\frac{L}{t}\right)^2. \tag{8}$$

The Cosmological Principle requires $\delta\Phi/\Phi$ to be small (< 1), but we notice that this does not require $\delta\rho/\rho$ to be small so long as the inhomogeneity is significantly smaller than the horizon. This, of course, is why the existence of galaxies like the Milky Way, within which $\delta\rho/\rho \sim 10^6$, in no way undermines the assumption of the Friedmann metric for the universe as a whole. Homogeneity of the observed distribution of luminous galaxies and clusters is not a necessary condition for the use of the Friedmann metric to a good approximation. In fact, it is the highly isotropic nature of the microwave background radiation that offers the primary evidence to support the Cosmological Principle and the use of the Friedmann metric, (Barrow,

1989). The temperature anisotropy, $\delta T/T$, of the microwave background probes the fluctuations in the gravitational potential directly:

$$\frac{\delta T}{T} \sim \frac{\delta \Phi}{\Phi}. \tag{9}$$

The fact that these fluctuations are so small ($\sim 10^{-5}$) over a range of angular scales is the primary evidence for the adoption of the Cosmological Principle. In fact, in the 1970s similar consideration of the metric perturbations associated with primordial vorticity perturbations enabled the cosmic turbulence theory of galaxy formation to be ruled out (Barrow, 1977). The gravitational potential perturbations associated with a (constant) velocity v over scale L is given by ($c = 1$)

$$\frac{\delta \Phi}{\Phi} \sim v^2 \left(\frac{L}{t}\right)^2. \tag{10}$$

Since $v = constant$ for angular momentum conserving rotational velocity perturbations, and $L \propto t^{1/2}$ during the radiation era we have $\delta\Phi/\Phi \to \infty$ as $t \to 0$ and the turbulence theory requires a chaotic non–Friedmannian beginning to the universe. In fact, it required chaotic conditions even at redshifts $10^9 - 10^{10}$ and so is ruled out by the observations of light elements and their accord with the standard picture of primordial nucleosynthesis (Barrow, 1977).

Equation (8) also reveals the significance of the famous Harrison–Zeldovich spectrum of inhomogeneities (Harrison, 1969; Zeldovich, 1972). If $\delta\rho/\rho \propto L^{-2}$ then the density inhomogeneities contribute a scale–independent spectrum of gravitational potential fluctuations. If we return to the steady–state model for a moment we can use it to understand why this spectrum arises inevitably in those inflationary universes which undergo a finite period of de Sitter inflation (Hawking, 1982; Barrow, 1988; Guth and Pi, 1982). During such an inflationary epoch the universe behaves as if the Perfect Cosmological Principle holds. That is, one must not be able to make observations of physical effects which enable the future to be distinguished from the past. If inhomogeneities are produced by any physical process during this period we must not be able to use their associated metric perturbation or (what amounts to the same thing), their density perturbation when they enter the horizon, to tell the time. This will not be possible unless the gravitational potential perturbations are the same on every scale and this will only be the case if $\delta\rho/\rho \propto L^{-2}$.

Having introduced significance of temperature anisotropies in the microwave background it is appropriate to highlight one of Dennis's many contributions to the study of the microwave background. In general there are three sources of temperature anisotropy in the microwave background radiation on the sky. We can resolve $\delta T/T$

in the following way:

$$\frac{\delta T}{T} = \frac{5}{3}(\Phi_o - \Phi_e) + \boldsymbol{n} \cdot (\boldsymbol{v}_o - \boldsymbol{v}_e) - 2\int_e^o \nabla\Phi.d\boldsymbol{x} \qquad (11)$$

where 'e' is the value at emission (the last scattering redshift) and 'o' the value at observation (now). Hence, using the expression for the total derivative of Φ under the integral term, we have

$$\frac{\delta T}{T} = \frac{1}{3}(\Phi_e - \Phi_o) + \boldsymbol{n} \cdot (\boldsymbol{v}_o - \boldsymbol{v}_e) + 2\int_e^o \frac{\partial \Phi(\boldsymbol{x}, t)}{\partial t}dt. \qquad (12)$$

The first term on the right–hand–side is the 'Sachs-Wolfe' effect (Sachs & Wolfe, 1967) – temperature anisotropy created by gravitational potential variations between us and the last scattering surface. The second term is the Doppler effect – temperature anisotropies created by non–Hubble velocities of material on the last scattering surface or of the observer. The final 'Rees–Sciama' term (Rees & Sciama, 1968) arises from explicit time–dependence in the potential Φ. It is the net work performed by a photon traversing a time–dependent potential well. There is no significant contribution from this term for linear inhomogeneities. They evolve with $\Phi \approx constant$ and so the photon climbs out of the same potential well as it fell into, giving it the same net redshift as a neighbouring photon that did not pass through the inhomogeneity. But when inhomogeneities are non–linear this term will contribute to $\delta T/T$ because the lump (or void) will have evolved significantly whilst the photon is traversing it. The energy change for a photon of frequency ν, ($\hbar = c = 1$), is $\delta E_\gamma \sim \nu\delta\Phi$, and this creates a temperature anisotropy $\delta T/T \sim \Phi(\delta\Phi/\Phi) \sim \Phi(t_c/t_{dyn})$ where t_c is the crossing time and t_{dyn} the dynamical time of the fluctuation respectively. For a virialized inhomogeneity $\Phi \sim v^2$, so

$$\frac{\delta T}{T} \sim \Phi v \sim \Phi^{3/2} \sim \left(\frac{GM}{R}\right)^{3/2} \qquad (13)$$

and this is of order a few $\times 10^{-6}$ in a rich cluster. It is worth noting that in a structure like a cosmic 'texture', (Spergel & Turok, 1990), the Rees–Sciama fluctuation is bigger because $t_c \sim t_{dyn}$ and so $\delta T/T \sim \Phi$ rather than $\Phi^{3/2}$.

4 CHAOTIC COSMOLOGY AND INFLATION

In the early 1970s the 'chaotic cosmology' programme initiated by Charles Misner (1967,1968), still formed a focus of attention for theoretical cosmologists. It sought to provide an explanation for the large scale uniformity and isotropy of the Universe irrespective of its initial conditions by exhibiting naturally arising dissipative mechanism – classical or quantum – which might eradicate anisotropies and inhomogeneities during the evolution of the Universe. In effect, its agenda was to prove the Cosmological Principle rather than simply assume it. The early attempts to implement this ambitious programme eventually encountered difficulties. Matzner and I, (1977), showed that the entropy production from dissipation in the very early universe was catastrophically high. If baryon number was conserved (as it was then thought to be), the second law of thermodynamics would produce an entropy per baryon far in excess of the observed value of 10^9 as a result of the dissipation of irregularities and shear anisotropy for any but highly isotropic initial conditions. Moreover, Collins and Hawking (1973) established that isotropic expansion is unstable under a wide range of conditions. These conditions are interesting because they offer a bridge into contemporary cosmological studies of the inflationary universe. It can be shown, in the context of spatially homogeneous universes, that so long as the total matter density $\rho > 0$ and the density and pressure obey $\rho + 3p > 0$, then isotropic expansion is not asymptotically stable (Barrow, 1982); that is $\sigma/H \nrightarrow 0$ as $t \rightarrow 0$. In fact, generic small perturbations around ever–expanding Friedmann models are stable; that is, asymptotically they have $\sigma/H \rightarrow constant$, but this results in a built up in microwave background anisotropy (Barrow & Sonoda, 1986; Barrow, 1986). These problems are severe enough but if one were to introduce inhomogeneity into the discussion then one would have to face the problem of explaining why the microwave background is highly isotropic over angular scales exceeding the horizon size ($\sim 2°$) at the last scattering redshift. Attempts to achieve this by exploiting the Mixmaster oscillations of some universes near the initial singularity failed, (Misner, 1969; Doroshkevich et al., 1971; Barrow, 1982a). All attempts to solve these problems within the context of conventional matter fields with $\rho > 0$ and $\rho + 3p > 0$ were ultimately unsuccessful but they resulted in a thorough study of the dynamics of spatially homogeneous cosmological models by a variety of mathematical methods and led to the extraction of new information about the possible magnitude of anisotropies in the Universe that would be compatible with the observed microwave background isotropy and cosmological nucleosynthesis (Doroshkevich at al., 1973; Barrow, 1976; Collins & Hawking, 1972; Barrow et al., 1985).

Late in the 1970s the new developments taking place in theoretical particle physics started to have an impact upon the study of the very early universe. The most important development was the discovery of asymptotic freedom in 1973, (Politzer, 1973; Gross & Perry, 1973). Prior to this studies of the strong interaction at very high energy were assumed to be intractable and the study of particle physics in the very early universe amounted to little more than deliberation about the equation of state at very high density. For example, if one looks at the relativity textbook I used as an undergraduate (Weinberg's *Gravitation and Cosmology*, 1972) one finds a detailed exposition of the Hagedorn statistical bootstrap theory and its application to cosmology (and certainly no mention of gauge theories or the Weinberg–Salam model!). The introduction of asymptotic freedom made discussion of high–energy particle physics in the very early universe a tractable problem. It guaranteed an asymptotic equation of state with $p = \rho/3$ for a discrete spectrum of spin states (Collins & Perry, 1975) and determined the high–energy interaction cross–section at temperature T as $\sigma \sim \alpha^2 T^{-2}$, where α is the coupling constant of the interaction. With a total effective number of spin states g the number density is $n \sim gT^3$ and so the mean–free path is $\lambda \sim (\sigma n)^{-1}$, whilst the inter–particle separation is $L \sim n^{-1/3}$. Hence,

$$\frac{L}{\lambda} \sim 0.07 g^{2/3}\alpha^2 \tag{14}$$

In general this ratio is much less than unity and so the ideal gas approximation $(L \ll \lambda)$ holds good in the early universe (except for short intervals during phase transitions). By comparing the interaction rate, σn, with the Hubble rate given by $3H^2 \sim GgT^4$, one finds the condition for these particle physics interactions to be in equilibrium $(\sigma n > H)$ as $T < g^{3/2}\alpha^2 m_p$, where $m_p = G^{-1/2} \sim 10^{19}$ GeV is the Planck mass. Thus asymptotically–free interactions will not be in equilibrium all the way up to the Planck temperature and the notion of temperature and thermal equilibrium may not be well-defined in the very early stages of the expansion.

The original goal of the chaotic cosmology programme was achieved by dropping the imposition $\rho + 3p > 0$ on the matter content of the universe. For if $\rho > 0$ and $\rho + 3p < 0$ then it is possible for isotropic expansion to become an asymptotically stable feature of general relativistic cosmological models as $t \to \infty$. The simplest way to see this is to consider the generalisation of Friedmann 's equations to the case of general anisotropic and inhomogeneous expansion. The scale factor $a(t)$ is now the geometric–mean scale factor, ω the vorticity scalar, σ the shear scalar, and A, the acceleration vector. For the acceleration equation we have Raychaudhuri's equation (Raychaudhuri, 1955; Ellis, 1971)

$$3\ddot{a}/a = -4\pi G(\rho + 3p) + 2(\omega^2 - \sigma^2) + \nabla \cdot A. \tag{15}$$

Integrating this with respect to time we obtain a generalisation of Friedmann's e-quation below which we have indicated the slowest possible decay (deduced from the remaining Einstein equations) of the shear, rotation, curvature and acceleration terms on the right–hand–side of the equation compared with the evolution of the density of a perfect fluid with equation of state $p = (\gamma - 1)\rho$, γ constant:

$$3\dot{a}^2/a^2 = 8\pi G\rho + \text{shear - rotation + acceleration - curvature}$$
$$a^{-3\gamma} \quad\;\; t^{-2} \quad\;\; t^{-2} \quad\quad\;\; t^{-2} \quad\quad\;\; a^{-2} \tag{16}$$

From this we can see why isotropy was unstable when $\rho + 3p > 0$, that is when $\gamma > 2/3$. In this case the isotropic stress contributed by the $8\pi G\rho$ term falls off faster than the anisotropic stresses and so they come to dominate at large $a(t)$ and

t. However, if $\gamma < 2/3$ then the isotropic stress falls off *slower* than the anisotropic stresses and so for large $a(t) \sim t^{2/3\gamma}$ and t the dynamics approach those of the isotropic, zero–curvature Friedmann universe (Barrow, 1988):

$$3\dot{a}^2/a^2 \to 8\pi G\rho. \tag{17}$$

Moreover, the horizon problem is automatically resolved. In general, the asymptote (17) requires $a \sim t^{2/3\gamma}$ and this will result in the scale factor overtaking the horizon scale increase $\propto t$ at early times if $2/3\gamma > 1$; that is, so long as $\gamma < 2/3$ – exactly the same condition that was required to produce approach to isotropy and local homogeneity.

Of course this simple model has an obvious defect, which it shares with the steady–state universe. A constant $\gamma < 2/3$ is no good. The universe does not expand with $\ddot{a} > 0$ today. However, by having a matter field with a time–varying equation of state the inflationary universe (Guth, 1981) manages to reap the benefits of a finite period of accelerated expansion with $\rho + 3p < 0$ during the early stages of the Universe without changing the late–time evolution or preventing the growth of density inhomogeneities into galaxies (which the accelerated expansion did in the steady–state universe). Given a scalar field φ with self–interaction potential $V(\varphi) \geq 0$, its energy momentum tensor corresponds to that of a fluid with pressure p and density ρ given by

$$p = \frac{1}{2}\dot{\varphi}^2 - V \qquad\qquad \rho = \frac{1}{2}\dot{\varphi}^2 + V. \tag{18}$$

Hence, we see that $\rho + 3p = 2(\dot{\varphi}^2 - V)$ and so when the field evolves fast and $\dot{\varphi}^2 \gg V$ then $p \approx \rho$; whilst for slow, potential–dominated evolution $p \approx -\rho$ and $\rho + 3p < 0$ so the universal expansion is accelerated. If the φ field subsequently decays then if, for example, V has a minimum with $V = m^2\varphi^n$, n even, then the averaged evolution of the φ field contributes an asymptotic stress with an effective equation of state

$$\langle p_\varphi \rangle \approx (\gamma_{ef} - 1)\langle \rho_\varphi \rangle \tag{19}$$

where

$$1 \leq \gamma_{ef} \leq 2n/(2+n) \leq 2. \tag{20}$$

The resolution of the isotropy and horizon problem by this means (Guth, 1981) carries with it the solution of another dilemma which is simply another manifestation of the horizon problem. The successful application of particle physics to the very early universe encountered a problem created by the existence of spontaneous symmetry breaking. The very small size of the horizon at early times means that the region which expands to become our entire visible universe today consisted of a very large number of causally disjoint regions at very early times. For illustration, our visible universe today is of order 10^{27} cm in size and the temperature of the background radiation is $T_o \sim 3K$. Since $T \propto a^{-1}$ if we follow the universe back to an early time $t \sim 10^{-35}$ s when the temperature was, say, $3 \times 10^{28} K$, then the scale factor must be reduced by a factor of 10^{28} and so our visible universe was then of size $10^{27}\,\mathrm{cm}/10^{28} = 0.1$ cm. However, at that time the horizon size over which causal connection can be maintained is only $\sim 3 \times 10^{10}\,\mathrm{cms}^{-1} \times 10^{-35}\,\mathrm{s} = 3 \times 10^{-25}$ cm in extent! At any symmetry breaking there will arise mismatches of gauge field alignments in group space over physical regions separated by more than the horizon size in space at the time of the symmetry breaking. Each one of these mismatches produces a topological defect (yielding essentially one per horizon volume) whose specific character depends upon the details of the transition involved. Some of these transitions produce defects (cosmic strings?, textures?) that might be of use in seeding the development of large scale structure later in the Universe. However, others (monopoles and domain walls) produce a cumulative density that is billions of times larger than observations of the universal expansion dynamics could permit. In order to overcome this problem one must expand the early universe faster so that the observable universe can arise from a much smaller (causally connected) region. This is what inflation does (Guth, 1981). It does not prevent the formation of lots of topological defects; it merely allows the present day universe to have expanded from a region which contained only ~ 1 defect at very early times. Likewise, whereas the chaotic cosmology programme sought to explain the homogeneity of the universe by dissipating primordial inhomogeneities, the inflationary resolution of this problem is to grow the entire observable universe from a single causally connected region which (it is *assumed*) is smooth up to quantum statistical fluctuations (no fractal structure down to zero length scales allowed). Inhomogeneity can still exist in the initial state but if any part of the universe can undergo accelerated expansion it will sweep the inhomogeneities outside its horizon.

5 THE PROBLEM WITH 'PRINCIPLES'

One final lesson can be drawn form these developments and the structure of the inflationary universe. They display the difficulty of using any global "Principle" to draw conclusions about the structure of the observed universe. There are several examples of such conditions – the 'no boundary condition' (Hawking, 1988), a minimum 'gravitational entropy' condition (Penrose, 1989) are two examples. However, these 'principles' apply to the whole of the initial data space whilst the observed part of the Universe tells us only about a tiny part of that initial data space. Since any principle governing initial conditions will have a quantum statistical aspect it may not be of much use to us unless we understand completely the probability of evolving complexity (of which our form of life is one possible manifestation) from those initial conditions. Without such information we cannot decide whether a particular observation of the properties of the observable part of the universe is at variance with some global principle. As an illustration we might recall the possibility that the coupling constants of physics and the cosmological constant, are determined statistically on connected pieces of space–time with large 4–volume by the concatenation of worm-hole connections. If, say, a log–normal probability distribution with a large variance but with a mean far from the observed value were predicted for the measured value of the fine structure constant, what should we conclude? If life can only evolve in universes with a fine structure constant close to our own we must conclude that we necessarily live in a somewhat improbable universe as universes go (Barrow, 1990a, 1992). General 'principles' can only tell us the most probable initial conditions. They may be of no interest to observers. Only when we understand the constraints imposed by the conditions required to evolve observers like ourselves will we be in a position to evaluate the relevance of such 'principles' to our understanding of that part of the Universe that we see.

ACKNOWLEDGEMENTS

I would like to thank Dennis Sciama for his unfailing help and guidance over many years and George Ellis, Antonio Lanza and John Miller for their invitation to partic-ipate in these Proceedings.

REFERENCES

J.D. Barrow, 1976. *Mon. Not. Roy. astr. Soc.* **175**, 359.

J.D. Barrow, 1977. *Mon. Not. Roy. astr. Soc.* **178**, 625.

J.D. Barrow, 1982. *Quart. Jl. Roy. astr. Soc.* **23**, 344.

J.D. Barrow, 1982a. *Phys. Rep.* **85**, 1.

J.D. Barrow, 1986. *Canadian J. Phys.* **64**, 152.

J.D. Barrow, 1988. *Quart. Jl. Roy. astr. Soc.* **29**, 101.

J.D. Barrow, 1989. *Quart. Jl. Roy. astr. Soc.* **30**, 163.

J.D. Barrow, 1990. Chapter 5 of *Modern Cosmology in Retrospect*, eds R. Bertotti, R. Balbinot, S. Bergia & A. Messina CUP, Cambridge.

J.D. Barrow, 1990a. *Theories of Everything: the quest for ultimate explanation*, OUP, Oxford.

J.D. Barrow, 1992. *Pi in the Sky: Counting, Thinking and Being*, OUP, Oxford.

J.D. Barrow, & R.A. Matzner, 1977. *Mon. Not. Roy. astr. Soc.* **181**, 719.

J.D. Barrow, & D. Sonoda, 1986. *Phys. Rep.* **139**, 1.

J.D. Barrow, & F.J. Tipler, 1986. *The Anthropic Cosmological Principle*, OUP, Oxford.

J.D. Barrow, R. Juszkiewicz & D. Sonoda, 1985. *Mon. Not. Roy. astr. Soc.* **213**, 917.

H. Bondi & T. Gold, 1948. *Mon. Not. Roy. astr. Soc.* **108**, 52.

R. Brandenberger, 1991. In *Relativistic Astrophysics, Cosmology, and Fundamental Physics*, eds J.D. Barrow, L. Mestel & P.A. Thomas, *Annals NY Acad. Sci.* **647** pp. 767-774.

C.B. Collins & S.W Hawking, 1972. *Mon. Not. Roy. astr. Soc.* **162**, 307.

C.B. Collins & S.W. Hawking, 1973. *Astrophys. J.* **180**, 317.

J. Collins & M. Perry, 1975. *Phys. Rev. Lett.* **34**, 1353.

A. Doroshkevich, V. Lukash & I.D. Novikov, 1971. *Sov. Phys. JETP* **33**, 649.

A. Doroshkevich, V. Lukash & I.D. Novikov, 1973. *Sov. Phys. JETP* **37**, 739.

G.F.R. Ellis, 1971. "Relativistic Cosmology" in *General Relativity and Cosmology*, ed. R.K. Sachs, Academic, NY.

D. Gross & F. Wilczek, 1973. *Phys. Rev. Lett.* **30**, 1343.

A. Guth, 1981. *Phys. Rev. D* **23**, 347.

A. Guth & S.Y. Pi, 1982. *Phys. Rev. Lett.* **49**, 1110.

E. Harrison, 1969. *Phys. Rev. D* **1**, 2726.

S.W. Hawking, 1982. *Phys. Lett. B* **115**, 295.

S.W. Hawking, 1983. *Phil. Trans. Roy. Soc. A* **310**, 303.

S.W. Hawking, 1988. *A Brief History of Time*, Bantam, NY.

F. Hoyle, 1948. *Mon. Not. Roy. astr. Soc.* **108**, 372.

F. Hoyle & J.V. Narlikar, 1963. *Proc. Roy. Soc. A* **273**, 1.

G. Lemaître, 1934. *Proc. Nat. Acad. Sci. Washington* **20**, 12.

W.H. McCrea, 1951. *Proc. Roy. Soc. A* **206**, 562.

C. Misner, 1967 *Nature* **214**, 40.

C. Misner, 1968 *Astrophys. J.* **151**, 431.

C. Misner, 1969 *Phys. Rev. Lett* **22**, 1071.

R. Penrose, 1989. *The Emperor's New Mind*, OUP, Oxford.

H.D. Politzer, 1973. *Phys. Rev. Lett.* **30**, 1346.

A. Raychaudhuri, 1955. *Phys. Rev.* **98**, 1123.

M.J. Rees, 1972. *Comments Astrophys. Sp. Sci.* **4**, 182.

M.J. Rees & D.W. Sciama, 1968. *Nature* **217**, 511.

R. Sachs & A. Wolfe, 1967. *Astrophys. J* **147**, 73.

D.W. Sciama, 1955. *Mon. Not. Roy. astr. Soc.* **115**, 3.

D.W. Sciama, 1971. *Modern Cosmology*, CUP Cambridge.

D. Spergel & N. Turok, 1990. *Phys. Rev. Lett.* **64**, 2736.

S. Weinberg, 1972. *Gravitation and Cosmology*, Wiley, NY.

H. Weyl, 1949. *Philosophy of Mathematics and Natural Science*, (translation), Princeton UP, Princeton.

Ya.B. Zeldovich, 1972. *Mon. Not. Roy. astr. Soc.* **160**, 1p.

Anisotropic and Inhomogeneous Cosmologies

MALCOLM A.H. MacCALLUM

1 INTRODUCTION

My first impressions of Dennis Sciama came from a short introductory astrophysics course he gave to undergraduates in 1964. Then in 1966-7 I took his Cambridge Part III course in relativity, in which he charitably ignored my inadvertent use of Euclidean signature in the examination (an error I spotted just at the very end of the allowed time) and gave me a good mark. In both these courses he showed the qualities of enthusiasm and encouragement of students with which I was to become more familiar later in 1967 when I began as a research student. A project on stellar structure had taught me that I did not want to work on that, and I began under Dennis with the idea of looking at galaxy formation. However, by sharing an office with John Stewart I came to read John's paper with George Ellis (Stewart and Ellis, 1968) and its antecedent (Ellis, 1967) and developed an interest in relativistic cosmological models, which led to George becoming my second supervisor.

I was still in Sciama's group, and I learnt a lot from the tea-table conversations, which seemed to cover all of general relativity and astrophysics. Dennis taught us by example that the field should not be sub–divided into mathematics and physics, or cosmological and galactic and stellar, but that one needed to know about all those things to do really good work. Of course he was not uncritically enthusiastic: his own opinions were strongly enough held that we used to joke that if we wanted to stop research all we need do was say loudly in the tea–room that we did not believe in Mach's principle. But it was a very supportive and stimulating atmosphere for which I will always be grateful.

This is a review of what we have learnt from the study of non–standard cosmologies in which I got involved 25 years ago. Only exact solutions will be considered: the perturbation theory will be left for others to discuss. In earlier reviews (MacCallum, 1979, MacCallum, 1984) I started from the mathematical classification of the solutions but here I want to take a different route and consider the application areas. So let me just quickly remind readers of the general groups of models to which I will later refer. They are:

[1] Spatially–homogeneous and isotropic models. In relativity these give the Friedman–Lemaître–Robertson–Walker cosmologies, and the "standard model".

[2] Spatially–homogeneous but anisotropic models. These are the Bianchi models, in general, the exceptions being the Kantowski–Sachs models with an $S2 \times R2$ topology.

[3] Isotropic but inhomogeneous models. These are spherically symmetric models, whose dust subcases, having been first discussed by Lemaître (1933), are called Tolman–Bondi models.

[4] Models with two ignorable coordinates, usually with a pair of commuting Killing vectors. These may be plane or cylindrically symmetric.

[5] Models with less symmetry than those above. Only a few special cases are known exactly.

In giving this review I only had time to mention and discuss some selected papers and issues, not survey the whole vast field. Thus the bibliography is at best a representative selection from many worthy and interesting papers, and authors whose work is unkindly omitted may quite reasonably feel it is *un*representative.

2 OBSERVATIONS, THE STANDARD MODEL AND ALTERNATIVES

What is it that a cosmological model should explain? There are the following main features:

[1] Lumpiness, or the clumping of matter. The evidence for this is obvious.

[2] Expansion, shown by the Hubble law.

[3] Evolution, shown by the radio source counts and more recently by galaxy counts.

[4] A hot dense phase, to account for the cosmic microwave background radiation (CMWBR) and the abundances of the chemical elements.

[5] Isotropy, shown to a high degree of approximation in various cosmological observations, but especially in the CMWBR.

[6] Possibly, homogeneity. (The doubt indicated here will be explained later.)

[7] The numerical values of parameters of the universe and its laws, such as the baryon number density, the total density parameter Ω, the entropy per baryon, and the coupling constants

[8] (Perhaps) such features as the presence of life.

The standard big–bang model at the time I started as a student was:

[1] Isotropic at all points and thus necessarily...

[2] Spatially–homogeneous, implying Robertson–Walker geometry.

[3] Satisfied Einstein's field equations

[4] At recent times (for about the last 10^{10} years) pressureless and thus governed

by the Friedman-Lemaître dynamics.

[5] At early times, radiation-dominated, giving the Tolman dynamics and a thermal history including the usual account of nucleogenesis and the microwave background.

To this picture, which was the orthodox view from about 1965-80, the last decade added the following extra orthodoxies:

[6] $\Omega = 1$. Thus there is dark matter, for which the Cold Dark Matter model was preferred.

[7] Inflation – a period in the early universe where some field effectively mimics a large cosmological constant and so causes a period of rapid expansion long enough to multiply the initial length scale many times.

[8] Non–linear clustering on galaxy cluster scales, modelled by the N–body simulations which fit correlation functions based on observations.

and also added, as alternatives, such concepts as cosmic strings, GUTs or TOEs[1] and so on.

The standard model has some clear successes: it certainly fits the Hubble law, the source count evolutions (in principle if not in detail), the cosmic microwave spectrum, the chemical abundances, the measured isotropies, and the assumption of homogeneity. Perhaps its greatest success was the prediction that the number of neutrino species should be 3 and could not be more than 4, a prediction now fully borne out by the LEP data.

However, the model still has weaknesses (MacCallum, 1987). For example, the true clumping of matter on large scales, as shown by the QDOT data (Saunders *et al.*, 1991) and the angular correlation functions of galaxies (Maddox *et al.*, 1990), is too strong for the standard cold dark matter account[2]. The uniformity of the Hubble flow is under question from the work of the "Seven Samurai" (Lynden–Bell *et al.*, 1988) and others. The question of the true value of Ω has been re–opened, partly because theory has shown that inflation does not uniquely predict $\Omega = 1$ and partly because observations give somewhat variant values. Some authors have pointed out that our knowledge of the physics valid at nucleogenesis and before is still somewhat uncertain, and we should retain some agnosticism towards our account of those early times.

Finally, we should recognize that our belief in homogeneity has very poor observa-

---

[1] Why so anatomical?
[2] These discoveries made it possible for disagreement with the 1980s dogmatism on such matters to at last be listened to.

tional support. We have data from our past light cone (and those of earlier human astronomers) and from geological records (Hoyle, 1962). Studying homogeneity requires us to know about conditions at great distances *at the present time*, whereas what we can observe at great distances is what happened a long time ago, so to test homogeneity we have to understand the evolution both of the universe's geometry and of its matter content[3]. Thus we cannot test homogeneity, only check that it is consistent with the data and our understanding of the theory. The general belief in homogeneity is indeed like the zeal of the convert, since until the 1950s, when Baade revised the distance scale, the accepted distances and sizes of galaxies were not consistent with homogeneity.

These comments, however, are not enough to justify examination of other models. Why do we do that? I think there are several reasons. Alternative models provide fully non-linear modelling of local processes. They may show whether characteristics thought to be peculiar to the standard model, and thus a test of it, can occur elsewhere. They may be used in attempted proofs that no universe could be anisotropic or inhomogeneous, by proving that any strong departures from the standard model decay away during evolution. They can be comparators in data analysis, to show that only standard models fit. Finally, they may even be advanced as replacements of the standard model.

Before starting to examine how the alternatives fare in these various rôles, I must point out two major defects in work up to now. One is that the matter content is almost always assumed to be a perfect fluid. Yet even in the simplest non-standard models, the Bianchi models, as soon as matter is in motion relative to the homogeneous surfaces (i.e. becomes 'tilted')[4] it experiences density gradients which should lead to heat fluxes (Bradley and Sviestins, 1984): similar remarks apply to other simple models. Attempting to remedy this with some other mathematically convenient equation of state is not an adequate response; one must try to base the description of matter on a realistic model of microscopic physics or thermodynamics, and few have considered such questions (Salvati *et al.*, 1987; Bona and Coll 1988; Romano and Pavon 1992). The other objection is that we can only explore the mathematically tractable subsets of models, which may be far from representative of all models.

---

[3] Local measures of homogeneity merely tell us that the spatial gradients of cosmic quantities are not too strong near us.

[4] Such models have recently been used to fit the observed dipole anisotropy in the CMWBR (Turner, 1992), though other explanations seem to me more credible.

3 MODELLING LOCAL NON-LINEARITIES

Cosmic strings have been modelled by cylindrically symmetric models, starting with the work of Gott, Hiscock and Linet in 1985. These studies have usually been done with static strings[5], and have considered such questions as the effects on classical and quantum fields in the neighbourhood of the string.

Similarly, exact solutions for domain walls, using plane symmetric models, usually static, have been considered (Vilenkin, 1983; Ipser and Sikivie, 1984; Goetz 1990; Wang 1991)[6].

Galactic scale inhomogeneities have frequently been modelled by spherically symmetric models, usually Tolman-Bondi. They have been used to study galaxy formation (e.g. Tolman (1934), Carr and Yahil (1990), and Meszaros (1991)), to estimate departures from the simple theory of the magnitude–redshift relations based on a smoothed out model[7] (e.g. Dyer (1976), Kantowski (1969b) and Newman (1979): note that these works show that the corrections depend on the choice of modelling), and as the simplest models of gravitational lenses[8]. Spherically symmetric inhomogeneities have also been used to model the formation of primordial black holes (Carr and Hawking, 1974).

On a larger scale still inhomogeneous spacetimes have been used to model clusters of galaxies (Kantowski, 1969b), variations in the Hubble flow due to the supercluster (Mavrides, 1977), the evolution of cosmic voids (Sato, 1984; Hausman et al., 1983; Bonnor and Chamorro, 1990 and 1991), the observed distribution of galaxies and simple hierarchical models of the universe (Bonnor, 1972; Wesson, 1978; Wesson, 1979; Ribeiro 1992a).

The references just cited are only the tip of the iceberg. For his mammoth survey of all inhomogeneous cosmological models which contain, as a limiting case, Friedman–Robertson–Walker models, Krasinski now has read about 1900 papers (as reported

---

[5] There is some controversy about whether these can correctly represent strings embedded in an expanding universe (Clarke et al., 1990).

[6] Note that since the sources usually have a boost symmetry in the timelike surface giving the wall, corresponding solutions have timelike surfaces admitting the (2+1)-dimensional de Sitter group.

[7] The point is that the beams of light we observe are focussed only by the matter actually inside the beam, not the matter that would be there in a completely uniform model.

[8] The very detailed modern work interpreting real lenses to study various properties of individual sources and the cosmos mostly uses linearized approximations.

at the GR13 conference in 1992[9]). As well as the issues mentioned above, these papers discuss many others including models for interactions between different forms of matter, generation of gravitational radiation, and the nature of cosmic singularities.

There is not enough space here to do all these arguments justice, and anyway it would be unfair to pre-empt Krasinski's conclusions. Moreover, I believe the issues for which I have given a few detailed references are (together with some appearing later in this survey) the most important astrophysically. So I will just mention two more points which have arisen. One is that some exact non-linear solutions obey exactly the linearized perturbation equations for the FLRW models (Goode and Wainwright 1982; Carmeli et al., 1983). The other is a *jeu d'esprit* in which it was shown that in a "Swiss cheese" model, made by joining two FLRW exteriors at the two sides of a Kruskal diagram for the Schwarzschild solution, one can have two universes each of which can receive (but not answer) a signal from the other (Sussman, 1985).

4 WHICH FLRW PROPERTIES ARE SPECIAL?

The earliest use of anisotropic cosmological models to study a real cosmological problem was the investigation by Lemaître (1933) of the occurrence of singularities in Bianchi type I models. The objective was to explore whether the big–bang which arose in FLRW models was simply a consequence of the assumed symmetry: it was of course found not to be.

A later similar investigation was to see if the helium abundance, as known in the 1960s, could be fitted better by anisotropic cosmologies than by FLRW models, which at the time appeared to give discrepancies. The reason this might happen is that anisotropy speeds up the evolution between the time when deuterium can first form, because it is no longer dissociated by the photons, and the time when neutrons and protons are sufficiently sparse that they no longer find each other to combine. Hawking and Tayler (1966) were pioneers in this effort, which continued into the 1980s but suffered some mutations in its intention.

First the argument was reversed, and the good agreement of FLRW predictions with data was used to limit the anisotropy during the nucleogenesis period (see e.g. Barrow (1976), Olson (1977)). Later still these limits were relaxed as a result of considering the effects of anisotropic neutrino distribution functions (Rothman and Matzner 1982) and other effects on reaction rates (Juszkiewicz et al., 1983). It has even been shown (Matravers et al., 1984; Barrow, 1984) that strongly anisotropic models, not obeying the limits deduced from perturbed FLRW models, can also produce correct element

---

[9] The survey is not yet complete and remains to be published, but interim reports have appeared in some places, e.g. (Krasinski, 1990).

abundances, though they may violate other constraints (Matravers and Madsen 1985; Matravers *et al.*, 1985).

The above properties turn out not to be special to FLRW geometry. One that might be thought to be is the exact isotropy of the CMWBR. To test this, many people in the 1960s and 70s computed the angular distribution of the CMWBR temperature in Bianchi models (e.g. Thorne (1967), Novikov (1968), Collins and Hawking (1972), and Barrow *et al.*, (1985)). These calculations allow limits to be put on small deviations from isotropy from observation, and also enabled, for example, the prediction of 'hot spots' in the CMWBR in certain Bianchi models, which could in principle be searched for, if there were a quadrupole component[10], to see if the quadrupole verifies one of those models.

Similar calculations, by fewer people, considered the polarization (Rees, 1968; Anile, 1974; Tolman and Matzner, 1984) and spectrum (Rees, 1968; Rasband, 1971). More recently still, work has been carried out on the microwave background in some inhomogeneous models (Saez and Arnau, 1990). It has been shown that pure rotation (without shear) is not ruled out by the CMWBR (Obukhov, 1992), but this result may be irrelevant to the real universe where shear is essential to non–trivial perturbations (Goode, 1983; Dunsby, 1992).

One property, the nature of the big–bang singularity, as distinct from its existence, has been so extensively discussed as to demand a section of its own.

5 THE ASYMPTOTIC BEHAVIOUR OF CLASSICAL COSMOLOGIES

One can argue that classical cosmologies are irrelevant before the Planck time, but until a theory of quantum gravity is established and experimentally verified (if indeed that will ever be possible) there will be room for discussions of the behaviour of classical models near their singularities.

In the late 1950s and early 60s Lifshitz and Khalatnikov and their collaborators showed (a) that singularities in synchronous coordinates in inhomogeneous cosmologies were in general 'fictitious' and (b) that a special subclass gave real curvature singularities (Lifshitz and Khalatnikov, 1963). From these facts they (wrongly) inferred that general solutions did not have singularities. This contradicted the later singularity theorems (for which see Hawking and Ellis (1973)), a disagreement which led to the belief that there were errors in LK's arguments. They themselves, in collaboration with Belinskii, and independently Misner, showed that Bianchi IX models

---

[10] Which, since the meeting this survey was given at, has been shown to exist in the COBE data.

gave a more complicated, oscillatory, behaviour than had been discussed in the earlier work, and Misner christened this the 'Mixmaster' universe after a brand of food mixer.

The detailed behaviour of the Mixmaster model has been the subject of still–continuing investigations: some authors argue that the evolution shows ergodic and chaotic properties, while others have pointed out that the conclusions depend crucially on the choice of time variable (Barrow 1982; Burd et al., 1990; Berger, 1991). Numerical investigations are tricky because of the required dynamic range if one is to study an adequately large time–interval, and the difficulties of integrating chaotic systems.

The extension of these ideas to the inhomogeneous case, by Belinskii, Lifshitz and Khalatnikov, has been even more controversial, though prompting a smaller literature. It was strongly attacked by Barrow and Tipler (1979) on a number of technical grounds, but one can take the view that these were not as damaging to the case as Barrow and Tipler suggested (Belinskii et al., 1980; MacCallum, 1982). Indeed the 'velocity–dominated' class whose singularities are like the Kasner (vacuum Bianchi) cosmology have been more rigorously characterized and the results justified (Eardley et al., 1971; Holmes et al., 1990. Sadly this does not settle the more general question, and attempts to handle the whole argument on a completely rigorous footing[11] have so far failed.

General results about singularity types have been proved. The 'locally extendible' singularities, in which the region around any geodesic encountering the singularity can be extended beyond the singular point, can only exist under strong restrictions (Clark, 1976), while the 'whimper' singularities (King and Ellis, 1973), in which curvature invariants remain bounded while curvature components in some frames blow up, have been shown to be non–generic and unstable (Siklos, 1978). Examples of these special cases were found among Bianchi models, and both homogeneous and inhomogeneous cosmologies have been used as examples or counter–examples in the debate.

A further stimulus to the study of singularities was provided by Penrose's conjecture that gravitational entropy should be low at the start of the universe and this would correspond to a state of small or zero Weyl tensor (Penrose, 1979; Tod 1992).

Studies of the behaviour of Bianchi models have been much advanced by the adoption of methods from the theory of dynamical systems. In the early 70s this began with

---

[11] One of them made by Smallwood and myself.

the discussion of phase portraits for special cases (Collins, 1971) and was extended in work in which (a) the phase space was compactified, (b) Lyapunov functions, driving the system near the boundaries of the phase space, were found and (c) analyticity together with the behaviour of critical points and separatrices was used to derive the asymptotic behaviour (Bogoyavlenskii, 1985). In the last decade these methods have been coupled with the parametrization of the Bianchi models using automorphism group variables (Collins and Hawking, 1973; Harvey, 1979; Jantzen, 1979; Siklos, 1980; Roque and Ellis, 1985; Jaklisch, 1987).

The automorphism group can be briefly described as follows. Writing the Bianchi metrics as

$$ds^2 = -dt^2 + g_{\alpha\beta}(t)(e^{\alpha}{}_{\mu}dx^{\mu})(e^{\beta}{}_{\nu}dx^{\nu})$$

where the corresponding basis vectors $\{e_{\alpha}\}$ obey

$$[e_{\alpha}, e_{\beta}] = C^{\gamma}{}_{\alpha\beta}e_{\gamma}$$

in which the C's are the structure constants of the relevant symmetry group, one uses a transformation

$$\hat{e}^{\alpha} = M^{\alpha}{}_{\beta}e^{\beta}$$

chosen so that the $\{\hat{e}_{\alpha}\}$ obey the same commutation relations as the $\{e_{\alpha}\}$. The matrices M are time-dependent and can be chosen so that the new metric coefficients $\hat{g}_{\alpha\beta}$ take some convenient form. The real dynamics is in these metric coefficients. The idea is present in earlier treatments which grew from Misner's methods for the Mixmaster case (Ryan and Shepley, 1975) but unfortunately the type IX case was highly misleading in that for Bianchi IX (and no others except Bianchi I) the rotation group is an automorphism group.

A long series of papers by Jantzen, Rosquist and collaborators (Jantzen, 1984; Rosquist *et al.*, 1990) have coupled these ideas with Hamiltonian treatments in a powerful formalism. Using a different, and in some respects simpler, set of variables, Wainwright has also attacked the asymptotics problem (Wainwright and Hsu 1989): his variables are well-suited for those questions because their limiting cases are physical evolutions of simpler models rather than singular behaviours.

The conclusions of these studies have justified the work of Belinskii *et al.* for the homogeneous case (but do not affect the arguments about the inhomogeneous cases) and have enabled new exact solutions to be found and some general statements about the occurrence of these solutions (which in general have self–similarity in time) to be made (Wainwright and Hsu, 1989), in particular showing their rôles as attractors of the dynamical systems.

The other class of models where techniques have improved considerably are the models with two commuting Killing vectors, even when these vectors are not hypersurface–orthogonal. Some studies have focussed on the mathematics, showing how known vacuum solutions can be related by solution–generating techniques (Kitchingham, 1984), while others have concentrated on the physics of the evolution of fluid models (not obtainable by generating techniques, except in the case of 'stiff' fluid, $p = \rho$) and interpretative issues (Wainwright and Anderson, 1984; Hewit et al., 1991). It emerges that the models studied are typically Kasner–like near the singularity (agreeing with the LK arguments), and settle down to self–similar or spatially homogeneous models with superposed high–frequency gravitational waves at late times. However, some cases have asymptotic behaviour near the singularity like plane waves, and others are non–singular (Chinea et al., 1991). The Penrose conjecture has been particularly developed, using exact solutions as examples, by Wainwright and Goode, who have given a precise definition to the notion of an 'isotropic singularity' (Goode et al., 1992; Tod, 1992).

Many authors have also considered the far future evolution (or, in closed models, the question of recollapse, whose necessity in Bianchi IX models lacked a rigorous proof until recently (Lin and Wald, 1991)). From various works (MacCallum, 1971; Collins and Hawking, 1973; Barrow and Tipler, 1978) one finds that the homogeneous but anisotropic models do not in general settle down to an FLRW–like behaviour but typically generate shears of the order of 25% of their expansion rates.

This last touches on an interesting question about our account of the evolution of the universe: is it structurally stable, or would small changes in the theory of the model parameters change the behaviour grossly? Several instances of the latter phenomenon, 'fragility', have recently been explored by Tavakol, in collaboration with Coley, Ellis, Farina, Van den Bergh and others (Coley and Tavkol, 1992).

6 DO NON-FLRW MODELS BECOME SMOOTH?

Attempts to smooth out anisotropies or inhomogeneities by any process obeying deterministic sets of differential equations satisfying Lipschitz–type conditions are doomed to fail, as was first pointed out by Collins and Stewart (1971) in the context of viscous mechanisms. The argument is simply that one can impose any desired amount of anisotropy or inhomogeneity now and evolve the system backwards in time to reach initial conditions at some earlier time whose evolution produces the chosen present–day values. This was one of the arguments which rebutted Misner's ingenious suggestion that viscosity in the early universe could explain the present level of isotropy and homogeneity regardless of the initial conditions.

The same argument also holds for inflationary models. Inflation in itself, without the use of singular equations or otherwise indeterminate evolutions, cannot wholly explain present isotropy or homogeneity, although it may reduce deviations by large factors (Sirousse–Zia, 1982; Wald, 1983; Moss and Sahni, 1986; Futamase *et al.*, 1989). Objections to some specific calculations have been given (Rothman and Ellis, 1986). Although one can argue that anisotropy tends to prolong inflation, this does not remove the difficulty.

Since 1981 I have been arguing a heretical view about one of the grounds for inflation, namely the 'flatness problem', on the grounds that the formulation of this problem makes an implicit and unjustified assumption that the *a priori* probabilities of values of Ω is spread over some range sufficient to make the observed closeness to 1 implausible. Unless one can justify the *a priori* distribution, there is no implausibility[12] (Ellis, 1991).

However, if one accepts there is a flatness problem, then there is also an isotropy problem, since at least for some probability distributions on the inhomogeneity and anisotropy the models would not match observation. Protagonists of inflation cannot have it both ways. Perhaps, if one does not want to just say "well, that's how the universe was born", one has to explain the observed smoothness by appeal to the 'speculative era', as Salam (1990) called it, i.e. by appeal to one's favourite theory of quantum gravity.

Incidentally, one may note that inflation does not solve the original form of the 'horizon problem', which was to account completely for the similarity of points on the last scattering surface governed by different subsets of the intial data surface. Inflation leads to a large overlap between these initial data subsets, but not to their exact coincidence. Thus one still has to assume that the non–overlap regions are not too different. While this may give a more plausible model, it does not remove the need for assumptions on the initial data.

A further interesting application of non-standard models has come in a recent attempt to answer the question posed by Ellis and Rothman (unpublished) of how the universe can choose a uniform reference frame at the exit from inflation when a truly de Sitter model has no preferred time axis. Anninos *et al.* (1991a) have shown by taking an inflating Bianchi V model that the answer is that the memory is retained and the

---

[12] One can however argue that only $\Omega = 1$ is plausible, on the grounds that otherwise the quantum theory before the Planck time would have to fix a length-scale parameter much larger than any quantum scale, only the $\Omega = 1$ case being scale-free. I am indebted to Gary Gibbons for this remark.

universe is never really de Sitter.

Finally, one may comment that if inflation works well at early times, then inflation actually enhances the chance of an anisotropic model fitting the data, and that since the property of anisotropy cannot be totally destroyed in general (because it is coded into geometric invariants which cannot become zero by any classical evolution) the anisotropy could reassert itself in the future!

7 ASTROPHYSICAL AND OBSERVABLE CONSEQUENCES OF NON-STANDARD MODELS

Galaxy formation in anisotropic models has been studied to see if they could overcome the well–known difficulties of FLRW models (without inflation), but with negative results (Perko *et al.*, 1972).

As mentioned above solutions with two commuting Killing vectors provide models for universes with gravitational waves. Aspects of these models have been considered by several authors, e.g. Carr and Verdaguer (1983), Ibanez and Verdaguer (1983), Feinstein (1988). There are in fact several mathematically related but physically distinct classes of solutions of the Einstein equations accessible by generating techniques: stationary axisymmetric spacetimes, colliding wave solutions (nicely summarized in Griffiths (1991) and Ferrari (1990)), and cosmological solutions, the differences arising from the timelike or spacelike nature of the surfaces of symmetry and the nature of the gradient of the determinant of the metric in those surfaces.

The generating techniques essentially work for forms of matter with characteristic propagation speed equal to the velocity of light, and use one or more of a battery of related techniques: Bäcklund transformation, inverse scattering, soliton solutions and so on. One interesting question that has arisen from recent work is whether solitons in relativity do or do not exhibit non–linear interactions: Boyd *et al.* (1991), in investigations of solitons in a Bianchi I background, found no non–linearity, while Belinskii (1991) has claimed there is a non–linear effect.

Work on the observable consequences of non–standard models has been done by many, as mentioned above. One intriguing possibility raised by Ellis *et al.* (1978) is that the observed sphere on the last scattering surface could lie on a timelike (hyper)cylinder of homogeneity in a static spherically symmetric model. This makes the CMWBR isotropic *at all points* not only at the centre, and although it cannot fit all the other data, the model shows how careful one must be, in drawing conclusions about the geometry of the universe from observations, not to assume the result one wishes to prove.

Recent work by Ribeiro (1992b), in the course of an attempt to make simple models of fractal cosmologies using Tolman–Bondi metrics, has reminded us of the need to compare data with relativistic models not Newtonian approximations. Taking the Einstein–de Sitter model, and integrating down the geodesics, he plotted the number counts against luminosity distances. At small distances, where a simple interpretation would say the result looks like a uniform density, the graph is irrelevant because the distances are inside the region where the QDOT survey shows things are lumpy (Saunders *et al.*, 1991), while at greater redshifts the universe ceases to have a simple power–law relation of density and distance. Thus even Einstein–de Sitter does not look homogeneous!

One must therefore ask in general "do homogeneous models look homogeneous?". Of course, they will if the data is handled with appropriate relativistic corrections, but to achieve such comparisons in general requires the integration of the null geodesic equations in each cosmological model considered, and, as those who have tried it know, even when solving the field equations is simple, solving the geodesic equations may not be.

Ultimately we will have to refine our understanding with the help of numerical simulations which can include fully three–dimensional variations in the initial data, and some excellent pioneering work has of course been done, e.g. Anninos *et al.* (1991b), but capabilities are still limited (for example Matzner (1991) could only use a space grid of 31^3 points and 256 time steps). Moreover before one can rely on numerical simulations one needs to prove some structural stability results.

8 IS THE STANDARD MODEL RIGHT?
While I do not think one can give a definitive answer to this question, I would personally be very surprised if anisotropic but homogeneous models turned out to be anything more than useful examples. However, the status of fully inhomogeneous models is less clear.

One argument is that while the standard models may be good approximations at present, they are unstable to perturbations both in the past and the future. The possible alternative pasts are quite varied, as shown in section 5, even without considering quantum gravity. Similarly, as also mentioned in section 5, the universe may not be isotropic in the far future. Moreover, there is the question of on what scale, if any, the FLRW model is valid. Its use implies some averaging, and is certainly not correct on small scales. Is it true on any scale? If so, on what scales? There may be an upper as well as a lower bound, since we have no knowledge of conditions outside our past null cone, where some inflationary scenarios would predict bubbles of differing FLRW universes, and perhaps domain walls and so on.

If the universe were FLRW, or very close to that, this means it is in a region, in the space of all possible models, which almost any reasonable measure is likely to say has very low probability (though note the remarks on assignments of probabilities in section 6). One can only evaluate, and perhaps explain, this feature by considering non–FLRW models. It is noteworthy that many of the "problems" inflation claims to tackle are not problems if the universe simply is always FLRW. Hence, as already argued above, one has a deep problem in explaining why the universe is in the unlikely FLRW state if one accepts the arguments about probabilities current in work on inflation.

Moreover, suppose we speculated that the real universe is significantly inhomogeneous at the present epoch (at a level beyond that arising from perturbations in FLRW). What would the objections be? There are only two relevant pieces of data, as far as I can see. One is the deep galaxy counts made by the automatic plate measuring machines, which are claimed to restrict variations to a few percent, and the other is the isotropy of the CMWBR. Although the latter is a good test for large lumps in a basically FLRW universe, one has to question (recalling the results of Ellis *et al.*) whether it really implies homogeneity.

There is a theorem by Ehlers, Geren and Sachs (1968) showing that if a congruence of geodesically–moving observers all observe an isotropic distribution of collisionless gas the metric must be Robertson-Walker. Treciokas and Ellis (1971) have investigated the related problem with collisions. Recently Ferrando *et al.* (1992) have investigated inhomogeneous models where an isotropic gas distribution is possible. These studies throw into focus a conjecture which is usually assumed, namely that an approximately isotropic gas distribution, at all points, would imply an approximately Robertson–Walker metric. (It is this assumption which underlies some of the arguments used, for example, by Barrow in his talk at this meeting.)

Whether the standard model is correct or not, I feel confident in concluding that one of the more outstanding inhomogeneities is the dedicatee of this piece, Dennis Sciama, and I hope some small part of his talents has been shown here to have been passed on to me. To show how it has influenced the subject, I have marked authors cited in the bibliography below who also appear in the Sciama family tree by an asterisk.

I would like to thank G.F.R. Ellis for comments on the first draft of this survey.

REFERENCES

Anile*, A.M. (1974). Anisotropic expansion of the universe and the anisotropy and linear polarization of the cosmic microwave background. *Astrophys. Sp. Sci.*, **29**, 415.

Anninos, P., Matzner, R.A., Rothman, T., and Ryan, M.P. (1991a). How does inflation isotropize the universe? *Phys. Rev. D*, **43**, 3821–3832.

Anninos, P., Matzner, R.A., Tuluie, R., and Centrella*, J.M. (1991b). Anisotropies of the cosmic background radiation in a "hot" dark matter universe. *Astrophys. J.*, **382**, 71–78.

Barrow*, J.D. (1976). Light elements and the isotropy of the universe. *Mon. Not. R.A.S.*, **175**, 359.

Barrow*, J.D. (1982). Chaotic behaviour in general relativity. *Phys. Repts.*, **85**, 1.

Barrow*, J.D. (1984). Helium formation in cosmologies with anisotropic curvature. *Mon. Not. R.A.S.*, **211**, 221.

Barrow*, J.D., Juszkiewicz, R., and Sonoda*, D.H. (1985). Reply to "the effect of "spottiness" in large-scale structure of the microwave background" by V.N. Lukash and I.D. Novikov. *Nature*, **316**, 48.

Barrow*, J.D and Tipler, F.J (1978). Eternity is unstable. *Nature*, **278**, 453.

Barrow*, J.D and Tipler, F.J. (1979). An analysis of the generic singularity studies by Belinskii, Lifshitz and Khalatnikov. *Phys. Repts.*, **56**, 371.

Belinskii, V.A., Lifshitz, E.M., and Khalatnikov, I.M. (1980). On the problem of the singularities in the general cosmological solution of the Einstein equations. *Phys. Lett. A*, **77**, 214.

Belinskii, V.A. (1991). Gravitational breather and topological properties of gravisolitons. *Phys. Rev D*, **44**, 3109–3115.

Berger, B.K. (1991). Comments on the calculation of Liapunov exponents for the Mixmaster universe. *Gen. Rel. Grav.*, **23**, 1385.

Bogoyavlenskii, O.I. (1985). *Methods of the qualitative theory of dynamical systems in astrophysics and gas dynamics*. Springer-Verlag, Berlin & Heidelberg. [Russian original published by Nauka, Moscow, 1980.].

Bona, C. and Coll, B. (1988). On the Stephani universes. *Gen. Rel. Grav.*, **20**, 297–303.

Bonnor, W.B. (1972). A non-uniform relativistic cosmological model. *Mon. Not. R.A.S.*, **159**, 261.

Bonnor, W.B. and Chamorro, A. (1990). Models of voids in the expanding universe. *Ap. J.*, **361**, 21–26.

Bonnor, W.B. and Chamorro, A. (1991). Models of voids in the expanding universe II. *Astrophys. J.*, **378**, 461–465.

Boyd, P., Centrella*, J.M., and Klasky, S. (1991). Properties of gravitational "solitons". *Phys. Rev. D*, **43**, 379–390.

Bradley, J.M. and Sviestins, E. (1984). Some rotating, time-dependent Bianchi VIII cosmologies with heat flow. *Gen. Rel. Grav.*, **16**, 1119–1133.

Burd*, A.B., Buric, N., and Ellis*, G.F.R. (1990). A numerical analysis of chaotic behaviour in Bianchi IX models. *Gen. Rel. Grav.*, **22**, 349–363.

Carmeli, M., Charach, C. and Feinstein, A. (1983). Inhomogeneous Mixmaster universes: some exact solutions. *Ann. Phys. (N.Y.)*, **150**, 392.]

Carr*, B.J. and Hawking*, S.W. (1974). Black holes in the early universe. *Mon. Not. R.A.S.*, **168**, 399.

Carr*, B.J. and Verdaguer, E. (1983). Soliton solutions and cosmological gravitational waves. *Phys. Rev. D*, **28**, 2995.

Carr*, B.J. and Yahil, A. (1990). Self-similar perturbations of a Friedmann universe. *Astrophys. J.*, **360**, 330–342.

Chinea, F.J., Fernandez-Jambrina, F., and Senovilla, J.M.M. (1991). A singularity-free spacetime. Madrid/Barcelona preprint FT/UCM/16/91.

Clarke*, C.J.S. (1976). Space-time singularities. *Comm. math. phys.*, **49**, 17.

Clarke*, C.J.S., Ellis*, G.F.R., and Vickers*, J. (1990). The large-scale bending of cosmic strings. *Class. Quant. Grav.*, **7**, 1–14.

Coley*, A.A. and Tavakol, R.K. (1992). Fragility in cosmology. QMW preprint.

Collins*, C.B. (1971). More qualitative cosmology. *Comm. math. phys.*, **23**, 137.

Collins*, C.B. and Hawking*, S.W. (1972). The rotation and distortion of the universe *Mon. Not. R.A.S.*, **162**, 307.

Collins*, C.B. and Hawking*, S.W. (1973). Why is the universe isotropic? *Astrophys. J.*, **180**, 37.

Collins*, C.B. and Stewart*, J.M. (1971). Qualitative cosmology. *Mon. Not. R.A.S.*, **153**, 419-434.

Dunsby*, P.K.S. (1992). *Perturbations in general relativity and cosmology.* Ph.D. thesis, Queen Mary and Westfield College, University of London.

Dyer, C.C. (1976). The gravitational perturbation of the cosmic background radiation by density concentrations. *Mon. Not. R.A.S.*, **175**, 429.

Eardley, D.M., Liang, E., and Sachs, R.K. (1971). Velocity-dominated singularities in irrotational dust cosmologies. *J. Math. Phys.*, **13**, 99.

Ehlers, J., Geren, P., and Sachs, R.K. (1968). Isotropic solutions of the Einstein-Liouville equation. *J. Math. Phys.*, **9**, 1344.

Ellis*, G.F.R. (1967). Dynamics of pressure-free matter in general relativity. *J. Math. Phys.*, **8**, 1171.

Ellis*, G.F.R. (1991). Standard and inflationary cosmologies. In Mann, R. and Wesson, P., editors, *Gravitation: a Banff summer institute*. World Scientific, Singapore.

Ellis*, G.F.R., Maartens*, R.A., and Nel*, S.D. (1978). The expansion of the universe. *Mon. Not. R.A.S.*, **184**, 439.

Feinstein, A. (1988). Late-time behaviour of primordial gravitational waves in expanding universe. *Gen. Rel. Grav.*, **20**, 183–190.

Ferrando, J., Morales, J., and Portilla, M. (1992). Inhomogeneous space-times admitting isotropic radiation. Valencia preprint.

Ferrari, V. (1990). Colliding waves in general relativity. In N. Ashby, D. Bartlett and Wyss, W., editors, *General Relativity and Gravitation, 1989*, pp. 3–20. Cambridge University Press, Cambridge, New York and Melbourne.

Futamase, T., Rothman, T., and Matzner, R. (1989). Behaviour of chaotic inflation in anisotropic cosmologies with nonminimal coupling. *Phys. Rev. D*, **39**, 405–411.

Goetz, G. (1990). Gravitational field of plane symmetric thick domain walls. *J. Math. Phys.*, **31**, 2683–2687.

Goode, S.W. (1983). *Spatially inhomogeneous cosmologies and their relation with the Friedmann-Robertson-Walker models*. Ph.D. thesis, University of Waterloo.

Goode, S.W., Coley*, A.A., and Wainwright, J. (1992). The isotropic singularity in cosmology. *Class. Quant. Grav.*, **9**, 445–455.

Goode, S. and Wainwright, J. (1982). Singularities and evolution of the Szekeres cosmological models. *Phys. Rev. D*, **26**, 3315.

Gott, J.R. (1985). Gravitational lensing effects of vacuum: exact solutions. *Astrophys. J.*, **288**, 422.

Griffiths, J.B. (1991). *Colliding plane waves in general relativity*, Oxford mathematical monographs. Oxford University Press, Oxford.

Harvey, A.L. (1979). Automorphisms of the Bianchi model Lie groups. *J. Math. Phys.*, **20**, 251.

Hausman, M., Olson, D.W., and Roth, B. (1983). The evolution of voids in the expanding universe. *Astrophys. J.*, **270**, 351.

Hawking*, S.W. and Ellis*, G.F.R. (1973). *The large-scale structure of space-time*. Cambridge University Press, Cambridge.

Hawking*, S.W. and Tayler, R.J. (1966). Helium production in an anisotropic big-bang cosmology. *Nature*, **209**, 1278.

Hewitt, C., Wainwright, J., and Glaum, M. (1991). Qualitative analysis of a class of inhomogeneous self-similar cosmological models II. *Class. Quant. Grav.*, **8**, 1505–1518.

Hiscock, W.A. (1985). Exact gravitational field of a string. *Phys. Rev. D*, **31**, 3288–90.

Holmes, G., Joly*, G.J., and Smallwood, J. (1990). On the application of computer algebra to velocity dominated approximations. *Gen. Rel. Grav.*, **22**, 749–764.

Hoyle, F. (1962). Cosmological tests of gravitational theories. In *Evidence for gravitational theories*, ed. C. Moller, Enrico Fermi Corso XX, Varenna, p.141. Academic Press, New York.

Ibanez, J. and Verdaguer, E. (1983). Soliton collision in general relativity. *Phys.*

Rev. Lett., **51**, 1313.

Ipser, J. and Sikivie, P. (1984). Gravitationally repulsive domain walls. *Phys. Rev. D*, **30**, 712–9.

Jaklitsch*, M. (1987). First order field equations for Bianchi types $II - VI_h$. Capetown preprint 87-6.

Jantzen, R. (1979). The dynamical degrees of freedom in spatially homogeneous cosmology. *Comm. math. phys.*, **64**, 211.

Jantzen, R. (1984). Spatially homogeneous dynamics: a unified picture. In Ruffini, R. and Fang L.-Z., editors, *Cosmology of the early universe*, pp. 233–305. Also in "Gamow cosmology", (Proceedings of the International School of Physics 'Enrico Fermi', Course LXXXVI) ed. R. Ruffini and F. Melchiorri, pp. 61-147, North Holland, Amsterdam, 1987.

Juszkiewicz, R., Bajtlik, S., and Gorski, K. (1983). The helium abundance and the isotropy of the universe. *Mon. Not. R.A.S.*, **204**, 63P.

Kantowski, R. (1969a). The Coma cluster as a spherical inhomogeneity in relativistic dust. *Astrophys. J.*, **155**, 1023.

Kantowski, R. (1969b). Corrections in the luminosity-redshift relations of the homogeneous Friedman models. *Astrophys. J.*, **155**, 59.

King*, A.R. and Ellis*, G.F.R. (1973). Tilted homogeneous cosmological models. *Comm. math. phys.*, **31**, 209.

Kitchingham*, D.W. (1984). The use of generating techniques for space-times with two non-null commuting Killing vectors in vacuum and stiff perfect fluid cosmological models. *Class. Quant. Grav.*, **1**, 677–694.

Krasinski, A. (1990). Early inhomogeneous cosmological models in Einstein's theory. In Bertotti, B., Bergia, S., Balbinot, R., and Messina, A., editors, *Modern cosmology in retrospect*. Cambridge University Press, Cambridge.

Lemaître, G. (1933). L'univers en expansion. *Ann. Soc. Sci. Bruxelles A*, **53**, 51.

Lifshitz, E.M. and Khalatnikov, I.M. (1963). Investigations in relativistic cosmology. *Adv. Phys.*, **12**, 185.

Lin, X.-F. and Wald, R.M. (1991). Proof of the closed universe recollapse conjecture for general Bianchi IX cosmologies. *Phys. Rev. D*, **41**, 2444.

Linet, B. (1985). The static metrics with cylindrical symmetry describing a model of cosmic strings. *Gen. Rel. Grav.*, **17**, 1109.

Lynden-Bell, D., Faber, S., Burstein, D., Davies, R., Dressler, A., Terlevich, R., and Wegner, G. (1988). Spectrosopy and photometry of elliptical galaxies V. Galaxy streaming toward the new supergalactic center. *Astrophys. J.*, **326**, 19.

MacCallum*, M.A.H. (1971). A class of homogeneous cosmological models III: asymptotic behaviour. *Comm. math. phys.*, **20**, 57–84.

MacCallum*, M.A.H. (1979). Anisotropic and inhomogeneous relativistic cosmologies. In Hawking, S.W. and Israel, W., editors, *General relativity: an Einstein* *centenary survey*, pp. 533–580. Cambridge University Press, Cambridge.

MacCallum*, M.A.H. (1982). Relativistic cosmology for astrophysicists. In de Sabbata, V., editor, *Origin and evolution of the galaxies*, pp. 9–33. World Scientific, Singapore. Also, in revised form, in "Origin and evolution of the galaxies", ed. B.J.T. and J.E. Jones*, Nato Advanced Study Institute Series, **B97**, pp. 9-39, D.Reidel and Co., Dordrecht, 1983.

MacCallum*, M.A.H. (1984). Exact solutions in cosmology. In Hoenselaers, C. and Dietz, W., editors, *Solutions of Einstein's equations: techniques and results (Retzbach, Germany, 1983)*, volume 205 of *Lecture Notes in Physics*, pp. 334–366. Springer Verlag, Berlin and Heidelberg.

MacCallum*, M.A.H. (1987). Strengths and weaknesses of cosmological big-bang theory. In W.R. Stoeger*, S.J., editor, *Theory and observational limits in cosmology*, pp. 121–142. Specola Vaticana, Vatican City.

Maddox, S., Efstathiou*, G., Sutherland, W., and Loveday, J. (1990). Galaxy correlations on large scales. *Mon. Not. R.A.S.*, **242**, 43P.

Matravers, D.T. and Madsen*, M.S. (1985). Baryon number generation in a class of anisotropic cosmologies. *Phys. Lett. B*, **155**, 43–46.

Matravers, D.T., Madsen*, M.S., and Vogel, D.L (1985). The microwave background and (m, z) relations in a tilted cosmological model. *Astrophys. Sp. Sci.*, **112**, 193–202.

Matravers, D.T., Vogel, D.L., and Madsen*, M.S. (1984). Helium formation in a Bianchi V universe with tilt. *Class. Quant. Grav.*, **1**, 407.

Matzner, R.A (1991). Three-dimensional numerical cosmology. *Ann. N.Y. Acad. Sci.*, **631**, 1–14.

Mavrides, S. (1977). Anomalous Hubble expansion and inhomogeneous cosmological models. *Mon. Not. R.A.S.*, **177**, 709.

Meszaros, A. (1991). On shell crossing in the Tolman metric. *Mon. Not. R. Astr. Soc.*, **253**, 619–624.

Moss*, I. and Sahni, V. (1986). Anisotropy in the chaotic inflationary universe. *Phys. Lett. B*, **178**, 159.

Newman, R.P.A.C. (1979). *Singular perturbations of the empty Robertson-Walker cosmologies*. Ph. D. thesis, University of Kent.

Novikov, I. (1968). An expected anisotropy of the cosmological radioradiation in homogeneous anisotropic models. *Astr. Zh.*, **45**, 538. Translation in Sov. Astr.-A.J. **12**, 427.

Obukhov, Y. (1992). Rotation in cosmology. *Gen. Rel. Grav.*, **24**, 121–128.

Olson, D.W. (1977). Helium production and limits on the anisotropy of the universe *Astrophys. J.*, **219**, 777.

Penrose, R. (1979). Singularities and time-asymmetry. In Hawking, S.W. and Israel, W., editors, *General relativity: an Einstein centenary survey*, pp. 581–638.

Cambridge University Press, Cambridge.

Perko, T., Matzner, R.A., and Shepley, L.C. (1972). Galaxy formation in anisotropic cosmologies. *Phys. Rev. D*, **6**, 969.

Rasband, S.N. (1971). Expansion anisotropy and the spectrum of the cosmic background radiation. *Astrophys. J.*, **170**, 1.

Rees*, M.J. (1968). Polarization and spectrum of the primeval radiation in an anisotropic universe. *Astrophys. J.*, **153**, 1.

Ribeiro*, M.B. (1992a). On modelling a relativistic hierarchical (fractal) cosmology by Tolman's spacetime. I. Theory. *Astrophys. J.*, **388**, 1.

Ribeiro*, M.B. (1992b). On modelling a relativistic hierarchical (fractal) cosmology by Tolman's spacetime. II. Analysis of the Einstein-De Sitter model. *Astrophys. J.*, (to appear).

Romano, V. and Pavon, D. (1992). Causal dissipative Bianchi cosmology. Catania/Barcelona preprint.

Roque*, W.L. and Ellis*, G.F.R. (1985). The automorphism group and field equations for Bianchi universes. In MacCallum, M., editor, *Galaxies, axisymmetric systems and relativity: essays presented to W.B. Bonnor on his 65th birthday*, pp. 54–73. Cambridge University Press, Cambridge.

Rosquist, K., Uggla, C., and Jantzen, R. (1990). Extended dynamics and symmetries in perfect fluid Bianchi cosmologies. *Class. Quant. Grav.*, **7**, 625–637.

Rothman, T. and Ellis*, G.F.R. (1986). Does inflation occur in anisotropic cosmologies? *Phys. Lett. B*, **180**, 19.

Rothman, T. and Matzner, R.A. (1982). Effects of anisotropy and dissipation on the primordial light isotope abundances. *Phys. Rev. Lett.*, **48**, 1565.

Ryan, M.P. and Shepley, L.C. (1975). *Homogeneous relativistic cosmologies*. Princeton University Press, Princeton.

Saez, D. and Arnau, J. (1990). On the Tolman Bondi solution of Einstein equations. Numerical applications. In E. Verdaguer, Garriga, J. and Cespedes, J., editors, *Recent developments in gravitation (Proceedings of the "Relativity Meeting – 89")*, pp. 415–422. World Scientific, Singapore.

Salam, A. (1990). *Unification of fundamental forces*. Cambridge University Press, Cambridge.

Salvati, G., Schelling, E., and van Leeuwen, W. (1987). Homogeneous viscous universes with magnetic field. II Bianchi type I spaces. *Ann. Phys. (N.Y.)*, **179**, 52–75.

Sato, H. (1984). Voids in the expanding universe. In *General relativity and gravitation: Proceedings of the 10th international conference on general relativity and gravitation* ed. B. Bertotti, F. de Felice and A. Pascolini, pp. 289-312, D. Reidel and Co., Dordrecht.

Saunders, W., Frenk, C., Rowan-Robinson, M., Efstathiou*, G., Lawrence, A.,

Kaiser*, N., Ellis, R., Crawford, J., and Parry, I. (1991). The density field of the local universe. *Nature*, **349**, 32.

Siklos*, S.T.C. (1978). Occurrence of whimper singularities. *Comm. math. phys.*, **58**, 255.

Siklos*, S.T.C. (1980). Field equations for spatially homogeneous spacetimes. *Phys. Lett. A*, **76**, 19.

Sirousse-Zia, H. (1982). Fluctuations produced by the cosmological constant in the empty Bianchi IX universe. *Gen. Rel. Grav.*, **14**, 751.

Stewart*, J.M. and Ellis*, G.F.R. (1968). On solutions of Einstein's equations for a fluid which exhibit local rotational symmetry. *J. Math. Phys.*, **9**, 1072.

Sussman*, R.A. (1985). Conformal structure of a Schwarzschild black hole immersed in a Friedman universe. *Gen. Rel. Grav.*, **17**, 251–292.

Thorne, K.S. (1967). Primordial element formation, primordial magnetic fields and the isotropy of the universe. *Astrophys. J.*, **48**, 51.

Tod*, K.P. (1992). Mach's principle and isotropic singularities. In this volume.

Tolman, B. and Matzner, R.A. (1984). Large scale anisotropies and polarization of the microwave background in homogeneous cosmologies. *Proc. Roy. Soc. A*, **392**, 391.

Tolman, R.A. (1934). Effect of inhomogeneity on cosmological models. *Proc. Nat. Acad. Sci. (Wash.)*, **20**, 169.

Treciokas*, R. and Ellis*, G.F.R. (1971). Isotropic solutions of the Einstein-Boltzmann equations. *Comm. math. phys.*, **23**, 1.

Turner, M. (1992). The tilted universe. *Gen. Rel. Grav.*, **24**, 1–7.

Vilenkin, A. (1983). Gravitational field of vacuum domain walls. *Phys. Lett. B*, **133**, 177–179.

Wainwright, J. and Anderson, P. (1984). Isotropic singularities and isotropization in a class of Bianchi VI_h cosmologies. *Gen. Rel. Grav.*, **16**, 609–24.

Wainwright, J. and Hsu, L. (1989). A dynamical systems approach to Bianchi cosmologies: orthogonal models of class A. *Class. Quant. Grav.*, **6**, 1409–1431.

Wald, R.M. (1983). Asymptotic behaviour of homogeneous cosmological models in the presence of a positive cosmological constant. *Phys. Rev. D*, **28**, 211.

Wang, A.-Z. (1991). Planar domain walls emitting and absorbing electromagnetic radiation. Ioannina preprint IOA-258/91.

Wesson*, P. (1978). General relativistic hierarchical cosmology: an exact model. *Astrophys. Sp. Sci.*, **54**, 489.

Wesson*, P. (1979). Observable relations in an inhomogeneous self-similar cosmology. *Astrophys. J.*, **228**, 647.

Mach's Principle and Isotropic Singularities

K. PAUL TOD

In this contribution, I review the work of Dennis Sciama and his collaborators on Mach's Principle, saying both what Mach's Principle is, and more generally what we should expect a 'Principle' to be and to do. Then I review the notion of an isotropic singularity, and the evidence for a connection between isotropic singularities and Mach's Principle. I suggest that a reasonable formulation of the cosmological part of Mach's Principle is that the initial singularity of space-time is an isotropic singularity, and that Mach's Principle may become a 'theorem' of quantum gravity.

1 WHAT IS MACH'S PRINCIPLE ?

Mach's Principle is the name usually given to a loose constellation of ideas according to which "the inertia of a body is due to the presence of all the other matter in the universe" (Milne 1952) and "the local inertial frame is determined by some average of the motion of the distant astronomical objects" (Bondi 1952). In Wheeler's aphorism "matter there governs inertia here" (Misner *et al.* 1973). The aim of Mach's Principle is to explain, without recourse to Absolute Space, the origin of inertia, inertial frames and the standard of non-rotation in Newtonian Mechanics, where the existence of these things is a basic assumption.

The proposal that *non-rotating* means *non-rotating with respect to the fixed stars* is already present in Newton's Principia (see e.g. Bradley 1971 Chapter 6). Bishop Berkeley went further, to suggest that motion and rest relative to the "heaven of the fixed stars" could replace the idea of absolute motion and rest, and thereby eliminate absolute space. The extra ingredient commonly attributed to Mach is the idea that the distant bodies actually exert a force which determines the inertial frames; put another way, inertial or fictitious forces, introduced into Newton's equations of motion when we work in non-inertial frames, are actually dynamical in origin and are due to the rest of the matter in the universe.

Mach's ideas had a great influence on Einstein. In his obituary of Mach (Einstein 1916) he traces both the special and general theories of relativity back to this influence. Elsewhere (Einstein 1948) he wrote that Mach's great merit lay in showing that the most important problems of physics were not mathematical-deductive in nature, but relate to *basic principles*. Evidently, *principle* is being used here in a special sense; a Principle is not thought of as just an axiom, to be used to set the mathematical-deductive machine in motion; so what is meant by a Principle in this sense?

Mach's *The Science of Mechanics: A critical and historical account of its development* (1883), whose influence Einstein specifically acknowledged (1949), is organized in terms of principles and their development. In the chapters on dynamics, one encounters among others d'Alembert's Principle, Maupertuis' Principle of Least Action, Hamilton's Principle and Gauss' Principle of Least Constraint. From a deductive point of view, these all follow from Newton's equations of motion and, if we are inclined to a mathematical-deductive approach, we might regard them rather as *theorems* than individual principles. Indeed Mach quotes Gauss as saying, of his own Principle of Least Constraint, "No essentially new principle can now be established in mechanics; but this does not exclude the discovery of new points of view, from which mechanical phenomena can be fruitfully contemplated".

In twentieth-century physics, there has been a whole range of Principles, for example Einstein's Equivalence Principle, Heisenberg's Uncertainty Principle, Pauli's Exclusion Principle. In each case, these originated as great physical insights which had enormous explanatory force. Later, as the relevant parts of physics developed and became more mathematical-deductive, each of these Principles has become a deduction or a theorem. In a sense, they have been tamed, and their force has become part of the strength of the theoretical edifice in which they are derived.

Even when a Principle becomes a theorem within a theory, it can be used independently of the theory to explain a phenomenon. A good example is the explanation of the gravitational red-shift from the Equivalence Principle as given for example in Dennis Sciama's *The Physical Foundations of General Relativity* (1969). The possibility of such an explanation has the logical consequence that the red-shift will be explained by any theory in which this part of the Equivalence Principle is a theorem. Also there may be more to a Principle than is captured in any specific mathematical formulation of it. For example, the Equivalence Principle can be used to guide work on quantum field theory in curved space or quantum gravity.

How does Mach's Principle fit into this picture? Since Einstein was guided by Mach in developing general relativity, is Mach's Principle a theorem of general relativity? Certainly some parts of it, or some mathematical formulations of it, are. In a famous discussion of Newton's rotating water bucket experiment (Mach 1883 p. 284) Mach says "No one is competent to say how the experiment would turn out if the sides of the vessel increased in thickness and mass until they were ultimately several leagues thick". However, this situation can be investigated in the context of general relativity (Lense and Thirring 1918, Brill and Cohen 1966, for a review see e.g. Raine and Heller 1981) and the conclusion is that the inertial frames are dragged into rotation by the massive rotating "vessel".

On the other hand, some parts of Mach's Principle, whatever the precise formulation, are generally agreed not to be incorporated into general relativity: it is possible to find rotating cosmological solutions of the Einstein field equations in which the local "compass of inertia" rotates with respect to the large-scale motion of the material content of the universe, and this would violate almost any formulation of Mach's Principle. The consensus which has emerged (foreshadowed in North 1965) is that Mach's Principle will hold only in certain cosmological solutions of the field equations of general relativity; some cosmological models are Machian and some are not. The question to be studied is then how to formulate and justify selection principles for Machian cosmological models.

In the next Section, I review the significant role which Dennis Sciama and his collaborators have played in the study of Mach's Principle since the 1950's, and the possible formulations of Mach's Principle which have emerged from this work.

In Section 3, I review the idea of an isotropic or conformally compactifiable cosmological singularity, and the evidence for a connection between isotropic singularities, Penrose's Weyl tensor hypothesis and Mach's Principle.

In an Appendix, I describe a "poor man's" version of Mach's Principle, where the requirement that locally non-rotating frames be fixed with respect to the fixed stars leads in an elementary way to the requirement that the universe be a Friedman-Robertson-Walker cosmology.

In the preparation of this contribution, I have followed the review of Derek Raine (Raine 1981) and I gratefully acknowledge his assistance. I am happy to be able to join in this celebration of Dennis' scientific career. No one has done more than Dennis Sciama to emphasise the explanatory power and significance of Mach's Principle, and to keep it alive for a generation of relativists.

2 THE WORK OF DENNIS SCIAMA AND HIS COLLABORATORS ON MACH'S PRINCIPLE

In his interview with Alan Lightman (Lightman and Brawer 1990), Dennis records an interest in Mach's Principle from the early 1950's and says he "probably picked up the idea from Bondi". He goes on "I found the idea *extremely* attractive....I like simple ideas with very great power in physics—the idea that centrifugal forces....are mainly due to galaxies." In the light of the discussion in Section 1, we recognize this as a liking for Principles.

In his early paper *On the Origin of Inertia* (1953) Dennis constructed a vector theory of gravity with the assumption that in the rest-frame of a body the total gravitational field at the body arising from all the other matter in the universe is zero. In this way, inertial forces on the body are attributed to the gravitational field of the moving universe, in other words one has a dynamical theory of inertial forces. The theory has various shortcomings, but from it Dennis deduces among other things the formula

$$G\rho\tau^2 \sim 1 \qquad (1)$$

where G is Newton's gravitational constant, ρ is the average density of matter in the universe and τ is the Hubble time.

In his book *The Physical Foundations of General Relativity* (1969) Dennis proposes an acceleration-dependent component of inertial force between two particles, to take the form

$$F = \frac{Km_1m_2a\Phi}{c^2r}$$

as well as a static interaction component

$$F = \frac{Km_1m_2}{r^2}$$

where K is a constant, m_1, m_2 are the masses of the particles, a is the relative acceleration and r is the separation, and Φ is an unspecified angular-dependence of order one. The inverse first power of separation in the acceleration-dependent component ensures that very distant objects still have a significant effect. By computing the effects of such an acceleration-dependent term in an expanding universe, and identifying K with G from the static term, Dennis again arrives at (1).

These represent two attempts to implement a version of Mach's Principle by constructing new dynamical theories. Interestingly, both lead to (1) which is in fact approximately true. Bondi (1952) gave as an implication of Mach's Principle that "the *magnitude* of the inertia of any body is determined by the masses of the universe and by their distribution" (my italics). In other words, Mach's Principle should provide a prediction for the magnitude of G, which is what (1) is.

A new approach to the study of Mach's Principle was provided by the *Generally-covariant integral formulation of Einstein's equations* (Sciama, Waylen and Gilman

1969). Following earlier ideas of Al'tshuler (1967) and Lynden-Bell (1967), these authors write down a linear field-equation corresponding to Einstein's equations for the perturbation of the metric with source equal to the perturbation in the stress-(energy-momentum)-tensor. Call this the SWG equation, then remarkably it is also satisfied by the unperturbed metric with source equal to the unperturbed stress-tensor. From this, it will follow that the Green's function of the SWG equation can be used to give a Kirchhoff integral formulation of the full Einstein equations.

To be specific, suppose that g_{ab} is a metric satisfying Einstein's equations with source T_{ab}, so that

$$R_{ab} - \frac{1}{2}g_{ab}R = -\kappa T_{ab} \tag{2}$$

where $\kappa = 8\pi G/c^2$, and R_{ab}, R are respectively the Ricci tensor and Ricci scalar of the metric g_{ab}. Suppose also that there is a perturbation in the contravariant metric:

$$\delta g^{ab} = \varepsilon \varphi^{ab} + O(\varepsilon^2) \tag{3}$$

due to a perturbation in the stress-tensor of the form:

$$\delta T_a^b = \varepsilon \left(\tau_a^b - \frac{1}{2}\delta_a^b \tau \right) \tag{4}$$

where ε is small and $\tau = \tau_a^a$, (thus δT_a^b is the "trace-reversed" form of $\varepsilon \tau_a^b$).

These forms for the perturbations are carefully chosen so that equation (5) below takes the form it does. Indices may be raised and lowered in the conventional way on φ^{ab} and τ_a^b, but care needs to be taken when performing these operations on the small quantities. For example, the perturbation in the covariant metric is

$$\delta g_{ab} = -\varepsilon \varphi_{ab} + O(\varepsilon^2)$$

The variation of (2) leads to the equation

$$\Box \varphi^{ab} - 2R^a{}_c{}^b{}_d \varphi^{cd} = 2\kappa g^{c(a}\tau_c^{b)} \tag{5}$$

where R_{abcd} is the Riemann tensor of g_{ab}, provided one imposes the gauge-condition

$$\psi^b \equiv \nabla_a \left(\varphi^{ab} - \frac{1}{2}g^{ab}\varphi \right) = 0 \tag{6}$$

where $\varphi = g_{ab}\varphi^{ab}$.

Equation (5) is the SWG equation (Sciama *et al.* 1969). Note that, if we substitute g^{ab} for φ^{ab} in (6), the equation is satisfied identically while making the same substitution in (5) leads back to the Einstein equations (2).

Sciama *et al.* next introduce the Green's function $E^{a'b'}{}_{cd}(x',x)$ for (5) which is a bitensorial object, and derive the Kirchhoff integral for the solution to (5):

$$\varphi^{ab}(x) = 2\kappa \int_\Omega E^{ab}{}_{c'd'} \tau^{c'd'} d^4x'$$
$$+ \oint_{\partial\Omega} (\varphi^{c'd'} \nabla_{e'} E^{ab}{}_{c'd'} - E^{ab}{}_{c'd'} \nabla_{e'} \varphi^{c'd'}) dS^{e'} \tag{7}$$

where Ω is a normal neighbourhood of the point x.

Now equation (5) holds up to terms of order ϵ^2 if φ^{ab} is replaced by $g^{ab} + \delta g^{ab}$ and τ_a^b is replaced by $T_a^b + \delta T_a^b$ with its trace reversed. Consequently (7) holds up to $O(\epsilon^2)$ with the same substitutions and an unchanged $E^{ab}{}_{c'd'}$. Imagine this substitution done and take the limit as ϵ tends to zero. The result is the integral expression:

$$g^{ab}(x) = 2\kappa \int_\Omega E^{ab}{}_{c'd'} \left(T^{c'd'} - \frac{1}{2} T g^{c'd'} \right) d^4x' + \oint_{\delta\Omega} \nabla_{e'} E^{ab}{}_{c'}{}^{c'} dS^{e'} \tag{8}$$

With (8), Sciama and his coworkers arrive at an explicit expression for the metric as the sum of an integral over the matter content of a space-time region Ω and a surface integral over $\partial\Omega$ not involving the matter. By requiring the Green's function to be a retarded one, they further ensure that the integrals are only over the intersection of Ω with the past of the point x.

This is already a fairly Machian way of writing general relativity: the metric, which determines inertial frames, is itself being determined explicitly by an integral over the matter, except for the appearance of the surface term.

Gilman (1970) went on to define a space-time as Machian if and only if the surface term in (8) is always zero, but this suggestion was criticised in the paper of Ellis and Sciama (1972). In this latter paper, the authors identify a number of cosmological problems and discuss them in the context of the integral formulation. It contains Dennis' most detailed exposition of Mach's Principle on the basis of the integral formulation. Their conclusion is that, rather than necessarily vanishing, the surface term should have no contribution from source-free terms, although it is difficult to see how these are to be defined. This point of view draws attention to another way of thinking of Mach's Principle, namely that there should be no source-free contributions to the metric, or put another way, no source-free Weyl tensor. This point will be taken up again in Section 3.

The study of Mach's Principle from the SWG point of view was carried on by Dennis' student, Derek Raine. He produced two Machian conditions (Raine 1975) which a space-time must satisfy if it is to be said to incorporate Mach's Principle. These were rather technical to state, and in a later work (Raine 1981) he suggested a single, much simpler condition. He remarks that with the SWG equation (5) and the gauge condition (6), there are constraints and gauge which make it difficult to identify the true degrees of freedom, and also that there need not be a unique retarded Green's function. These difficulties also confuse the identification of the source-free contribution to the surface term in (8). One would like to impose the gauge condition directly on the Green's function, and also thereby reduce the freedom in the Green's function, by requiring

$$\bigtriangledown_a M^{ab}{}_{c'd'} = 0; \qquad \bigtriangledown^{c'} M^{ab}{}_{c'd'} = 0 \qquad (9)$$

where $M^{ab}{}_{c'd'} = E^{ab}{}_{c'd'} - \frac{1}{2} g^{ab} g_{ef} E^{ef}{}_{c'd'}$. However, this cannot be done in general. Instead, Raine's suggestion is that a space-time be deemed Machian if the surface integral vanishes, i.e. the Gilman criterion holds, but for what he calls a *Mach-Green's function*, by which he means a Green's function satisfying (9) on some initial surface.

To make a bridge to Section 3, consider what happens to the integral formula (8) in a cosmological model, when Ω is such that the integrals extend back to the initial singularity. The various Machian conditions must become conditions on the behaviour of the Green's function or the Mach-Green's function on the approach to the singularity, and therefore must become conditions on the nature of the singularity. The natural culmination of this line of argument is that Mach's Principle implies restrictions on the initial singularity. For a proper understanding of Mach's Principle from this approach one needs to know what these restrictions are.

According to Gilman (1970), most and probably all Friedman-Robertson-Walker (FR-W) cosmologies are Machian, but the Gödel universe and the Kantowski-Sachs models are not. According to Raine (1975) all the FRW solutions are Machian [in particular there is no distinction between the values of the usual parameter k which characterises open and closed models, and thus no analogue of (1)]; other spatially-homogeneous cosmological models, and particularly rotating ones, are not; while the spherically-symmetric dust-filled cosmologies of Bondi are Machian if the initial singularity is of "FRW-type", but not if it is of "Heckman-Schücking type". These different "types" of singularity arise naturally in the study of these models, and we shall return to them in Section 3.

3 ISOTROPIC SINGULARITIES

Consider a cosmological solution of the Einstein equations with a perfect-fluid source and an initial singularity. Focus attention on a small (strictly speaking, infinitesimal) comoving blob of fluid as it moves backwards in time towards the singularity. The mass of the blob, in the sense of volume-times-density, will stay constant or diverge; the volume will drop to zero, and the density will diverge to infinity. Since the density is a scalar invariant of the Ricci tensor, this signals a curvature singularity.

Now imagine rescaling the metric, or equivalently blowing up the length scale, at such a rate that the volume of the blob stays constant in time. The blob may still suffer arbitrarily large distortions as it approaches the singularity: it may collapse in two directions and stretch out infinitely in the third; it may spread out infinitely in two directions and collapse in the third; or, as may be generic, it may oscillate between such states with the directions of collapse and stretching rotating and being permuted chaotically.

Alternatively, once the volume has been rescaled to constancy, the blob may suffer only finite distortions on its way back to the singularity. In this case, the singularity can be said to be *isotropic* (Goode and Wainwright 1985), since the rates of collapse are the same (up to finite corrections) in all directions.

Changes of shape of the fluid blob are due to shear of the fluid velocity field, and shear is driven by Weyl curvature. Thus the infinite-distortion singularities are, roughly-speaking, Weyl tensor singularities. Rescaling the metric has rescaled the density back to a finite quantity and so, again roughly speaking, has removed the Ricci tensor singularity. Thus an isotropic singularity is one at which the Ricci tensor diverges but, once this divergence has been removed by conformally rescaling the metric, the Weyl tensor is finite, and therefore all of the curvature is finite. For this reason these singularities may also be conveniently called *conformally-compactifiable singularities* (Tod 1987).

The recent interest in isotropic singularities (see e.g. Goode and Wainwright 1985, Goode, Coley and Wainwright 1992, Tod 1987, 1991) has largely been motivated by Penrose's *Weyl tensor hypothesis* (Penrose 1979). To say what this is, I quote from Penrose (1979 p. 630):

> "I propose that there should be a complete lack of chaos in the initial geom-etry.....This restriction on the early geometry should be something like: *the Weyl curvature C_{abcd} vanishes at any initial singularity.*"

Penrose is led to make this hypothesis by the need for some kind of low-entropy constraint on the initial state of the universe, at the same time as the matter content of the universe is in thermal equilibrium. He argues that it is simply an observational fact that a low-entropy constraint is needed on the geometry, and that low-entropy in the geometry is tied to constraints on the Weyl curvature.

In attempting to study the Weyl tensor hypothesis, one seeks to define singularities at which the Weyl tensor is finite but the Ricci tensor diverges. Now the Weyl tensor with its indices arranged as $C^a{}_{bcd}$ is invariant under conformal rescalings of the metric:

$$g^{ab} \rightarrow \hat{g}^{ab} = \Omega^2 g^{ab}.$$

(This is deliberately different from the usual convention for rescaling.) Consequently, one may obtain a space-time with an initial singularity of the desired type by rescaling a regular metric with a conformal factor Ω which vanishes on a space-like surface. Therefore cosmological models with isotropic or conformally-compactifiable initial singularities form a simply-defined class for which the Weyl tensor $C^a{}_{bcd}$ is finite at the initial singularity. The extent to which this class includes all models with finite initial Weyl tensor is a question under study.

In a cosmological model with an isotropic singularity, one can speak of the *singularity surface* as a regular space-like surface Σ at which Ω vanishes, added by conformal-rescaling. One can then seek to pose an initial-value-problem for the model with data given on Σ (Goode and Wainwright 1985, Tod 1987, 1990, 1991, Newman 1991, 1992). A study of the problem suggests that, for a perfect-fluid source with an equation of state, the data for the initial-value-problem consist of just the 3-metric h_{ij} of Σ. For the particular case of the radiation equation of state, $p = \rho/3$, Newman (1991, 1992) has shown that the Einstein equations can be written as a symmetric hyperbolic system, and that solutions are unique given h_{ij}.

In their analysis, Goode and Wainwright (1985) showed that the initial value of the electric part of the Weyl tensor, E_{ij} is proportional to the trace-free part of the Ricci tensor r_{ij} of h_{ij}:

$$E_{ij} = \lambda[r_{ij} - (r/3)h_{ij}] \tag{10}$$

where $\lambda = 3\gamma/(3\gamma + 2)$ and γ is the limiting value of the polytropic index on the approach to the singularity:

$$p/\rho \rightarrow \gamma - 1$$

which necessarily exists and is constant. The magnetic part of the Weyl tensor necessarily vanishes initially, while the matter density, at least for the exactly polytropic equations of state (Tod 1991), takes the form:

$$\rho = \frac{A}{Z^m}[1 + BrZ^2 + O(Z^3)] \tag{11}$$

where Z is a geometrically-defined time-coordinate which vanishes at the singularity Σ, A and B are numerical factors, functions of the polytropic index γ, r is the Ricci scalar of Σ, and

$$m = \frac{6\gamma}{3\gamma - 2}$$

Now let us assume that this initial-value-problem is well-posed and that one has existence and uniqueness of perfect-fluid cosmological models given an equation of state and the 3-metric of the singularity surface (this is a large assumption, but the issue is a straightforward mathematical question). Inhomogeneities in the matter density arise via (11) from the 3-Ricci scalar, while the finite initial Weyl tensor is determined by the trace-free part of the 3-Ricci tensor by (10). If the initial Weyl tensor vanishes, then the initial 3-metric is a metric of constant curvature and uniqueness of solution forces the cosmology to be an FRW model: if there is no initial Weyl tensor then there is never any Weyl tensor. On the other hand, the initial 3-metric is free data, and therefore so is its 3-Ricci tensor; the initial Weyl tensor is constrained only by the Bianchi identity on Σ:

$$\nabla_i E_j^i = (\lambda/6) \, \nabla_j \, r \tag{12}$$

To make the connection with Mach's Principle, we look again at the cosmologies considered at the end of Section 2. The FRW cosmologies are uniformly agreed to be Machian, and they certainly have isotropic singularities since they are actually conformally flat. Rotating cosmologies are generally agreed not to be Machian, and it is a result of Goode (1987) that these cannot have isotropic singularities. The spatially homogeneous cosmologies and the Kantowski-Sachs cosmologies usually have Weyl tensor singularities, although exceptional ones have isotropic singularities with finite and non-zero initial Weyl tensor; however, the arguments which deem these cosmologies to be non-Machian may not apply to these exceptional ones. Finally, among the spherically symmetric, dust-filled cosmologies of Bondi, those with FRW-type singularities, which are Machian according to Raine (1975), have isotropic singularities and those with Heckman-Schücking type, which are non-Machian, do not.

In all cases where it has proved possible to classify cosmologies as Machian or non-Machian, with the possible exception of some homogeneous, anisotropic models, the classification coincides with the classification of the singularities into isotropic and non-isotropic.

In a sense, this is not surprising since non-isotropic singularities are Weyl tensor singularities. Typically the Weyl tensor grows more rapidly than the Ricci tensor on the approach to a singularity, and the non-Machian character can be attributed to the presence of large amounts of source-free Weyl-tensor. At an isotropic singularity, there is at least only a finite amount of Weyl tensor and it is tied to the matter via (11,12), but it is not *determined* by the matter. Now one could try to formulate Machian selection principles on (12), requiring that the Weyl tensor be determined by the matter and thereby constraining the original 3-metric. The most Machian requirement which one could make in the sense of "no free initial Weyl tensor" would lead, as pointed out above, uniquely to the FRW cosmologies, at least for a perfect-fluid source.

Given the above, one naturally conjectures that there is a simple connection, in a cosmological model, between the vanishing of the surface term in (8) for a Mach-Green's function in the sense of Raine (1981), and the presence of an isotropic singularity in the model. This remains to be investigated. Since the presence of an isotropic singularity seems to be the hallmark of finite initial Weyl tensor, one might sidestep this investigation by the formulation: Mach's Principle is the requirement for an isotropic singularity.

The final point that I want to make brings me back to the question of Principles. I have suggested that cosmological models with isotropic singularities are precisely the Machian ones. Furthermore they provide some, if not all, examples of cosmologies with finite initial Weyl tensor which certainly includes all the ones satisfying Penrose's Weyl tensor hypothesis. It may even be that the Machian ones are obliged to have zero initial Weyl tensor. Let us assume that this is so, then Penrose's Weyl tensor hypothesis is equivalent to the cosmological part of Mach's Principle.

Now Penrose (1979) justified his Weyl tensor hypothesis by an appeal to observation, and to the existence of the second law of thermodynamics. However, he went on to say that initial singularities must have this form, while final singularities or singularities formed in collapse need not, because of some time-asymmetric quantum theory of gravity which has yet to be found. In other words, there must be a time-asymmetric dynamical mechanism which forces initial singularities to be regular. If this is true, then when this theory comes to be found, the cosmological part of Mach's Principle will change its character and become a deduction or a theorem of the theory.

APPENDIX

In Section 3, we were being led to the FRW solutions as the only cosmologies with no free initial Weyl tensor, and therefore as the only Machian ones. There is an elementary or "poor man's" approach to Mach's Principle which leads to this same conclusion, and I would like to give it here.

One cannot imagine doing relativistic cosmology without introducing a time-like congruence of fundamental observers, the "fixed stars". Suppose the generator of the congruence is the unit vector-field u^a. For the covariant derivative of u^a we necessarily have

$$\nabla_a u_b = u_a A_b + \omega_{ab} + \sigma_{ab} + (\theta/3)h_{ab} \qquad (A.1)$$

where A_b is the acceleration of the congruence; ω_{ab}, which is skew-symmetric, is its rotation; σ_{ab}, which is symmetric and trace-free, is its shear; θ is its expansion; and h_{ab} is the projection orthogonal to u^a (or equivalently the 3-metric on the space of observers). Each of these quantities is orthogonal to u^a.

Concentrate on a particular observer O, with world-line Γ. A vector-field X^a defined along T is a connecting-vector to an infinitesimally neighbouring observer if it is Lie-dragged along T:

$$\pounds_u X^a \equiv u^b \nabla_b X^a - X^b \nabla_b u^a = 0 \qquad (A.2)$$

The vector-field X^a cannot in general remain orthogonal to u^a under Lie-dragging, so we will take its projection W^a orthogonal to u^a. This we will take to define a spatial direction "fixed with respect to the fixed stars". Now we ask whether this direction is rotating as it is carried along Γ, where the criterion for non-rotating will be defined shortly in terms of the Fermi-derivative of W^a along Γ:

$$D_F W^a \equiv u^b \nabla_b W^a + (W^b A_b)u^a \qquad (A.3)$$

Substituting from (A.1,2) into (A.3) we find

$$D_F W^a = W^b(\sigma_b^a + \omega_b^a) + (\theta/3)W^a \qquad (A.4)$$

Now we shall claim that the direction defined by W^a is locally non-rotating if and only if its Fermi-derivative along Γ is proportional to itself. From (A.4), all spatial directions fixed with respect to the fixed stars are non-rotating according to this definition if and only if the rotation and the shear of the fundamental congruence vanish.

Finally, it follows from a result of Collins (1986) that a perfect-fluid cosmology with an equation of state and an initial singularity, for which the fluid flow is shear-free and rotation-free, is necessarily an FRW cosmology.

REFERENCES

Al'tshuler, B. L., 1967, *Sov. Phys. JETP*, **24**, 766.

Bondi, H., 1952, *Cosmology* (Cambridge University Press: Cambridge).

Bradley, J., 1971, *Mach's Philosophy of Science* (Athlone Press: London).

Brill, D. R. and Cohen, J. M., 1966, *Phys. Rev.*, **143**, 1011.

Collins, C. B., 1986, *Can. Jour. Phys.*, **64**, 191.

Einstein, A., 1916, *Phys. Zeit.*, **17**, 101.

—. 1948, letter to M. Besso.

—. 1949, in: *Albert Einstein: Philosopher-Scientist*, ed. P. A. Schilpp, (Evanston).

Ellis, G. F. R. and Sciama, D. W., 1972, *Global and non-global problems in Cosmology*, in: *General Relativity: papers in honour of J. L. Synge*, ed. L. O'Raifeartaigh (Clarendon Press: Oxford).

Gilman, R. C., 1970, *Phys. Rev. D*, **2**, 1400.

Goode, S. W., 1987, *Gen. Rel. Grav.*, **19**, 1075.

Goode, S. W., Coley, A. A. and Wainwright, J., 1992, *Class. Quant. Grav.*, **9**, 445.

Goode, S. W. and Wainwright, J., 1985, *Class. Quant. Grav.*, **2**, 99.

Lense, J and Thirring, W., 1918, *Phys. Zeit.*, **19**, 156.

Lightman, A. and Brawer, R., 1990, *Origins: the lives and worlds of modern cosmologists* (Harvard University Press: Cambridge, Mass).

Lynden-Bell, D., 1967, *Mon. Not. R. astr. Soc.*, **135**, 413.

Mach, E., 1883, *The Science of Mechanics*, English edition 1960 (Open Court: Lasalle, Illinois).

Milne, E. A., 1952, *Modern Cosmology and the Christian Idea of God* (Oxford University Press: Oxford).

Misner, C. W., Thorne, K. S. and Wheeler, J. A., 1973, *Gravitation* (W. H. Freeman: San Francisco).

Newman, R. P. A. C., 1991, *Conformal Singularities*, to appear in Rendiconti del Seminario Matematico.

—. 1992, *On the structure of conformal singularities in classical general relativity*, submitted to *Proc. Roy. Soc.*.

North, J. D., 1965, *The Measure of the Universe* (Oxford University Press: Oxford; 1990 Dover: New York).

Penrose, R., 1979, in: *General Relativity: an Einstein centennial volume* eds. W. Israel and S. W. Hawking, (Cambridge University Press: Cambridge).

Raine, D. J., 1975, *Mon. Not. R. astr. Soc.*, **171**, 507.

Raine, D. J., 1981, *Rep. Prog. Phys.*, **44**, 1151.

Raine, D. J. and Heller, M. 1981, *The Science of Space-time* (Pachart: Tucson).

Sciama, D. W., 1953, *Mon. Not. R. astr. Soc.*, **113**, 34.

—. 1957, *Scientific American*, Feb., 99.

—. 1959, *The Unity of the Universe* (Faber and Faber: London).

—. 1969, *The Physical Foundations of General Relativity* (Heineman: London).

Sciama, D. W., Waylen, P. C. and Gilman, R. C., 1969, *Phys. Rev.*, **187**, 1762.

Tod, K. P., 1987, *Class. Quant. Grav.*, **4**, 1457.

—. 1990, *Class. Quant. Grav.*, **7**, L13.

—. 1991, *Class. Quant. Grav.*, **8**, L77.

—. 1992, *Isotropic Singularities*, to appear in *Rendiconti del Seminario Matematico*.

Implications of Superconductivity in Cosmic String Theory

BRANDON CARTER

Although Kibble's original toy cosmic string model is characterised by longitudinal Lorentz invariance, it is argued that the tacit assumption that this feature would be preserved in a realistic treatment is rather naive. Strict longitudinal Lorentz invariance is incompatible with equilibrium, but its violation allows closed string loops to survive in centrifugally supported states instead of radiating all their energy away. Following the explicit suggestion by Witten of a superconductivity mechanism whereby such a violation would be achieved, it was pointed out by Davis and Shellard that although the ensuing distribution of centrifugally supported string loops would be cosmologically admissible in a "lightweight" (electroweak transition) string scenario, it would imply a highly excessive cosmological mass density ratio, $\Omega \gg 1$ in a "heavyweight" (G.U.T. transition) string scenario of the kind postulated to account for galaxy formation. In order to salvage such scenarios, it might be hoped that Witten type superconductivity does not occur, except perhaps as an ephemeral phenomenon subject to decay by quantum tunnelling. However such optimism overlooks the point that the Witten mechanism is just one particularly simple example, and that even if it fails to apply, experience shows that there are many other ways by which Lorentz symmetry breaking in extended material systems is usually achieved.

1 THE CONCEPT OF A COSMIC STRING

Following the general acceptance of the Higgs mechanism as a plausible model for spontaneous symmetry breaking in renormalisable field theories, attention was drawn by Kibble (1977) to the potential cosmological importance of the consequent formation of *cosmic strings*, meaning topological defects of the vacuum of *local vortex* type, for which the relevant stress-momentum energy distribution is effectively *confined* to the neighbourhood of a *two dimensional* timelike worldsheet (so that at the level of macroscopic approximation it will be describable in terms of 2-surface supported Dirac distributions).

It is to be emphasized that this usage of the term *string* (to which the present discussion is restricted) excludes the qualitatively different case of unconfined (*global*) vortex defects for which quantities such as the energy per unit length are divergent. The description as a string in this strict (*local*) sense thus implicitly requires that the long range (inherently unconfined) gravitational and (if present) electromagnetic fields attributable to the vortex (for which an extended continuum as opposed to Dirac distributional treatment is necessary) must be representable separately from the string itself as extrinsic perturbations. In order to be applicable with a satisfactory degree of accuracy such an approximate description in terms of (by definition confined) string structures requires that the relevant dimensionless gravitational and electromagnetic coupling constants should be sufficiently small, as is certainly the case in the analogous string models that are relevant in the more familiar contexts exemplified by musical instrumentation or ordinary telephone wiring, and as may likewise be safely presumed to be the case also in the cosmological applications that were envisaged first by Kibble (1977) and later in more realistic detail by Witten (1985), who was the first to raise the question of electromagnetic effects.

2 LONGITUDINAL LORENTZ INVARIANCE AND ITS VIOLATION
As an explicit model for the representation of cosmic string behaviour at a macroscopic level, the simplest possibility is specifiable by a variation principle of the well known Goto-Nambu type, which means that it is determined by an action given as the integral over the two-dimensional string world sheet of a scalar Lagrangian function L that is simply a constant. Explicitly, in units with $\hbar = c = 1$, the appropriate Lagrangian will be given by an expression of the form

$$L = -m_x{}^2 \tag{1}$$

where m_x is a fixed mass scale whose precise value (as derived by working out the corresponding Nielsen & Olesen (1973) type cylindrically symmetric equilibrium state) will be dependent on the details of the particular underlying field theory that is supposed to be relevant. As was already pointed out by Kibble (1977) in the earliest discussion of the potential cosmological importance of the phenomenon, the string mass scale m_x may be expected to be of the same order of magnitude as the mass scale characterizing the Higgs boson responsible for the spontaneous symmetry breaking, while the corresponding Compton wavelength $1/m_x$ may be expected to provide an estimate of the string thickness within which the vortex core may be considered to be effectively confined.

In most of the early discussions of the subject it was postulated that the strings under consideration were of the "heavyweight" variety for which the relevant symmetry breaking was that of grand unification, as characterised by $Gm_x{}^2 \approx 10^{-6}$, but [as has recently been shown explicitly by Peter (1992c)] it is also possible to envisage "lightweight" cosmic string formation for which the relevant symmetry breaking is that of electroweak unification, as characterised by $Gm_x{}^2 \approx 10^{-32}$.

In the particular case of a Goto Nambu type model, as characterised by a trivially constant Lagrangian of the form (1) both the tension T and the energy per unit length U of the string are also constant, having exactly the *same* magnitude $T = U = m_x{}^2$. Such "degenerate" string models are characterised by the very special property of longitudinal (two-dimensional) Lorentz invariance, which distinguishes them from more general "non-degenerate" string models of the kind needed for application in the terrestrially familiar contexts exemplified by musical instrumentation or telephone wiring for which (in relativistic units) the surface stress momentum energy density tensor has distinct eigenvalues related by the strict inequality $T < U$.

In view of the fact that such "non-degeneracy" characterizes all of the "terrestrial string" models that are appropriate in experimentally known cases, it is remarkable that it was not until as recently as 1985 that someone [it was of course Witten (1985)] at last put forward the first suggestion that there might be a mechanism whereby longitudinal Lorentz invariance can also be broken in cosmic strings models. As a specific example, on the basis of a particular toy field theoretical model only slightly more complicated than the kind originally considered by Kibble, Witten demonstrated the possibility of the string core being inhabited by a confined boson condensate of the kind that can act as a vehicle for superfluidity or superconductivity.

To provide a classical macroscopic description of a Witten type superconducting cosmic string the "degenerate" Goto-Nambu type Lagrangian (1) is no longer adequate, but it suffices (Carter 1989) to replace it by a "non-degenerate" (variable) Lagrangian that is given as an appropriate function of the magnitude of the covariant derivative of a scalar variable φ representing the phase of the condensate on the string worldsheet. Wherever the current is spacelike (the only case considered in the earliest discussions) the Lagrangian function is identifiable simply with the negative string tension T itself, i.e. the replacement for (1) takes the form

$$L = -T , \qquad (2)$$

while the corresponding energy density function U will be given by an expression of the form

$$U = T + \mu\nu \qquad (3)$$

where ν is the magnitude of the phase gradient (in terms of which T is given) which means that it is interpretable as the *number density* of particles (as defined in terms of phase rotations) associated with the flux, and μ its dynamical conjugate, which is to be expected to have the same order of magnitude,

$$\nu^2 \approx \mu^2 \lesssim m_x{}^2 , \qquad (4)$$

both being subject to an upper "current saturation" that on dimensional grounds is expected not to exceed the fixed mass value m_x to which both U and T tend in the "chiral" limit, $U \to m_x{}^2$, $T \to m_x{}^2$, at which the current vector becomes null so that $\nu \to 0$. Beyond this limit, where the current is timelike, the formula (3) will still be applicable, but the roles of ν and its dynamical conjugate μ will be interchanged, the gradient magnitude being interpretable directly as a chemical potential, and U must be used instead of T in (2). (It is to be noticed that such a model still admits the local existence of Lorentz invariant Kibble type states characterised by uniformity of the phase variable φ as a special case within the intermediate "chiral" case.)

The practical relevance of longitudinal Lorentz symmetry breaking by the formation of a Witten type current depends of course on whether there is a sufficiently long timescale τ over which it possible to neglect quantum dissipation effects so that a classical superfluidity or superconductivity description of the type provided by (2) remains valid. Dimensional analysis, as developed particularly by Davis (1988) suggests that for a string loop with current characterised by a mean phase winding number N, the relevant timescale τ may be expected to be given in terms of the basic Higgs mass scale m_x characterizing the typical dimensions of the core of the string vortex, and a second mass scale m_σ characterizing the free state of the relevant carrier boson field, by a crude order of magnitude estimate of the form

$$\ln(\tau\, m_x) \approx N \left(\frac{m_\sigma}{m_x}\right)^3 . \tag{5}$$

This would imply that, provided the carrier mass scale m_σ is not too small compared with the Higgs mass scale m_x, the current survival timescale can indeed be long even by cosmological standards for quite moderate and therefore entirely plausible values of the effective mean winding number N.

3 CATASTROPHIC CONSEQUENCES OF LONGITUDINAL LORENTZ SYMMETRY BREAKING

In the absence of any other sufficiently satisfactory explanation of galaxy formation, Kibble's original suggestion that "heavyweight" (G.U.T.) strings, as characterised by $Gm_x{}^2 \approx 10^{-6}$ might be responsible for galaxy formation was received with immediate enthusiasm by empirical cosmologists (Zel'dovich 1980, Vilenkin 1981). In this climate of opinion the further suggestion (Ostriker et al 1986) that electromagnetic effects arising from Witten's superconductivity phenomenon in such strings might also account for intergalactic voids was also favorably received, while comparatively little attention was paid to the warning that the whole scenario might be invalidated by a perhaps less desirable but not so easily avoidable side effect arising from the concomitant violation of longitudinal Lorentz symmetry, namely the production of a cosmological mass density excess $\Omega \gg 1$ in the form of stationary loop relics.

The point whose importance still seems to be commonly underestimated is that whereas the simple Goto Nambu type model given by (1) is characterised by longitudinal (2-dimensional) Lorentz invariance, which implies that a string loop of this type has no stationary equilibrium state and so must oscillate until all its energy has been lost by gravitational radiation (and possibly other processes as well) on the other hand in generic string models for which the longitudinal Lorentz invariance will be broken, and in particular for a Witten type model as governed by an action of the form (2), various kinds of stationary equilibrium state can exist, whose survival – as was first emphasized by Davis and Shellard (1988, 1989) – strongly threatens to produce a cosmological mass density excess $\Omega \gg 1$ at least if they are of the "heavyweight" (GUT) variety that is relevant for galaxy formation scenarios, though probably not (Davis & Shellard 1989, Carter 1991a, b) if they are of the "lightweight" (electroweak) variety whose existence as a real cosmological phenomenon would therefore seem to be relatively more plausible.

4 INEFFICACY OF THE MAGNETOSTATIC SUPPORT MECHANISM

Prior to the epoch making intervention of Davis and Shellard (1988), the first kind of equilibrium state to have been envisaged for string loops was the magnetostatically supported variety (Ostriker *et al* 1986) in which the tendency to contraction due to the local string tension is balanced by the the effect of a globally extended dipolar magnetostatic field due to an electromagnetically coupled current in the loop. It was however recognised at the outset (Ostriker *et al* 1986), and also implicitly in subsequent discussions (Copeland *et al* 1987 and 1988, Haws *et al* 1988) that such a magnetostatic support mechanism was unlikely to be sufficiently strong to be of practical importance. This can be seen from the fact that the required electric current magnitude I would need to have a rather large magnitude given roughly (modulo a numerical factor with logarithmic dependance on the phase winding number N) by

$$I^2 \approx T \ . \tag{6}$$

Since the current magnitude will be given simply by $I^2 = e^2 \mu^2$, it is evident (from the smallness of the electromagnetic coupling constant $e^2 = 1/137$) that it will be difficult (though, depending on the particular model, not necessarily impossible) to reconcile this magnetic support condition (6) with the expected upper limit (4) on the maximum value at which "saturation" or "quenching" occurs.

Although there was broad qualitative agreement on this point, the follow up discussions (Copeland *et al* 1987 and 1988, Haws *et al* 1988) of this magnetic dipole effect nevertheless engendered a certain amount of confusion due to their failure to clearly distinguish this essentially *global* support mechanism (due to the long range effect of the external magnetostatic field arising from the charge coupled current) from the qualitatively different mechanism (which can also be considered to account for the phenomenon of current saturation, and which operates even in the absence of any electromagnetic coupling) whereby the *local* string tension T is reduced below its Kibble limit value $T = m_x{}^2$ by the mechanical effect of the current in the immediate neighbourhood of the vortex core. This confusion was embodied in the use of the potentially misleading term "spring" to describe states in which the effective tension was reduced to zero so as to allow static equilibrium without an external supporting force. More detailed examination (Hill *et al* 1988, Babul *et al* 1988) of the local effect of the current appeared to confirm that in certain models the string tension T actually could in principle reach zero and even negative values, but it was pointed out that in a local (as opposed to global) string state negative tension automatically implies instability (Carter 1989b, 1990a and 1992b) and it has been shown more recently using improved numerical methods (Peter 1992a, b) that although it can be considerably reduced, the local string tension will in fact remain strictly positive, consistently with the qualitative prediction that magnetic configurations – for which the current is spacelike – will satisfy

$$T \gtrsim \mu^2 , \qquad (7)$$

in all the kinds of cosmic string model that have been examined so far, which, since $I^2 = e^2 \mu^2$, confirms the difficulty of satisfying (6).

5 EFFICACY OF THE CENTRIFUGAL SUPPORT MECHANISM
As far as cosmological implications are concerned, the questions about whether magnetically supported equilibrium states might be of any practical importance (and about the sense, if any, in which locally confined – as opposed to global – "spring" states can be said to exist at all) were effectively relegated to obsolescence when attention was drawn by Davis & Shellard (1988) to the obviously very much greater (though hitherto overlooked) importance of the essentially quite different (and simpler, since inherently *local*) mechanism of *centrifugal* support, which works even in the absence of any electromagnetic coupling of the relevant current.

As a reflection of the *duality* (Carter 1990b, Carter *et al* 1991) that survives the longitudinal Lorentz symmetry breaking, the presence of a superconduction current in a closed loop determines not just one but two independent conserved numbers, namely a phase winding number N and a charge counting number C (which is well defined even if the charge in question is not electromagnetically coupled). For chosen values of this pair of numbers there will be centrifugally supported stationary (but not static) "vorton" (Davis & Shellard 1988, 1989) equilibrium states, of minimum but (unless both numbers vanish) non zero energy M. Instead of (6), the equilibrium condition in the centrifugally supported case is that the longitudinal circulation velocity v should be given (exactly) by

$$Uv^2 = T \ . \tag{8}$$

Among such stationary solutions, the simplest (Carter 1990a, b), though not the only type (Carter *et al* 1991, Carter 1992a), are circular ring configurations with angular momentum $J = CN$. It is not of course to be supposed that equilibrium will be possible for all values of C and N since in some cases the current would build up towards a saturation value at which "quenching" will take place, but there will be at least a range of allowed values (of the order of unity) for the ratio C/N in the neighbourhood of the "chiral" limit (Davis & Shellard 1989) where the current vector is approximately null, so that although the equilibrium frame current components may be large, the absolute current magnitude as measured in the comoving frame will remain small compared with the saturation limit.

6 HOPES OF SALVAGING HEAVYWEIGHT STRING FORMATION SCENARIOS

Although a definitive analysis will require much more thorough work on the (classical and quantum) stability properties of these (not too far from chiral) ring shaped and more general "vorton" states, and also much more detailed investigations of the numbers and size distribution with which they are likely to be formed, it is apparent (Davis & Shellard 1988 and 1989, Carter 1990b and 1991a) that in a "heavyweight" string scenario there would be a very serious cosmological mass excess problem even if the efficiency of relic loop formation and preservation were extremely low. The persistent popularity of such scenarios has however encouraged a perhaps unreasonable faith that they might be saved by some loophole in the argument, the only obvious possibility being that the various conceivable kinds of "vortons" should all turn out to be effectively unstable after all.

A glimmer of hope that an adequate instability mechanism may exist has recently been raised by Peter's discovery (Peter 1992a, b) that (contrary to what was implied by the linear relation between tension and energy density that was effectively assumed in most early discussions such as those cited above (Copeland *et al* 1987 and 1988, Haws *et al* 1988)) the longitudinal perturbation velocity is generally lower (at least in all the cases examined so far) than the transverse perturbation velocity which is known (as a general theorem – Carter 1990a, b and 1992a, Carter *et al* 1991) to be the same as the circulation speed v of the stationary "vorton" states, as given by (8). This means that there will be forward circulating but nevertheless relatively retrograde longitudinal modes which might be expected to be subject to an instability analogous to the well known Friedman-Schutz type instability (Schutz 1987) that occurs generically in rotating perfect fluid star models. This glimmer of hope is however dimmed by the consideration that the characteristic timescale of the instability has generally turned out to be far too long for it to be of any practical consequence in the stellar case: it may plausibly be conjectured (but still ought to be checked) that the same applies to the "vortonic" case, at least in the relevant "chiral" limit where the difference between the longitudinal and transverse perturbation speeds will be very small.

If (as I would conjecture) Peter's instability mechanism is insufficient to save the situation, there remains – as the only other plausible suggestion for setting up a "heavyweight" cosmic string formation theory that avoids the formation of a disastrous cosmological mass excess – the imposition of the requirement (Davis & Shellard 1988, 1989) that the underlying field be such that the vortex core admits no current carrying condensate except perhaps for fields characterised by a mass scale m_σ that is negligibly small compared with the basic symmetry mass scale m_x so that the corresponding quantum instability timescale (5) is cosmologically short. However unless the advocates of "heavyweight" cosmic string scenarios are able and willing to resort to the use of what has been referred to in the related context of axion theory as "the last refuge of a scoundrel" (Dowrick & McDougal 1988) meaning what I have called the "strong" anthropic principle (in order to distinguish it from the ordinary honest though not quite incontestable (Carter 1992c) "weak" version) the requirement

$$m_\sigma \ll m_x \qquad (9)$$

would seem too much like special pleading to be satisfactory, since it must be applied not just to a single particular mass scale but to every mass scale m_σ characterizing any possible current carrier in the theory. To appreciate the severity of such a requirement, one needs only to notice that the further one goes towards replacing simplified toy models (such as those on which the discussion was historically founded (Kibble 1977, Witten 1985)) by more realistic field theories (with a much richer content) the greater the range of possibilities for the creation of superconducting condensates.

The natural violation of the condition (9) is illustrated in the most recent and phenomenologically realistic example (Nielsen & Olesen 1973) which gives rise to a current carrying condensate whose mass scale $m_\sigma \approx m_W$ is that of the W meson, in a string whose symmetry breaking mass scale $m_x \approx m_{WS}$ is that of the Weinberg Salam phase transition which is commonly presumed to be only of the same order of magnitude, i.e. $m_{WS} \approx m_W$. Although this value is sufficiently small ($G m_{WS}{}^2 \approx 10^{-32}$) to ensure that this model is not itself threatened by a cosmological overdensity disaster, it strongly suggests that an analogous "Grand Unified" model with $m_x \approx m_{GU}$ (where $G m_{GU}{}^2 \approx 10^{-32}$) would be likely to give rise to an analogous current carrying condensate characterised by a comparable mass scale, $m_\sigma \approx m_{GU}$, whose cosmological consequences would be fatal.

Regardless of particular examples, it is to be emphasized that since phenomena such as superfluidity and superconductivity are familiar as common generic features of ordinary matter at sufficiently low temperatures, particularly at high densities, it would seem rather unreasonable *a priori*, to expect that they would not also turn up generically in cosmic strings. Moreover it is to be further emphasized that these are not the only mechanisms for the breaking of Lorentz invariance which is all that is needed for centrifugally supported string loops to be able to survive in equilibrium. Even at zero temperature, all the extended material systems that are experimentally known involve spontaneous Lorentz symmetry breaking, typically by the formation of an elastic lattice structure. Although such structures are not so easy to derive directly from first principles of quantum field theory, this theoretical difficulty of working them out is not a valid reason for doubting that they should exist in cosmic strings just as in experimentally known material systems.

Such phenomenological considerations all tend to indicate the naivety of the assumption (implicit in so much of the work on cosmological effects of cosmic strings) that the longitudinal Lorentz symmetry characterizing Kibble's original toy string model would be preserved in a realistic treatment. The implication is that the warning by Davis and Shellard that "heavyweight" string scenarios are vulnerable to cosmological catastrophe should be taken more seriously.

I wish to thank Dennis Sciama for introducing me to cosmology in general, and to thank Rick Davis, Patrick Peter, Tsvi Piran, and Alex Vilenkin for many enlightening discussions about cosmic strings in particular.

REFERENCES

Babul, A., Piran, T. and Spergal, D. N. 1988, *Phys. Lett. B*, **202**, 307.

Carter, B. 1989a, *Phys. Lett. B*, **224**, 61.

—. 1989b, *Phys. Lett. B*, **228**, 446.

—. 1990a, "Covariant Mechanics of Simple and Conducting Strings and Membranes", in: *The Formation and Evolution of Cosmic Strings*, eds. G. Gibbons, S. Hawking, T. Vachaspati, pp. 143-178 (Cambridge University Press).

—. 1990b, *Phys. Lett. B*, **238**, 166.

—. 1991a, *Ann. N. Y. Acad. Sci.*, **647**, 758.

—. 1991b, in: *Early Universe and Cosmic Structures (Xth Moriond Astrophysics Meeting, 1990)* eds. A. Blanchard *et al.*, pp. 213-221 (Editions Frontières, Gif-sur-Yvette).

—. 1992a, "Basic Brane Mechanics", in: *Relativistic Astrophysics and Gravitation (10th Potsdam Seminar, Oct., 1991*, eds. S. Gottloeber, J. P. Muecket, V. Mueller (World Scientific, Singapore).

—. 1992b, *Class. and Quantum Grav.*, **9**, 19.

—. 1992c, in: *The Anthropic Principle (2nd Venice Conf. on Cosmology and Philosophy, November 1988)*, eds. F. Bertola, V. Curi (Cambridge University Press).

Carter, B., Frolov, V. P. and Heinrich O. 1991, *Class. and Quantum Grav.*, **8**, 135.

Copeland, E., Haws, D., Hindmarsh, M. and Turok N. 1988, *Nucl. Phys. B*, **306**, 908.

Copeland, E., Hindmarsh, M., Turok, N. 1987, *Phys. Rev. Lett*, **58**, 1910.

Davis, R. L. 1988, *Phys. Rev. D*, **38**, 3722.

Davis, R. L. and Shellard, E. P. S. 1988, *Phys. Lett. B*, **209**, 485.

—. 1989, *Nucl. Phys. B*, **323**, 209.

Dowrick, N. and McDougal, N. A. 1988, *Phys. Rev. D*, **38**, 3619.

Haws, D., Hindmarsh, M. and Turok, N. 1988, *Phys. Lett. B*, **209**, 225.

Hill, C., Hodges, H. and Turner, M. 1988, *Phys. Rev. D*, **37**, 263.

Kibble, T. W. B. 1977, *J. Phys. A*, **9**, 1387.

Nielsen, H. B. and Olesen, P. 1973, *Nucl. Phys. B*, **61**, 45.

Ostriker, J., Thompson, C. and Witten, E. 1986, *Phys. Lett. B*, **180**, 231.

Peter, P. 1992a, *Phys. Rev. D*, **45**, 1091.

—. 1992b, *Phys. Rev. D*, to be published.

—. 1992c, *Phys. Rev. D*, to be published.

Schutz, B. 1987, "Relativistic Gravitational Instabilities", in: *Gravitation in Astrophysics, (NATO A.S.I., Cargèse 1986)*, eds. B. Carter, J. B. Hartle, pp. 123 - 153 (Plenum, New York).

Vilenkin, A. 1981, *Phys. Rev. Lett.*, **46**, 1169; erratum: ibid. 1496.

Witten, E. 1985, *Nucl. Phys. B*, **249**, 557.

Zel'dovich, Ya. B. 1980, *Mon. Not. R.A.S.*, **192**, 663.

The Formation and Evaporation of Primordial Black Holes

BERNARD J. CARR

It is a great pleasure to speak at this meeting since it gives me a chance to acknowledge the great influence Dennis Sciama has had on my life. It was Dennis who first introduced me to relativity as an undergraduate at Cambridge in 1968 and it was through a popular lecture he gave to the Cambridge University Astronomical Society in that year that I first learnt about the microwave background radiation. I well recall his remark that he was "wearing sackcloth and ashes" as a result of his previous endorsement of the Steady State theory. This made a great impression on me and was an important factor in my later choosing to do research in Big Bang cosmology. When I was accepted as a PhD student by Stephen Hawking, I was therefore delighted to become Dennis' academic grandson. (Incidentally since Stephen has related how he had originally wanted to do his PhD under Fred Hoyle, having never heard of Dennis, I must confess - with some embarrassment - that, when I applied for a PhD, I had never heard of Stephen!) The subject of my PhD thesis was primordial black holes, so it seems appropriate that I should talk on this topic at this meeting, especially as Dennis was my PhD examiner.

1 HISTORICAL REVIEW

It was first pointed out by Zeldovich & Novikov (1967) and Hawking (1971) that black holes could have formed in the early Universe as a result of density inhomogeneities. Indeed this is the only time when black holes smaller than a solar mass could form since a region of mass M must collapse to a density $\rho_{BH} \simeq 10^{18}(M/M_\odot)^{-2}$ g/cm^3 in order to fall within its Schwarzschild radius and only in the first moments of the Big Bang could the huge compression required arise naturally. In order to collapse against the background pressure, overdense regions would need to have a size comparable to the particle horizon size at maximum expansion. On the other hand, they could not be much bigger than that else they would be a separate closed universe rather than part of our universe, so primordial black holes (PBHs) forming at time t would need to have of order the horizon mass $M_H \simeq c^3 G^{-1} t \sim 10^5 (t/s) M_\odot$. Thus PBHs could span an enormous mass range: those forming at the Planck time (10^{-43} s) would have the Planck mass (10^{-5} g), whereas those forming at 1 s would be as large as the holes thought to reside in galactic nuclei.

For a while the existence of PBHs seemed unlikely since Zeldovich & Novikov (1967) had pointed out that they might be expected to grow catastrophically. This is because a simple Newtonian argument suggest that, in a radiation–dominated universe, the mass of a black hole should evolve according to

$$M(t) = M_H(t) \left\{ 1 + \frac{t}{t_o} \left[\frac{M_H(t_o)}{M_o} - 1 \right] \right\}^{-1} \simeq \left\{ \begin{matrix} M_o & \text{for } M_o \ll M_H(t_o) \\ M(t) & \text{for } M_o \sim M_H(t_o) \end{matrix} \right. \quad (1)$$

where M_o is the mass of the hole at some initial time t_o. This implies that holes much smaller than the horizon cannot grow much at all, whereas those of size comparable to the horizon could continue to grow at the same rate as the horizon ($M \propto t$) throughout the radiation era. Since we have seen that a PBH *must* be of order of the horizon size at formation, this suggests that all PBHs could grow to have a mass of order $10^{15} M_\odot$ (the horizon mass at the end of the radiation era). There are strong observational limits on how many such giant black holes could exist in the Universe, so the implication seemed to be that very few PBHs ever formed.

The Zeldovich–Novikov argument was questionable since it neglected the cosmological expansion and this would presumably hinder the black hole growth. Indeed the notion that PBHs could grow at the same rate as the horizon was disproved by myself and Hawking: we demonstrated that there is no spherically symmetric similarity solution which contains a black hole attached to a Friedmann background via a pressure wave (Carr & Hawking, 1974). Since a PBH must therefore soon become much smaller than the horizon, at which stage cosmological effects become unimportant and eq. (1) *does* pertain, one concludes that a PBH cannot grow very much at all.

The realization that small PBHs could exist after all prompted Hawking to consider their quantum properties and this led to his famous discovery that black holes should radiate thermally with a temperature $T \simeq 10^{-7}(M/M_\odot)^{-1} K$ and evaporate completely in a time $\tau \simeq 10^{10}(M/10^{15} \text{ g})^3$ y (Hawking 1975). Dennis played an important part in this story since he was one of the organizers at the Second Quantum Gravity conference at Rutherford, where the result was first announced. The claim was met with scepticism in certain quarters but Dennis was quick to appreciate and promulgate its profound importance. Indeed he and his student Philip Candelas made an important contribution in the area by connecting black hole radiance to the "fluctuation–dissipation" theorem (Candelas & Sciama, 1977).

Despite the conceptual importance of Hawking's result (it illustrates that it is sometimes usefully to study something even if it does not exist'), it was rather bad news for PBH enthusiasts. For since PBHs with mass of 10^{15} g would have a temperature of order 100 MeV and radiate mainly at the present epoch, the observational limit on the gamma–ray background density at 100 MeV immediately implied that the density of such holes could be at most 10^{-8} in units of the critical density (Chapline, 1975; Page & Hawking, 1976). Not only did this exclude PBHs as solutions of the dark matter problem, but it also implied that there was little chance of detecting black hole explosions at the present epoch (Porter & Weekes, 1979).

Despite this negative conclusion, it was realized that PBH evaporations could still have interesting cosmological consequences and the next five years saw a spate of papers focusing on these. In particular, people were interested in whether PBH evaporations could generate the microwave background (Zeldovich & Starobinsky, 1976) or modify the standard cosmological nucleosynthesis scenario (Novikov et al., 1979; Lindley, 1980) or account for the cosmic baryon asymmetry (Barrow, 1980). On the observational front, people were interested in whether PBH evaporations could account for the unexpectedly high fraction of antiprotons in cosmic rays (Kiraly et al., 1981; Turner, 1982) or the interstellar electron and positron spectrum (Carr, 1976) or the annihilation line radiation coming from the Galactic centre (Okeke & Rees, 1980). Renewed efforts were also made to look for black hole explosions after the realization that - due to the interstellar magnetic field - these might appear as radio rather than gamma-ray bursts (Rees, 1977).

In the 1980s attention switched to several new formation mechanisms for PBHs. Originally it was assumed that they would need to form from primordial density fluctuations but it was soon realized that PBHs might also form very naturally if the equation of state of the Universe was ever soft (Khlopov & Polnarev, 1980) or if there was a cosmological phase transition leading to bubble collisions (Kodama et al., 1979; Hawking et al., 1982; Hsu, 1990). In particular, the formation of PBHs during an inflationary era (Naselsky & Polnarev 1985) or at the quark-hadron era (Crawford & Schramm 1985) received a lot of attention. More recently, people have considered the production of PBHs through the collapse of cosmic strings (Polnarev & Zembovicz 1988, Hawking 1989) and the issue of forming Planck mass black holes through thermal fluctuations has also been of interest (Gross et al. 1982, Kapusta 1984, Hayward & Pavon 1989). All these scenarios are constrained by the quantum effects of the resulting black holes.

Recently work on the cosmological consequence of PBH evaporations has been revitalized as a result of calculations by my former PhD student Jane MacGibbon (Dennis' academic great grand–daughter). Previous approaches to this problem (including my own) had been rather simplistic, merely assuming that the relevant particles are emitted with a black–body spectrum as soon as the black hole temperature exceeds their rest mass. However, if one adopts the conventional view that all particles are composed of a small number of fundamental point–like constituents (quarks and leptons), it would seem natural to assume that it is these fundamental particles rather than the composite ones which are emitted directly once the temperatures goes above the QCD confinement scale of 140 MeV. MacGibbon therefore envisages a black hole as emitting relativistic quark and gluon jets which subsequently fragment into the stable leptons and hadrons (i.e. photons, neutrinos, gravitons, electrons, positrons, protons and antiprotons). On the basis of both experimental data and Monte Carlo simulations, one now has a fairly good understanding of how quark jets fragment. It is therefore straightforward in principle to convolve the thermal emission spectrum of the quarks and gluons with the jet fragmentation function to obtain the final particle spectra (MacGibbon & Webber, 1990; MacGibbon, 1991; MacGibbon & Carr, 1991). As discussed here, the results of such a calculation are very different from the simple direct emission calculation, essentially because each jet generates a Brehmstrahlung tail of decay products, with energy extending all the way down to the decay product's rest mass.

Lately attention has turned to the issue of Planck mass relics. Most early work assumed that PBHs evaporate completely. However, this is by no means certain and several people have argued that evaporation could discontinue when the black hole gets down to the Planck mass (Bowick et al., 1988; Coleman et al., 1991). In this case, one could end up with stable Planck mass objects. MacGibbon (1987) pointed out that such relics would naturally have around the critical density if 10^{15} g PBHs have the density required to explain the gamma–ray background and Barrow et al. (1992) have considered the constraints associated with such relics in more general circumstances. Another possibility, presented at this conference by Hawking himself, is that black hole evaporation could end by generating a "thunderbolt" - a naked singularity which is never seen because it travels at the speed of light.

2 THE FORMATION OF PRIMORDIAL BLACK HOLES

(1) Initial inhomogeneities. If the PBHs form directly from primordial density perturbations, then the fraction of the Universe undergoing collapse at any epoch is just determined by the root–mean–square amplitude of the horizon–scale fluctuations at that epoch ϵ and the equation of state $p = \gamma\rho$ $(0 < \gamma < 1)$. We have seen that an overdense region must be larger than the Jeans length at maximum expansion and this is just $\sqrt{\gamma}$ times the horizon size. This requires the density fluctuation to exceed γ at the horizon epoch, so one can infer that the fraction of regions of mass M which form a PBH is (Carr, 1975)

$$\beta(M) \sim \epsilon(M)\exp(-\gamma^2/2\epsilon(M)^2) \qquad (2)$$

where $\epsilon(M)$ is the value of ϵ when the horizon mass is M. This assumes that the fluctuations have a Gaussian distribution and are spherically symmetric. The PBHs can have an extended mass spectrum only if the fluctuations are scale–invariant (i.e. with ϵ independent of M) and, in this case, the number density of PBHs is given by (Carr, 1975)

$$dn/dM = (\alpha - 2)(M/M_*)^{-\alpha} M_*^{-2} \Omega_{PBH} \rho_{crit} \qquad (3)$$

where $M_* \simeq 10^{15}$g is the current lower cut–off in the mass spectrum due to evaporations, Ω_{PBH} is the total density of the PBHs in units of the critical density ρ_{crit} (which itself depends on β), and the exponent α is uniquely determined by the equation of state:

$$\alpha = (1 + 3\gamma)/(1 + \gamma) + 1. \qquad (4)$$

If one has a radiation equation of state ($\gamma = 1/3$), as is consistent with the Elementary Particle picture, then $\alpha = 5/2$. This means that the integrated PBH mass spectrum falls off as $M^{-1/2}$, so most of the PBH density is contained in the smallest ones. If $\epsilon(M)$ decreases with M, then the spectrum falls off exponentially with M and PBHs can form only around the Planck time ($t_{Pl} \sim 10^{-43}$ s) if at all; if $\epsilon(M)$ increases with M, the spectrum rises exponentially with M and PBHs would form very prolifically at sufficiently large scales but the microwave anisotropies would then be larger than observed. Fortunately, many scenarios for the cosmological density fluctuations do predict that ϵ is scale–invariant, so eq. (3) represents the most likely mass spectrum.

(2) Soft equation of state. The pressure may be reduced for a while ($\gamma \ll 1$) if the Universe's mass is ever channeled into particles which are massive enough to be non–relativistic (Khlopov & Polnarev, 1980). In this case, the effect of pressure in stopping collapse is unimportant and the probability of PBH formation depends upon the fraction of regions which are sufficiently spherical to undergo collapse; this can be shown to be (Polnarev & Khlopov, 1981)

$$\beta = 0.02\epsilon^{13/2} \qquad (5)$$

The holes should have a mass which is smaller than the horizon mass at formation by a factor $\epsilon^{3/2}$, so the period for which the equation of state is soft directly specifies their mass range. In this case, the value of β and hence Ω_{PBH} is not as sensitive to ϵ as in case (1).

(3) Inflationary period. In the standard inflationary scenario, the amplitude of the density fluctuations increases logarithmically with mass and the normalization required to explain galaxy formation would then preclude the fluctuations being large enough to give PBHs on a smaller scale. One therefore has to involve a "double inflation" scenario, in which there is a second period of inflation associated with a

larger value of the self–coupling constant λ (Naselsky & Polnarev, 1984). Since the amplitude of the resulting fluctuations scales as $\lambda^{1/2}$, one needs fine–tuning of λ to get an interesting value of Ω_{PBH}. Note that λ also determines the duration of the inflationary period since this should go as λ^{-1} (Polnarev & Khlopov, 1981). Thus, if ϵ is to be as high as 0.05 (as required for $\Omega_{PBH} \sim 1$), inflation can persist for at most a factor of 10^3 in time and this implies that the PBH spectrum can only extend over three decades.

(4) Bubble collisions. Even if the Universe starts off perfectly smooth, bubbles of broken symmetry might arise at a spontaneously broken symmetry epoch and it has been suggested that black holes could form as a result of bubble collisions (Kodama et al., 1982; Hawking et al., 1982; La & Steinhardt, 1989). In fact, this happens only if the bubble formation rate is finely tuned: if it is too large, the entire Universe under goes the phase transition immediately; if it is too small, the bubbles never collide. In consequence, the holes should again have a mass of order the horizon mass at the phase transition. For example, PBHs forming at the Grand Unification epoch $(10^{-35}$ s$)$ would have a mass of order 10^3 g. Only a phase transition before 10^{-23} s would be relevant in the context of evaporating PBHs.

(5) Collapse of cosmic loops. A typical cosmic loop will be larger than its Schwarzschild radius by the inverse of the factor $G\mu$ which represents the mass per unit length. In the favoured scenario, $G\mu$ is of order 10^{-6}. However, Hawking (1989) and Polnarev & Zemboricz (1983) have shown that there is still a small probability that a cosmic loop will get into a configuration in which every dimension lies within its Schwarzschild radius. Hawking estimates this to be

$$\beta \sim (G\mu)^{-1}(G\mu x)^{2x-2} \qquad (6)$$

where x is the ratio of the loop length to the correlation scale. If one takes x to be 3, Ω_{PBH} exceeds 1 for $G\mu > 10^{-7}$, so he argues that one overproduces PBHs. However, Ω_{PBH} is very sensitive to x and a slight reduction would give a rather interesting value. Note that spectrum (3) still applies since the holes are forming at every epoch.

In all these scenarios, the value of Ω_{PBH} associated with PBHs which form at a redshift z or time t is related to β by

$$\Omega_{PBH} = \beta\Omega_R(1+z)^{-1} = 10^6\beta(t/s)^{-1/2} \qquad (7)$$

where $\Omega_R \sim 10^{-4}$ is the density of the microwave background. Since t is very small, the constraint $\Omega_{PBH} < 1$ implies that β must be tiny over all mass ranges. This is because the radiation density falls off as $(1+z)^4$, whereas the PBH density falls off as $(1+z)^3$. If the PBHs form at a phase transition, as in cases (2) to (4), then they have a very narrow mass spectrum and t is just the time of the transition. If they have a continuous mass spectrum, as in cases (1) and (5), then the dominant contribution to Ω_{PBH} comes from the holes with mass $M \sim 10^{15}$ g evaporating at the present epoch. These form at $t \sim 10^{-23}$ s and so eq. (7) implies $\beta \sim 10^{-17}\Omega_{PBH}$.

3 PRIMORDIAL BLACK HOLE EVAPORATIONS

A black hole of mass M will emit particles in the energy range $(Q, Q+dQ)$ at a rate (Hawking, 1975)

$$dN = \frac{\Gamma dQ}{2\pi\hbar}\{\exp\left(\frac{Q}{kT}\right) \pm 1\}^{-1} \tag{8}$$

where T is the black hole temperature, Γ is the absorption probability and the $+$ and $-$ signs refer to fermions and bosons respectively. This assumes that the hole has no charge or angular momentum. This is a reasonable assumption since charge and angular momentum will also be lost through quantum emission but on a shorter timescale than the mass (Page, 1977). Γ goes roughly like $Q^2 T^{-2}$, though it also depends on the spin of the particle and decreases with increasing spin. Thus a black hole radiates roughly like a black–body with temperature

$$T = \frac{\hbar c^3}{8\pi GkM} = 10^{26}\left(\frac{M}{g}\right)^{-1} K = \left(\frac{M}{10^{13} \text{ g}}\right) \text{ GeV} \tag{9}$$

This means that it loses mass at a rate

$$\dot{M} = -5 \times 10^{25} M^{-2} f(M) \text{ g s}^{-1} \tag{10}$$

where the factor $f(M)$ depends on the number of particle species which are light enough to be emitted by a hole of mass M, so the lifetime is

$$\tau(M) = 6 \times 10^{-27} f(M)^{-1} M^3 \text{ s} \tag{11}$$

The factor f is normalized to be 1 for holes larger than 10^{17} g and such holes are only able to emit "massless" particles like photons, neutrinos and gravitons. Holes in the mass range 10^{15} g $< M < 10^{17}$ g are also able to emit electrons, while those in the range 10^{14} g $< M < 10^{15}$ g emit muons which subsequently decay into electrons and neutrinos. The latter range includes, in particular, the critical mass for which τ equals the age of the Universe. This can be shown to be (MacGibbon & Webber 1990)

$$M_* = 4.4 \times 10^{14} h^{-0.3} \text{ g} \tag{12}$$

where h is the Hubble parameter in units of 100 km/s/Mpc and we have assumed that the total density parameter is $\Omega = 1$.

Once M falls below 10^{14} g, the hole can also begin to emit hadrons. However, hadrons are composite particles made up of quarks held together by gluons. For temperatures exceeding the QCD confinement scale of $\Lambda_{QCD} \simeq 250 - 300$ GeV, one would therefore expect these fundamental particles to be emitted rather than composite particles. Only pions would be light enough to be emitted below Λ_{QCD}. Since there are 12 quark degrees of freedom per flavour and 16 gluon degrees of freedom, one would also expect the emission rate (i.e. the value of f) to increase dramatically once the QCD temperature is reached.

The physics of quark and gluon emission from black holes is simplified by a number of factors. Firstly, since the spectrum peaks at an energy of about $5kT$, eq. (9) implies that most of the emitted particles have a wavelength $\lambda \simeq 2.5M$ (in units with $G = c = 1$), so the particles have a size comparable to the hole. Secondly, one can show that the time between emissions is $\Delta\tau \simeq 20\lambda$, which means that short range interactions between successively emitted particles can be neglected. Thirdly, the condition $T > \Lambda_{QCD}$ implies that $\Delta\tau$ is much less than $\Lambda_{QCD}^{-1} \simeq 10^{-13}$ cm (the characteristic strong interaction range) and this means that the particles are also unaffected by gluon interactions. The implication of these conditions is that one can regard the black hole as emitting quark and gluon jets of the kind produced in collider events. The jets will decay into hadrons over a distance T/m^2 and, since this is much larger than M for $T \gg m$, gravitational effects can be neglected. The hadrons will themselves decay into protons, antiprotons, electrons, positrons and photons on a somewhat longer timescale.

To find the final spectrum of stable particles emitted instantaneously from a black hole one must convolve the Hawking emission spectrum given by eq. (8) with the jet fragmentation function. This gives

$$\frac{dN_X}{dE} = \sum_j \int_{Q=0}^{Q=\infty} \frac{\Gamma_j(Q,T)}{2\pi\hbar} \left(\exp\frac{Q}{T} \pm 1\right)^{-1} \frac{dg_{jx}(Q,E)}{dE} dQ \qquad (13)$$

Here x and j label the final particle and the directly emitted particle, respectively, and the last factor specifies the number of final particles with energy in the range $(E, E + dE)$ generated by a jet of energy Q. For hadrons this can be represented by

$$\frac{dg_{jh}}{dE} = \frac{1}{E}\left(1 - \frac{E}{Q}\right)^{2m-1} \theta(E - km_h c^2) \qquad (14)$$

where m_h is the hadron mass, k is a constant of order 1, and m is 1 for mesons and 2 for baryons. The fragmentation function therefore has an upper cut–off at Q, a lower cut–off around m_h and an E^{-1} Brehmstrahlung tail in between. It also peaks around m_h. By examining the dominant contribution to the Q–integral one obtains

$$\frac{dN}{dE} \sim \begin{cases} E^2 e^{-E/T} & \text{for } E \gg T \;\; (Q \sim E) \\ E^{-1} & \text{for } T \sim E \gg m_h \;\; (Q \sim T) \\ dg/dE & \text{for } E \sim m_h \ll T \;\; (Q \sim m_h) \end{cases} \qquad (15)$$

where the terms in parentheses indicate the value of Q which dominates. This equation enables one to understand qualitatively the form of the instantaneous emission spectrum shown in Figure 1 for a $T = 1$ GeV black hole (MacGibbon & Webber, 1990). The direct emission just corresponds to the small bumps on the right. All the particle spectra show a peak at 100 MeV due to pion decays; the electrons and neutrinos also have peaks at 1 MeV due to neutron decays.

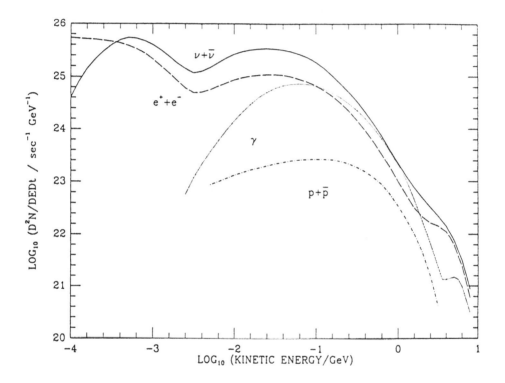

Figure 1. This shows the istantaneous emission from a black hole with a temperature of 1 GeV, taken from MacGibbon & Webber (1990). The neutrino emission is summed over all neutrino species

4 COSMIC RAYS FROM PRIMORDIAL BLACK HOLES

In order to determine the present day background spectrum of particles generated from PBH evaporations, we must integrate first over the lifetime of each hole of mass M and then over the PBH mass spectrum (MacGibbon, 1991). In doing this, we must allow for the fact that smaller holes will evaporate at an earlier cosmological epoch, so that the particles they generate will be redshifted in energy by the present epoch. If the holes are uniformly distributed throughout the Universe, the background spectra should have the form indicated in Figure 2. All the spectra have rather similar shapes: an E^{-3} fall-off for $E > 100$ MeV, due to the final phases of evaporations at the present epoch and an E^{-1} tail at $E < 100$ MeV due to the fragmentation of jets produced at the present and earlier epochs. Note that the E^{-1} effect masks any effect associated with PBH mass spectrum: in the absence of jets, the spectra would rise like $E^{2-\alpha}$ as one goes to lower energies (Carr, 1976) but this is shallower than E^{-1} for $\alpha < 3$, so the E^{-1} tail dominates the integral.

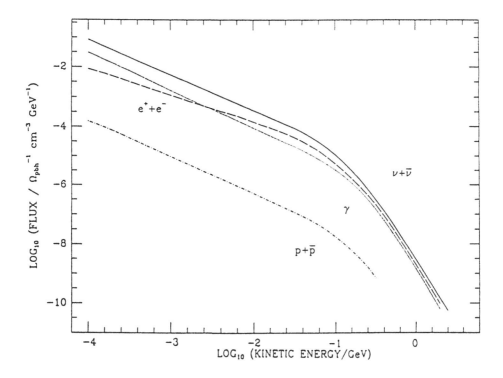

Figure 2. This shows the background produced by a distribution of PBHs emitting over the lifetime of the Universe. We assume $\Omega = 1$ and $h = 0.5$. All interactions are neglected apart from redshift effects.

The situation is more complicated if the PBHs evaporating at the present epoch are clustered inside our own Galactic halo (as is most likely). In this case, any charged particles emitted after the epoch of galaxy formation will have their flux enhanced relative to the photon spectra by a factor ζ which depends upon the halo concentration factor and the time for which the particles are trapped inside the halo by the Galactic magnetic field. Assuming the particles are uniformly distributed throughout a halo of radius R_h, one finds

$$\zeta = \left(\frac{\tau_{leak}}{\tau_{gal}}\right)\left(\frac{\rho_{halo}}{\Omega\rho_{crit}}\right) = 10^6 h^{-2}\left(\frac{\tau_{leak}}{\tau_{gal}}\right)\left(\frac{R_h}{10kpc}\right)^{-2} \tag{16}$$

The ratio of the leakage time to the age of the Galaxy, is rather uncertain and also energy dependent. At 100 MeV we take τ_{leak} to be about 10^7 y for electrons or positrons ($\zeta \sim 10^3$) and 10^8 y for protons or antiprotons ($\zeta \sim 10^4$). The total background of charged particles should therefore consist of the superposition of two components: one produced before galaxy formation and the other after it. The latter contribution, shown in Figure 3, corresponds to just a narrow range of masses below M_* (a factor of 2 if galaxies form at a redshift ~ 10).

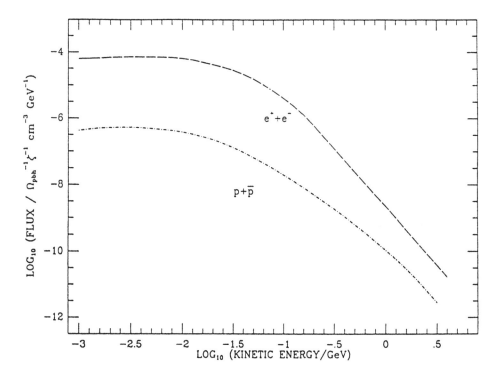

Figure 3. This shows the postgalactic $e^+ + e^-$ and $p + \bar{p}$ emission from PBHs clustered in the Galactic halo. We assume $\Omega = 1$, $h = 0.5$ and $\tau_{gal} = 1.2 \times 10^{10}y$. The enhancement factor ζ is in the range $10^3 - 10^4$.

For comparison with the observed cosmic ray spectra, one needs to determine the amplitude of the spectra at 100 MeV. This is because the observed fluxes all have slopes between E^{-2} and E^{-3}, so the strongest constraint comes from measurements at 100 MeV. The amplitudes all scale with the density parameter of the holes Ω_{PBH} and are found to be

$$\frac{dF}{dE} = \begin{cases} 1.5 \times 10^{-5} h^2 \Omega_{PBH} & \text{GeV}^{-1}\text{cm}^{-3} & (\text{photons}) \\ 9.5 \times 10^{-3} h^2 \Omega_{PBH}(\zeta/10^3) & \text{GeV}^{-1}\text{cm}^{-3} & (e^+, e^-) \\ 4.5 \times 10^{-4} h^2 \Omega_{PBH}(\zeta/10^4) & \text{GeV}^{-1}\text{cm}^{-3} & (p, \bar{p}) \end{cases} \quad (17)$$

We now apply this result to examining whether PBH evaporations could contribute appreciably to the observed spectra of these particles.

(1) Photons. Since the observed γ-ray background spectrum (Fichtel et al., 1975) goes like $E^{-2.4}$ at around $E \sim 100$ MeV, which is much steeper than the Brehmstrahlung tail from the jets, the dominant constraint on Ω_{PBH} comes from measurements of the background at 100 MeV itself. This gives an upper limit (MacGibbon & Carr, 1991)

$$\Omega_{PBH} \leq 7.6(\pm 2.6) \times 10^{-9} h^{-2} \quad (18)$$

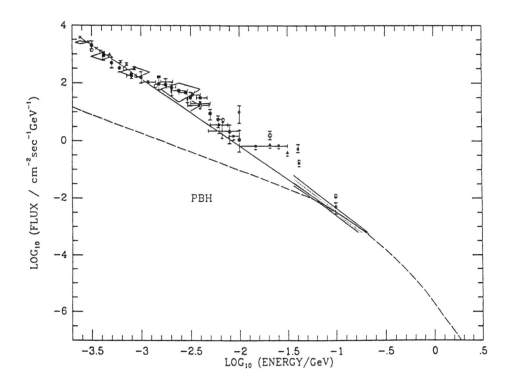

Figure 4. This compares the γ-ray background observations with the maximum PBH background (broken line) which is permitted by the Fichtel et al. data (dotted line).

as illustrated in Figure 4. In principle, PBH emission could be the dominant contribution to the photon flux above 50 MeV, in which case one has a clear prediction for the
spectrum. The only observations above 600 MeV come from the *EGRET* experiment
but there is the problem of separating the Galactic and extragalactic components.
Note that eq. (18) corresponds to a limit of $\beta(M_*) < 10^{-26}$. If PBHs form from initial
inhomogeneities, eq. (2) implies that the corresponding limit on their amplitude is
$\epsilon < 0.03$. It should be stressed that photons emitted prior to a redshift $z_{free} \simeq 400$
will be degraded due to pair–production off background nuclei but this will only affect
the present–day spectrum at energies below 1 MeV. This contrasts to the situation
which would pertain if one only had direct emission of photons because, in this case,
the spectrum would be modified up to 10 MeV (Page & Hawking, 1976). Eq. (18)
implies that the frequency of black hole explosions at the present epoch could be at
most 0.1 pc^{-3}y^{-1} even if they are clustered inside the Galactic halo, and there is then
little chance of their being detected (Halzen et al., 1991).

(2) Electrons and positrons. If the PBHs are not clustered within galaxies, the electrons and positrons they generate should have the spectrum indicated by Figure 2.
However, all the ones generated pregalactically would have been degraded through
inverse scattering off the microwave background photons, so one could only observe
the ones produced recently and the flux would then be uninteresting. However, we
have seen that the flux would be enhanced if the PBHs were currently clustered inside the Galactic halo. In this case, electrons and positrons with $E < 10$ MeV would
be degraded by ionization losses, while those above 10 GeV would be degraded by
inverse Compton losses. Thus the PBH spectrum should be dominated by 10 MeV
to 10 GeV particles produced within the last $\tau_{leak} \sim 10^7$y. An interesting feature
of the observations is that the electron and positrons have comparable fluxes at 100
MeV, even though the electrons are much more numerous at higher energies. This
feature is unexplained in most cosmic ray models but it is a natural consequence of
the PBH scenario since electrons and positrons are emitted in equal numbers. It is
also interesting that the positron spectrum falls off like E^{-3} above a few GeV, as
expected in the PBH model. It is difficult to estimate the value of Ω_{PBH} required to
generate *all* the observed positrons accurately but comparison with the interstellar
positron flux at 300 MeV (Ramaty & Westergaard, 1976) indicates that it should be
about

$$\Omega_{PBH} \simeq 2 \times 10^{-8}(\tau_{leak}/10^7\mathrm{y})^{-1}(R_h/10\mathrm{kpc})^{-2} \qquad (19)$$

(MacGibbon &, Carr 1991). This is comparable to the γ–ray limit given by eq. (18)
for reasonable values of τ_{leak} and R_h.

(3) Annihilation line radiation. If PBHs are clustered inside the Galactic halo, their
density should be even more enhanced towards the Galactic centre. One would there-

fore expect an especially strong emission of positrons from that direction. Some of these positrons should annihilate, producing a 0.511 MeV line, so it is relevant that such a line has indeed been detected from the Galactic centre (Leventhal et al., 1989). The intensity of the line corresponds to $3 - 10 \times 10^{42}$ annihilations s^{-1}. Okeke & Rees (1980) discussed whether these annihilations could be generated by PBH positrons. For relativistic particles, the optical depth of the Galaxy to annihilation is only about 0.1. However, the annihilation cross–section scales as the inverse speed of the particle, so annihilations can still be important if the antiprotons are slowed down by ionization losses. Assuming one has mainly molecular hydrogen at the Galactic centre, then positrons will be slowed sufficiently to annihilate providing their energy is less than $E_{slow} \simeq 13$ MeV. In order to produce the observed line, one would then require (MacGibbon & Carr, 1991)

$$\Omega_{PBH} \simeq 0.5 - 17 \times 10^{-6} h^{-2} (R_c/\text{kpc})^2 (\theta/20°)^{-3} \qquad (20)$$

where R_c is the halo core radius and θ is the angle subtended by the region generating the 0.511 MeV line. This is well above the γ–ray limit, so we conclude that PBHs are not a plausible explanation.

(4) Antiprotons. The protons and antiprotons generated in the final explosive phase of PBH evaporations should contribute to the cosmic ray background. However, since the observed $p : \bar{p}$ ratio is less than $10^{-3} - 10^{-4}$ over the energy range $0.1 - 10$ GeV, whereas PBHs should produce the particles in equal numbers, PBHs could only contribute appreciably to the antiprotons. It is usually assumed that antiproton cosmic rays are secondary particles, produced by the spallation of the interstellar medium by primary cosmic rays. However, Buffington et al. (1981) claimed that the observed $\bar{p}$ flux at $130 - 320$ MeV exceeds the predicted secondary flux by a factor of 100 and this prompted Kiraly et al. (1981) and Turner (1982) to examine whether PBH evaporations could explain the antiproton cosmic rays. In fact, more recent observations (Streitmatter et al., 1990) around 100 MeV give an upper limit which is a factor of 10 below Buffington's claim but it still exceeds the expected secondary flux by an order of magnitude and it includes two possible detections. It is therefore interesting to redetermine the expected antiproton spectrum on the basis of the jet calculations. If the PBHs are uniformly distributed throughout the Universe, the antiproton flux is too small to be interesting. However, if the PBHs are clustered in halos, the spectrum would be dominated by the antiprotons produced within our halo in the last $\tau_{leak} \sim 10^8$ y. In order to compare with observations, one must allow for the effects of ionization (which are important below 50 MeV) but, if the $p : \bar{p}$ ratio has the value $\sim 10^{-5}$ indicated by Streitmatter et al., one gets a rough fit with the interstellar proton flux at 1 GeV (MacGibbon & Carr, 1991) for

$$\Omega_{PBH} \simeq 0.6 - 0.4 \times 10^{-9} h^{-2} (\tau_{leak}/10^8 \text{y})^{-1} (R_h/10\text{kpc})^{-2} \qquad (21)$$

This is somewhat less than the value of Ω_{PBH} required to explain the positron and γ-ray observations but within an order of magnitude of it.

5 CONCLUSIONS

The jet calculations described above suggest that PBH evaporations could contribute appreciably to photons, positrons and antiprotons in the energy range above 100 MeV. Indeed it is rather remarkable that the value of Ω_{PBH} is of order 10^{-8} in all three cases. However, PBH evaporations could not contribute appreciably to the 0.5 MeV line from the Galactic centre and it will be hard to detect black hole explosions. If cosmic ray positrons and antiprotons really do derive from PBHs, then their spectra could yield vital information about particle physics. However, it should be stressed that one would expect the same signature for any other process which produces jets [eg. the annihilation of supersymmetric particles (Rudaz & Stecker, 1988)].

If $\Omega_{PBH} \sim 10^{-8}$, then the fraction of the Universe going into PBHs is $\beta \sim 10^{-26}$ at their formation epoch. If the holes form from initial inhomogeneities, this requires fine-tuning: the horizon-scale fluctuations need to have an amplitude of about 3%. If the form from the collapse of cosmic loops, then the string parameter μ must be finely tuned, although the precise value required is uncertain. Note that if PBHs leave Planck mass relics and if the PBH spectrum is given by eq. (3), then one expects the Planck relics to have a density which is $(M_*/M_{Pl})^{1/2} \sim 10^9$ times higher than Ω_{PBH}. As MacGibbon (1987) has pointed out, this is intriguingly close to the critical density if M_* holes have the density required to explain the cosmic ray observations.

REFERENCES

Barrow, J.D., *Mon.Not.R.Astron.Soc.* 1980, **192**, 127.

Barrow, J.D., Copeland, E.J. & Liddle, A.R. 1992, *Phys.Rev.D.*,

Bowick, M.J. et al. 1988, *Phys.Rev.Lett.*, **61**, 2823.

Buffington, A., Schindler, S.M. & Pennypacker, C.R. 1981, *Ap.J.*, **248**, 1179.

Candelas, P. & Sciama, D.W. 1977, *Phys.Rev.Lett.*, **38**, 1372.

Carr, B.J. & Hawking, S.W. 1974, *Mon.Not.R.Astron.Soc.*, **168**, 399.

Carr, B.J. 1975, *Ap.J.*, **201**, 1.

Carr, B.J. 1976, *Ap.J.*, **206**, 8.

Chapline, G.F. 1975, *Nature*, **253**, 251.

Coleman, S., Preskill, J. & Wilczek, F. 1991, *Mod.Phys.Lett.A.*, **6**, 1631.

Crawford, M. & Schramm, D.N. 1982, *Nature*, **298**, 538.

Fitchel, C.E. et al. 1975, *Ap.J.*, **198**, 163.

Gross, D.J., Perry, M.J. & Yaffe, L.G. 1982, *Phys.Rev.D.*, **25**, 230.

Halzen, F., Zas, E., MacGibbon, J.H. & Weekes, T.C. 1991, *Nature*, **353**, 807.

Hayward, G. & Pavon, D. 1989, *Phys.Rev.D.*, **40**, 1748.

Hawking, S.W. 1971, *Mon.Not.R.Astron.Soc.*, **152**, 75.
Hawking, S.W. 1975, *Comm.Math.Phys.*, **43**, 199.
Hawking, S.W. 1989, *Phys.Lett.B.*, **231**, 237.
Hawking, S.W., Moss, 1. & Stewart, J. 1982, *Phys.Rev.D.*, **26**, 2681.
Hsu, S.D.U. 1990, *Phys.Lett.B.*, **251**, 343.
Kapusta, J.I. 1984, *Phys.Rev.D.*, **30**, 831.
Kiraly, P. et al. 1981, *Nature*, **293**, 120.
Khlopov, M.Yu. & Polnarev, A.G. 1980, it Phys.Lett.B., **97**, 383.
Kodama, H., Sasaki, M. & Sato, K. 1982, *Pro.Theor.Phys.*, **68**, 1979.
La, D. & Steinhardt, P.J. 1989, *Phys.Lett.B*, **220**, 375.
Leventhal, M. et al. 1989, *Nature*, **339**, 36.
Lindley, D. 1980, *Mon.Not.R.Astron.Soc.*, **196**, 317.
MacGibbon, J.H. 1987, *Nature*, **329**, 308.
MacGibbon, J.H.& Webber, B.R. 1990, *Phys.Rev.D.*, **41**, 3052.
MacGibbon, J.H. 1991, *Phys.Rev.D.*, **44**, 376.
MacGibbon, J.H. & Carr, B.J. 1991, *Ap.J.*, **371**, 447.
Naselsky, P.D. & Polnarev, A.G. 1985, *Sov.Astron.*, **29**, 487.
Novikov,I.D., Polnarev, A.G., Starobinsky A.A. & Zeldovich, Ya.B. 1979, *Astron.Ap.*,
    80, 104.
Okeke, P.N. & Rees, M.J. 1980, *Astron.Ap.*, **81**, 263.
Page, D.N. 1977, *Phys.Rev.D.*, **16**, 2402.
Page, D.N. & Hawking, S.W. 1976, *Ap.J.*, **206**, 1.
Polnarev, A.G. & Khlopov, M.Yu. 1981, *Astron.Zh.*, **58**, 706.
Polnarev, A.G. & Zemboricz, R. 1988, *Phys.Rev.D.*, **43**, 1106.
Porter, N.A. & Weekes, T.C. 1979, *Nature*, **277**, 199.
Ramaty, R. & Westergaard, N.J. 1976, *Astrophys.Sp.Sci.*, **45**, 143.
Rees, M.J. 1977, *Nature*, **266**, 333.
Rudaz, S. & Stecker, F.W. 1988, *Ap.J.*, **325**, 16.
Streitmatter, R.E. et al. 1990, *Bull.Am.Phys.Soc.*, **35**, 1066.
Turner, M.S. 1982, *Nature*, **297**, 379.
Zeldovich, Ya.B. & Starobinsky, A.A. 1976, *JETP Lett.*, **24**, 571.
Zeldovich, Ya.B. & Novikov, I.D. 1967, *Sov.Astron.A.J.*, **10**, 602.

Evaporation of Two Dimensional Black Holes

STEPHEN W. HAWKING

1 INTRODUCTION

In 1973, I was studying quantum fields in the curved space background of a black hole. To my great surprise, and that of everyone else, I found that black holes weren't completely black. Instead, a black hole formed by collapse, should give off radiation exactly as if it were a hot body. In the case of a non rotating uncharged black hole, the temperature in Planck units would be

$$T = \frac{1}{8\pi M}.$$

There were two important consequences.

First, the radiation would carry energy away from the black hole. The black hole would therefore presumably lose mass. In the case of a non rotating, uncharged black hole, this would cause the temperature to go up. It would therefore radiate energy at a faster rate. Eventually, in a time of order M^3 Planck units, the black hole would radiate away all its original rest mass energy. Although this is not of much practical significance for stellar mass black holes, because the time scale is much longer than the age of the universe, it raises an important question of principle: What is the final fate of an evaporating black hole? Does it settle down to a stable non radiating remnant? Does it disappear completely? Or does the singularity spread out to infinity, and prevent the solution being continued into the future, more than a finite time?

The second issue raised by black hole radiation, is the question of the loss of quantum coherence. In the semi classical approximation, the radiation that is given off, has no relation to what went into the black hole. Instead, the radiation is thermal, and is in a mixed state. One would expect this to be true even beyond the semi classical approximation. Unless the radiation is exactly thermal, and in a mixed state, the whole analogy between the laws of black hole mechanics, and the laws of thermodynamics, breaks down. This is so beautiful, it can't be just an accident, or an artifact of the semi classical approximation. But if the emitted radiation is in a mixed state, and if the black hole evaporates completely, there must be loss of quantum coherence. Some people have found this idea so objectionable, they have tried a number of ways to preserve quantum coherence. But none have been very convincing. Personally, I don't see any good reason why quantum coherence should not be lost. It is not violating the laws of quantum mechanics, as is sometimes claimed. It is just taking into account, that spacetime can have a non trivial topology.

It is difficult to answer these questions, because quantum gravity in four dimensions is not a renormalizable theory. However, gravity in two dimensions is renormalizable. Recently, Callan, Giddings, Harvey, and Strominger, have suggested an interesting two dimensional theory, with the metric,

$$ds^2 = e^{2\rho} dx_+ dx_-,$$

coupled to a dilaton field, and N minimal scalar fields. The Lagrangian,

$$L = \frac{1}{2\pi} \sqrt{-g} [e^{-2\phi} (R + 4(\nabla \phi)^2 + 4\lambda^2) - \frac{1}{2} \sum_{i=1}^{N} (\nabla f_i)^2],$$

is similar to that obtained from string theory, but this is meant to be considered as a field theory in its own right. The classical field equations are,

$$\partial_+ \partial_- f_i = 0,$$

$$2\partial_+ \partial_- \phi - 2\partial_+ \phi \partial_- \phi - \frac{\lambda^2}{2} e^{2\rho} = \partial_+ \partial_- \rho,$$

$$\partial_+ \partial_- \phi - 2\partial_+ \phi \partial_- \phi - \frac{\lambda^2}{2} e^{2\rho} = 0.$$

These equations have a solution

$$\phi = \rho = -\frac{1}{2} \log(\lambda^2 x_+ x_-).$$

By a change of coordinates one obtains a flat metric and a linear dilaton field that is linear in the radial coordinate. This solution is known as the linear dilaton. It is the maximally symmetric solution admitted by the equations, and is therefore naturally interpreted as the ground state.

These equations also admit a black hole solution,

$$\phi = \rho - c = -\frac{1}{2} \ln(M\lambda^{-1} - \lambda^2 e^{2c} x_+ x_-).$$

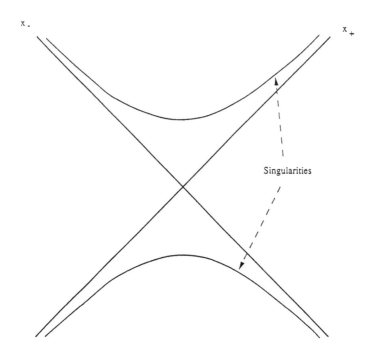

Figure 1. The conformal diagram of the black hole solution

It has a conformal diagram like the Schwarzschild solution (see Figure 1). The horizons are at $x_+ = 0$, and at $x_- = 0$. The singularities are at the top and bottom, and the right and left regions are asymptotic to the linear dilaton solution.

This black hole solution is periodic in the imaginary time, with period $2\pi/\lambda$. One would therefore expect it to have a temperature,

$$T = \frac{\lambda}{2\pi},$$

and to emit thermal radiation. This is confirmed by Callan et al. They considered a black hole, formed by sending in a thin shock wave of one of the scalar fields, from the right (see Figure 2). One can calculate the
energy momentum tensors of the scalar fields, using the conservation and trace anomaly equations. One imposes the boundary condition that there is no incoming energy momentum, apart from the shock wave. One finds that at late retarded times, there is a steady out flow of energy in each field, at a rate independent of the mass of the black hole:

$$\text{Energy Flux} = \frac{\lambda^2}{48}.$$

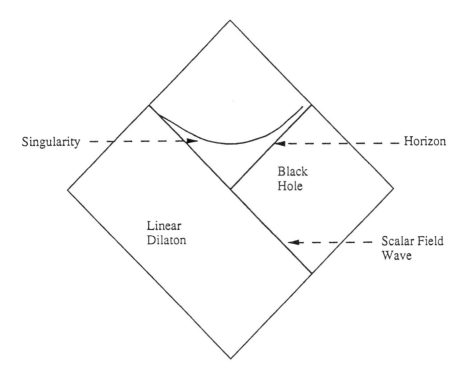

Figure 2. The black hole formed by a thin shock wave

If this radiation continued indefinitely, the black hole would radiate an infinite amount of energy, which seems absurd. One might therefore expect that the back reaction would modify the emission, and cause it to stop, when the black hole has radiated away its initial mass. A fully quantum treatment of the back reaction seems very difficult, even in this two dimensional theory. But Callan et al suggested that in the limit of a large number, N, of scalar fields, one could neglect the quantum fluctuations of the dilaton and the metric. One could treat the back reaction to the radiation in the scalar fields semi–classically, by adding a trace anomaly term to the action. The semi classical effective Lagrangian then becomes.

$$L = \frac{1}{2\pi}\sqrt{-g}[e^{-2\phi}(R + 4(\nabla\phi)^2 + 4\lambda^2) - \frac{1}{2}\sum_{i=1}^{N}(\nabla f_i)^2]$$
$$- \frac{N}{96\pi}\int\int d^2x d^2y\sqrt{-g}R(x)\frac{1}{\nabla^2}R(y)$$

The evolution equations that result from this action are a fairly simple set of non linear hyperbolic equations in two dimensions,

$$\partial_+\partial_-\phi = (1 - \frac{N}{24}e^{2\phi})\partial_+\partial_-\rho,$$

$$2(1 - \frac{N}{12}e^{2\phi})\partial_+\partial_-\phi = (1 - \frac{N}{24}e^{2\phi})(4\partial_+\phi\partial_-\phi + \lambda^2 e^{2\rho}).$$

Even these semi classical equations seem too difficult to solve in closed form. Callan *et al* suggested that a black hole, formed from a scalar field wave, would evaporate completely, without there being any singularity. The solution would approach the linear dilaton, at late retarded times, and there would be no horizons. They therefore claimed that there would be no loss of quantum coherence, in the formation and evaporation of a two dimensional black hole: the radiation would be in a pure quantum state, rather than in a mixed state.

The Stanford and Rutgers groups have shown that this scenario could not be correct. The solution would develop a singularity on the incoming wave, at the point where the dilaton field reached a critical value,

$$\phi_0 = -\frac{1}{2}\ln\frac{N}{12}.$$

This singularity will be spacelike near the incoming wave. Thus at least part of the final quantum state will end up on the singularity. This implies that the radiation at infinity, in the weak coupling region, will not be in a pure quantum state.

The outstanding question is: How does the spacetime evolve, to the future of the thin shock wave. There seem to be three main possibilities:

Possibility 1. The singularity remains hidden behind an event horizon. One can continue an infinite distance into the future, outside the horizon, without ever seeing the singularity. For this to be possible, the rate of radiation would have to go to zero.

Possibility 2. The singularity is naked. A naked singularity is one that is visible from the weak coupling region, at a finite time to the future of the incoming wave. Any evolution of the solution after this would not be uniquely determined by the semi–classical equations and the initial data.

Possibility 3. The singularity is what I call a thunder-bolt. This is a singularity that comes out and hits you, on an asymptotically null path, after a finite time. Such a singularity is not naked. It approaches at the speed of light, or faster, so one wouldn't see it before it hit you, and wiped you out.

In this seminar, I shall present evidence that possibility 1 can not be correct. The singularity that the semi–classical equations predict, can not remain hidden for all time, behind an event horizon. Some computer calculations that John Stewart and I have done, indicate that what you get is a thunder-bolt, rather than a naked singularity. However, one might expect the semi–classical approximation to break down near the singularity, when the dilaton field approaches the critical value.

2 STATIC BLACK HOLES

Suppose the solution were to evolve, without a thunder-bolt or naked singularity. Then it would presumably approach a static state, in which a singularity was hidden behind an event horizon. This motivates a study of static black hole solutions of the semi–classical equations,

$$\phi'' + \frac{1}{r}\phi' = \left(1 - \frac{N}{24}e^{2\phi}\right)\left(\rho'' + \frac{1}{r}\rho'\right),$$

$$\left(1 - \frac{N}{12}e^{2\phi}\right)\left(\phi'' + \frac{1}{r}\phi'\right) = 2\left(1 - \frac{N}{24}e^{2\phi}\right)\left((\phi')^2 - \lambda^2 e^{2\rho}\right).$$

One can express the equations for a static black hole in terms of a radial coordinate

$$r^2 = -x_- x_+.$$

The horizon is then at $r = 0$. The two other papers that have been written on these static solutions, have used the radial coordinate, $\sigma = x_+ - x_-$. In this coordinate, the horizon is at $\sigma = -\infty$, so it is difficult to set the boundary conditions. On the other hand, in the r coordinate, the boundary conditions for a regular horizon are that the first derivatives of ϕ and ρ are zero on the horizon. A static black hole solution, is therefore determined by the values of ϕ and ρ on the horizon. The value of ρ, however, can be changed by a constant, by rescaling the coordinates. The physical distinct static solutions with a horizon are therefore characterized by the value of the dilaton on the horizon.

The horizon value, ϕ_h, has to be less than the critical value, ϕ_0, to get a solution that is asymptotic to the linear dilaton. At large r, the back reaction terms proportional to N can be neglected. The solution will depend on four parameters, b, c, K, and L,

$$\rho = \phi + c = -\ln r + \ln \frac{2b}{\lambda} - \frac{K + L \ln r}{r^{4b}} + \dots$$

The parameters b and c correspond to coordinate freedom in the linear dilaton, that the solution approaches at large r. If the parameter L is non zero, the ADM mass M of the solution will be infinite. This is what one might expect for a black hole in thermal equilibrium, because there will be in–coming and out–going radiation, all the way to infinity. However, a black hole formed by sending in a shock wave, would not have the in–coming radiation, and presumably only a limited amount of out–going radiation.

If the value of the dilaton on the horizon is much less than the critical value, the back reaction terms will be small at all values of r. So the solutions of the semi–classical equations will be almost the same as the classical black holes. This means the horizon value of ϕ will be approximately

$$\phi_h = -\frac{1}{2} \ln \frac{M}{\lambda}.$$

Jonathan Brenchley integrated these equations, using Mathematica. As expected, for horizon dilaton fields less than the critical value, the solutions were almost the same as the black holes without back reaction.

Consider a situation in which a black hole of large mass is created by sending in a scalar field wave. One could approximate the subsequent evolution by a sequence of static black hole solutions, with steadily increasing values of the dilaton, ϕ, on the horizon. However, when the value of ϕ on the horizon approaches the critical value, ϕ_0, the back reaction will become important, and will change the black hole solutions significantly.

As the dilaton field on the horizon approaches the critical value, ϕ_0, the coefficient of the second derivative of ϕ will become very small. This will drive the second derivative of ϕ to be very large, until ϕ' approaches $-\lambda e^\rho$, in a very short distance. The second derivative of ρ will also be very large, for a short distance. A power series solution, and numerical calculations carried out by Jonathan Brenchley, confirm that as the dilaton on the horizon tends to the critical value, the solution tends to a limiting form. This limiting solution appears to have non–zero gradients of ϕ and ρ on the horizon, but the gradients are actually zero there.

The limiting black hole is regular everywhere outside the horizon, but has a fairly mild singularity on the horizon. At large values of r, the solution will tend to the linear dilaton. One or both of the asymptotic parameters, K and L, must be non zero, because the solution is not exactly the linear dilaton.

Suppose the singularity inside the black hole were to remain hidden at all times, as in possibility 1 that I described earlier. Then the temperature and rate of evolution of the black hole would have to approach zero, as the dilaton field on the horizon approached the critical value. However, this is not what happens. The fact that the black holes tend to a limiting solution means that the period in imaginary time, and hence the temperature, will remain finite. The energy momentum tensor of one of the scalar fields can be calculated from the conservation equations and the trace anomaly. One imposes the boundary conditions that there is no incoming radiation, and that the energy momentum tensor is regular on the future horizon. One finds that there should be a steady rate of energy outflow.

3 CONCLUSIONS

What I have shown, is that the temperature and rate of emission of the limiting black hole do not go to zero. This implies that the solution cannot settle down to a static state, in which the singularity is hidden behind an event horizon. Of course, this does not tell us what the semi–classical equations will predict. But it makes it very plausible, that they will lead either to a naked singularity, or a thunder-bolt. To check this, and to find whether the singularity was naked or a thunder-bolt, John Stewart and I have integrated the semi–classical equations numerically. The program was similar to one we developed some time ago, for bubble collisions in the inflationary universe. We consider the integration of two coupled hyperbolic equations, in $1 + 1$ dimensions, in a region bounded by two null lines (see Figure 3). The null line on

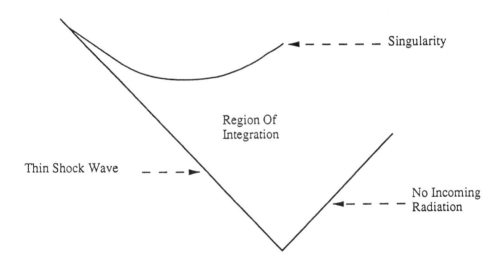

Figure 3. The region of numerical integration

the left, represents the path of a thin shock wave in one of the N scalar fields, that is sent into the linear dilaton solution from the region of weak coupling. The boundary condition here is that ϕ and ρ have the same values as they have in the linear dilaton. The null line on the right of the diagram, represents an out–going null line in the weak coupling region. The boundary condition here, is that the in–coming energy momentum crossing this line should be zero, apart from a delta function on the shock wave, of strength a. With these boundary conditions, one can integrate the differential equations numerically. To check the stability and accuracy of the program, we first tried it in the case $N = 0$ (see Figure 4). In other words, without

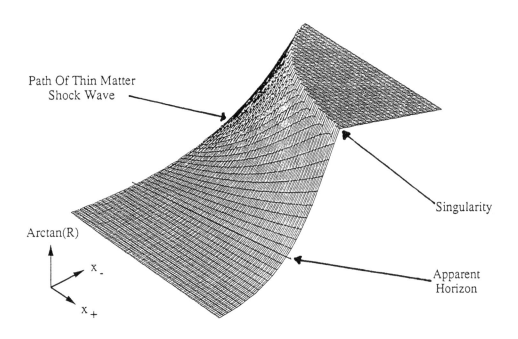

Figure 4. Curvature scalar for a black hole without back reaction ($a = 0.9$, $N = 0$)

any back reaction. In this case, we know the solution analytically. It is just the two dimensional black hole solution, discovered by Witten. In order to avoid things going off to infinity, we have plotted the arctan of the curvature scalar, R. You can see that there is a singularity of infinite R, on a space–like curve at the top. Beyond the singularity, R has been set to infinity, but this region has no physical significance. Also shown is what is called, the apparent horizon, where the gradient of ϕ is null. In this case, the solution is static, and the apparent horizon coincides with the global event horizon. But in more general non–static solutions, the position of the apparent horizon can be determined locally, while the position of the event horizon depends on the behavior of the solution in the future.

The $N = 0$ numerical integration gives good agreement with the analytic solution. We were therefore encouraged to proceed to the case of non–zero N. The actual

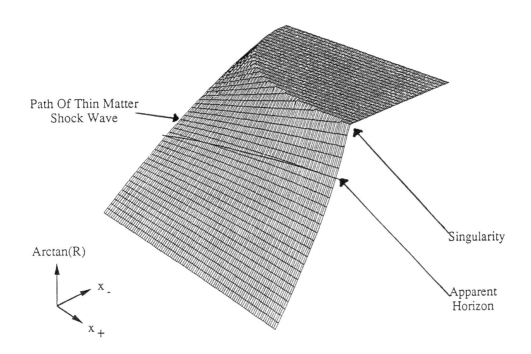

Path Of Thin Matter
Shock Wave

Arctan(R)

x_-

x_+

Singularity

Apparent
Horizon

Figure 5. Curvature scalar when back reaction is included ($a = 0.9$, $N = 12$)

non–zero value of N can be scaled out of the problem. For convenience, we chose $N = 12$ (see Figure 5). Again, you can see we get a singularity on a space like, asymptotically null curve. The apparent horizon is not now an out–going null line. Rather, it moves in towards the singularity. This is what one would expect, as the energy of the black hole created by the shock wave is given off to infinity. Similar results have been obtained by Susskind and Thorlacius.

At first sight, the solution with back reaction looks qualitatively similar to the solution without. But there is an important difference. Consider going out an affine distance, t, on an out–going null line, and then turning and going in to the singularity, on an in going null line. The affine distance s to the apparent horizon will increase asymptotically linearly with the distance, t (see Figure 6), in the case of the black hole solution without back reaction. Another way of saying this, is that an observer can travel arbitrarily far away from the horizon. Our numerical calculations in the $N = 0$ case show this effect: s increases linearly with t. However, in the non zero N case, s increases linearly at first, but seems to approach an upper bound (see Figure 7). What this means is that even if you move away at the speed of light, you can never get very far from the horizon, and the singularity behind it. If you move at

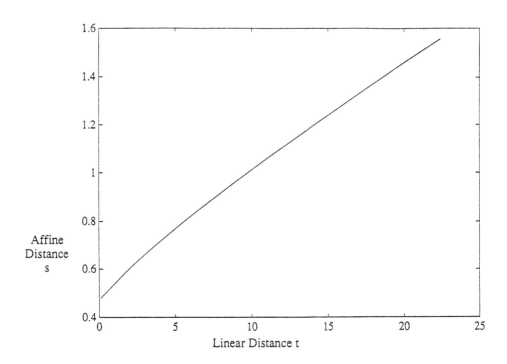

Figure 6. Linear variation of affine parameter distance to apparent horizon without back reaction

any speed less than that of light, you will be over taken by the singularity. Thus the singularity that the semi–classical equations predict is a thunder-bolt, rather than a naked singularity.

The semi classical evolution of these two–dimensional black holes is very similar to that of magnetically charged black holes in four dimensions, with a dilaton field. Because there are no fields in the theory that can carry away the charge, the steady loss of mass would suggest that the black hole would approach an extreme state. However, unlike the case of the Reissner-Nordstrom solutions, the extreme black holes with a dilaton have a finite temperature and rate of emission. So one obtains a similar contradiction. If the solution were to evolve to a state of lower mass but the same charge, the singularity would become naked.

There seems no way of avoiding naked or thunder-bolt singularities, in the context of the semi–classical theory. If spacetime is described by a semi–classical Lorentz metric, a black hole cannot disappear completely, without there being some sort of naked or thunder-bolt singularity. But there seem to be zero temperature non radiating black holes only in a few cases. For example, magnetically charged black holes with no

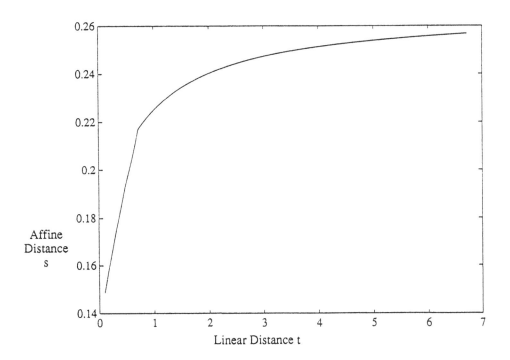

Figure 7. Asymptotic variation of affine parameter distance to apparent horizon when back reaction is included

dilaton field. In other cases, there is no remnant black hole solution that the system could settle down to.

What seems to be happening is that the semi classical approximation is breaking down, in the strong coupling regime. In conventional general relativity, this break-down occurs only when the black hole gets down to the Planck mass. But in the two and four dimensional dilatonic theories, it can occur for macroscopic black holes, when the dilaton field on the horizon approaches the critical value. When the coupling becomes strong, the semi classical approximation will break down. Quantum fluctuations of the metric will become important, and can no longer be represented by a single Lorentzian metric. How to treat the strong coupling regime is not clear. One possibility is that some sort of Euclidean wormhole can occur, which could carry away the particles that went in to form the black hole, and bring in the particles to be emitted. These wormholes could be in a coherent state, described by α-parameters. These parameters might be determined, by the minimization of the effective gravitational constant, G. In this case, there would be no loss of quantum coherence, if a black hole were to evaporate and disappear completely. Or the α-parameters might be different moments of a quantum field, α, on superspace. In this case, there would be effective loss of quantum coherence, but it might be possible, to measure all the α-parameters involved in the evaporation of a black hole of a given mass. In that case, there would be no further loss of quantum coherence.

I have a bet with Kip Thorne and John Preskill, about naked singularities. Maybe we should have included thunder-bolt singularities, but we didn't think of them as distinct. If it can be shown that the theory predicts naked singularities, I have to buy them clothes to cover their nakedness. So I would have lost my bet, except that I was careful to insert the words, in classical general relativity, in the bet. I believe I'm right, in the strictly classical theory, but I don't know how to prove it.

Topology and Topology Change in General Relativity

GARY W. GIBBONS

1 INTRODUCTION

The organisers have asked us to review the progress of some aspect of general relativity and cosmology in which they have a particular interest and to introduce their remarks by describing its relation to their interaction with Dennis Sciama. It is a great pleasure for me to do so and also to pay tribute to the inspiration that he, and his style of doing physics, has been to me over the years. In particular I have tried to follow his example by asking simple physical questions and trying to answer them with the simplest appropriate tools available. For that reason in what follows I shall not give extensive mathematical details but refer the reader to the references. Moreover because of the personal nature of the review I have made no attempt to include in those references every paper on the subject, especially where the story is widely known and can be read up in standard textbooks. For the same reason I have perhaps erred in including too many papers of my own.

I became a Research student of Dennis Sciama in October 1969 after being enthralled by his marvelously lucid and exciting Part III lectures on General Relativity. When Dennis left for Oxford a year later I transferred to Stephen Hawking, himself a former student of Dennis. I can thus be said to be the answer to a riddle in that I am the academic son of my father's son. The problem that Dennis originally suggested to me was to give an upper bound to the proportion of the rest mass of a gravitating system's rest which could be lost in gravitational radiation - a question which was at that time very topical because of Weber's claims to have observed substantial amounts of gravitational radiation coming from the galactic centre.

I failed to make much headway on that particular problem, unlike my next supervisor who spectacularly solved one aspect of it - that involving black holes, about which I knew almost literally nothing at that time - with his celebrated Area Theorem. However, as all good Ph.D. problems should, it did set me thinking about numerous things including my subject today, topology and topology change in general relativity, and I would like to describe some of my thoughts over the years on this subject, at the same time trying to put them in a more general context.

I shall begin by considering the classical theory and then pass on to the necessarily more speculative quantum theory. I will restrict myself to the case of four spacetime dimensions, and I will concentrate on two particular theories: pure general relativity (with no matter) and Einstein-Maxwell theory. Much of what I have to say can be readily extended to more complicated theories, as I shall indicate briefly at the end, but these two simple theories already exhibit in essence the basic ideas and at the same time indicate the great richness of possible physical effects one can expect.

2 THE CLASSICAL THEORY

2.1 The Einstein-Rosen Bridge
The most direct way of answering the question Dennis had posed seemed at first to be to examine the motion of 'point particles' à la E.I.H. I read many papers on this subject and understood very little of them. However the one thing that did strike me was that the point particle picture seemed rather inappropriate in general relativity because it involved extrapolating the gravitational field of the particle inside its Schwarzschild radius. I found far more appealing the view, espoused in the famous paper of Einstein and Rosen (1935) in which they first described the "Einstein-Rosen Bridge" in the Schwarzschild solution, that one should look for everywhere non-singular solutions of the non-linear field equations representing a particle. Nowadays we call such solutions "solitons" and their use has become commonplace: the topological defects such as magnetic monopoles we now consider in cosmology are of course just solutions of this sort. Einstein and Rosen were greatly impressed by the non-trivial (spatial) topology of their throat, which seemed to be responsible for the existence of the solution and absence of singularities. Sadly however the singularities were still there lurking, as it were, in the full Kruskal spacetime rather than in space at one time. Indeed this conflict between the Hamiltonian and Lagrangian viewpoints, i.e. between space and spacetime is characteristic of our present subject, as it is of a great deal else in gravity.

2.2 Geometrodynamics
Einstein and Rosen's idea was taken up by Wheeler and his group at Princeton and their main results were summarised in that marvelously inspiring book 'Geometro-dynamics' (1961). From it I came away with three ideas which are important for our story.

1) That there exist very many non-singular initial data sets which are topologically non-trivial both for pure general relativity and for Einstein-Maxwell theory.

2) In the latter case the non-trivial topology can be responsible for the phenomenon

of "Charge without Charge".

3) The complete duality symmetry between electric and magnetic fields, a symmetry which can even be extended to massless spinor fields using chiral rotations involving γ_5.

2.3 Causal Selection Rules

The existence of topologically non-trivial initial data sets of this sort naturally raises questions about their evolution in time. Can there be transitions for example between spaces of different spatial topologies? One may ask such questions at the most basic kinematical level: what kind of evolutions are possible?, and at the dynamical level: what kind of evolutions will arise if field equations are imposed?

Perhaps the best known results at the purely kinematical level are those of Geroch (1967) which despite their great elegance and beauty have had the somewhat unfortunate effect until very recently of discouraging work on topology change. Let us assume the existence of an everywhere non-singular time-orientable and space-orientable Lorentzian metric g_L on a compact 4-manifold M with a closed but not necessarily connected spacelike boundary $\partial M = \Sigma$. Reinhart and later Geroch had shown that there exist many such spacetimes $\{M, g_L\}$ for every boundary Σ but Geroch also showed that if Σ consists of two components, Σ_i and Σ_f in the past and in the future respectively, and if Σ_i and Σ_f do not have the same topology then the spacetime M must contain closed timelike curves. If on the other hand M does not have closed timelike curves it is topologically the product of a closed time interval and some 3-manifold of fixed topology. Because people have, until recently, been loathe to consider closed timelike curves as physically possible, Geroch's theorem has usually been interpreted as a selection rule forbidding topology change at the classical level. But like most No Go Theorems in physics, its basic assumptions have come to be questioned, and a number of physicists have begun to take seriously closed timelike curves and time machines and even to ask whether the laws of physics forbid in principle the construction of wormholes in the laboratory and their subsequent use as time machines.

2.4 Spinorial Selection Rules

In my opinion a potentially much worse pathology for a spacetime is for it not to admit an $SL(2, C)$ spinor structure, since it is difficult according to our present understanding of matter to imagine a world without fermions described by fundamental spinor fields. If spacetime M is a product of a closed time interval and a compact 3-manifold it must necessarily admit an $SL(2, C)$ spinor structure but if it is not a product it need not. Bichteler (1968) and Geroch (1968,1970) pointed out that the necessary and sufficient condition for the existence of a spinor structure in this case is the vanishing of the second Stiefel Whitney class of the 4-manifold M. Stephen Hawking and I (1992a, 1992b) have recently obtained an interesting extension of Geroch's results. We find that the Lorentzian spacetime M will admit an $SL(2,C)$ spinor structure if and only if a certain mod 2 topological invariant $u(\Sigma)$ of the spacelike boundary vanishes. Since the invariant is additive under disjoint union:

$$u(\Sigma_1 \cup \Sigma_2) = u(\Sigma_1) + u(\Sigma_2) \quad \text{mod } 2$$

one may interpret our result as giving a topological conservation law for 3-manifolds. The invariant takes the value one for S^3 and zero for the wormhole, $S^1 \times S^2$. It follows that it is not possible even in principle to construct a wormhole in the laboratory in order, for example, to build a time machine. However under connected sum our invariant satisfies:

$$u(\Sigma_1 \# \Sigma_2) = u(\Sigma_1) + u(\Sigma_2) + 1 \quad \text{mod } 2.$$

Thus it takes the value zero for the connected sum of two wormholes. Our selection rule therefore allows the construction of a pair of wormholes in the laboratory. On the other hand it does not allow the birth of a single S^3 universe from nothing or a trouser-leg spacetime in which one S^3 universe splits into two, even though there are perfectly good Lorentzian spacetimes of this sort without spinor structure . An example would be CP^2 with three points removed.

Our invariant may be calculated in terms of the mod 2 homology groups of the boundary manifold Σ. It is called by mathematicians the mod 2 Kervaire semi-characteristic. The formula is:

$$u(\Sigma) = \dim_{Z_2}(H_0(\Sigma; Z_2) \oplus H_1(\Sigma; Z_2)) \quad \text{mod } 2.$$

The first term counts the connected components of Σ and the second term counts, roughly speaking, the number of wormholes or handles.

2.5 Singularities

So far we have been ignoring the field equations. If we take them into account we soon encounter more difficulties. The topologically non-trivial initial data sets we have been considering will in general possess apparent horizons and closed trapped surfaces. In the time-symmetric case for example, it is easy to see (Gibbons 1972) that the minimal 2–spheres (they must be topologically spheres if they are truly minimal by a simple application of the formula for the second variation of the area and the Gauss-Bonnet theorem) are also apparent horizons. Thus one expects, by the Singularity Theorems of Penrose and Hawking, that the future evolution will contain spacetime singularities. If our usual beliefs about Cosmic Censorship and Black Hole

Physics are true these singularities should be hidden inside event horizons outside of which the metric should settle down to one of the stationary solutions allowed by the Black Hole Uniqueness Theorems. If this is true, and I have no reason to doubt it, it means that dynamically, apart from providing the apparent horizons, the topology has no real effect on the dynamics. After all the *same* black holes could have been formed from a topologically trivial initial data set just consisting of gravitational waves or from some topologically extremely exotic initial data set, even a non-orientable initial 3-manifold or one allowing "spin without spin". According to our present views all details of any complicated non-trivial topology should in the classical theory of general relativity ultimately be lost in the singularity, and certainly hidden from our view by the event horizon. All of this is in marked contrast to conventional flat space theories admitting solitons where the influence of topology can play a decisive role in the dynamics.

At the classical level therefore it seems that we can conclude at least that:
1) It is difficult to incorporate topology change into the classical theory even at the purely *kinematic* level.
2) If the Einstein Equations hold then dynamics is no respecter of topology.

3 THE QUANTUM THEORY

3.1 Hawking Evaporation and Central Charges
These preliminary conclusions would seem to be reinforced when we turn to the (semi-classical) quantum theory. The classically stable Black Hole final states are no longer stable in the quantum theory but rather undergo Hawking evaporation, ultimately shrinking down to a size below which we cannot reasonably calculate because of our ignorance of the correct theory of quantum gravity. If it is really true that neutral black holes evaporate completely leaving no remnant then it seems reasonable to conclude that topological invariants, unless associated with some conserved charge, are simply not "good quantum numbers" . If the black holes are electrically or magnetically charged however the situation may be different and this is where Wheeler's idea of charge without charge comes in. If a black hole carries a charge which is not carried by the fundamental fields of the theory it can't be radiated away by Hawking evaporation, and it may thus stabilise the hole. In flat space theories with solitons charges of this kind, usually topological in origin, are often called "central" because in the quantum theory they commute with the other observables of the theory.

3.2 Supersymmetry
In supersymmetric theories with N spinorial supercharges Q_i, $i = 1, 2, ..N$, they may be central charges in the technical sense that central charges appear in the

supersymmetry algebra:

$$\{\overline{Q}_i, Q_j\} = \gamma_\mu P^\mu \delta_{ij} + Z_{ij}$$

where $Z_{ij} = -Z_{ji}$ are central charges, and P^μ is the total energy momentum vector of the system. In the simplest case, $N = 2$ when $Z_{12} = -Z_{21} = Z$, it follows that the mass M and central charge Z satisfy the Bogomolnyi bound:

$$M \geq |Z|$$

with equality if and only if the soliton is partially supersymmetric in the sense that it is invariant under half the maximum number of supersymmetries.

The obvious model theory to which to apply these ideas is $N = 2$ supergravity theory, that is the supersymmetric extension of Einstein Maxwell theory. The fundamental degrees of freedom make up a supermultiplet consisting of the graviton, the photon and two (electrically neutral) Majorana gravitini. These three particles carry neither electric nor magnetic charge so either type of "charge without charge" is a candidate as a "central charge" (Gibbons, 1981). Moreover the $N = 2$ supergravity theory shares the duality invariance of its purely bosonic sector provided one extends the symmetry to the spin $\frac{3}{2}$ gravitini by means of precisely the type of chiral rotation to be found in "Geometrodynamics" applied to the spin $\frac{1}{2}$ neutrino field. Gratifyingly the whole story works out rather well. Witten's proof of the Positive Energy Theorem may be extended to the Einstein-Maxwell theory (Gibbons and Hull 1982, Gibbons, Hawking, Horowitz and Perry 1983) to bound from below the mass M of any initial data set regular outside an apparent horizon and carrying total electric charge Q and magnetic charge P consistent with the Cosmic Censorship Hypothesis:

$$M \geq \frac{\sqrt{(Q^2 + P^2)}}{\sqrt{(4\pi G)}}.$$

If one has equality, then any regular solution must be one of the static multi-black hole solutions of Papapetrou and Majumdar admitting 4 out of a possible maximum of 8 Majorana Killing spinors, that is (considered as a Supergravity solution) it has half the maximum number of supersymmetries. It is illuminating to view this from the point of view of the supersymmetric ground states or vacua of the $N = 2$ supergravity theory (Gibbons 1984)

A minimal requirement of a vacuum is that it be spacetime homogeneous, and to be fully supersymmetric it should admit the maximum number of supersymmetries, i.e. 8 Majorana Killing spinors. For the $N = 2$ theory there are two such ground states:
  1) Flat spacetime with vanishing electromagnetic field.

2) The Robinson–Bertotti solution, which is hoinogeneous but anisotropic and whose electromagnetic field is homogeneous in space and time.

The Robinson-Bertotti solution may be thought of as a compactified state of the theory, since it is a metric product of the 2-sphere and 2-dimensional anti-de-Sitter spacetime. Both of these ground states appear to be stable against small perturbations subject to suitable boundary conditions. Now the typical behaviour of a supersymmetric soliton is that it admits half the maximum number of supersymmetries and spatially interpolates between two distinct vacua or ground states of the theory. The extreme Reissner-Nordstrom solution does just that. It is asymptotically flat but in the extreme limit the Einstein-Rosen throat becomes infinitely long and the geometry near the horizon is well approximated by the Robinson-Bertotti solution. In common with other supersymmetric solitons the extreme black holes may be fitted into a so-called shortened supermultiplet (Gibbons 1981, Gibbons and Hull 1982, Gibbons 1982a). One also has solutions in which the asymptotically flat region is replaced by another asymptotically Robinson-Bertotti like region (Whitt 1985). Brill (1992) has recently interpreted these as mediating the splitting of black holes, but in view of the stability of Robinson-Bertotti ground state this interpretation may encounter difficulties, particularly in a supersymmetric theory. The creation of pairs of black holes from a Melvin background, which will be described later, seems to me physically more plausible.

3.3 Sol;tons and Moduli
As mentioned above, in addition to the single soliton solution, i.e. the extreme Reissner-Nordstrom solution, there are multi-soliton solutions. These describe an arbitrary number, k, say of extreme black holes placed at k arbitrary positions and having k arbitrary masses. The moduli space is given by k distinct points in R^3, possibly quotiented by the symmetric group on k objects if the masses are all equal and the black holes are regarded as being identical. The black holes are in equipoise because, at rest, their gravitational attraction is just balanced by their electromagnetic repulsion. If they are given a small amount of kinetic energy the subsequent motion may be described as a one–parameter family of static solutions or as a curve in the moduli space. The curve is in fact a geodesic with respect to a certain non-flat metric on the moduli space (Gibbons and Ruback 1986, Eardley and Ferrell 1987, 1989) which has been calculated fully in the case $k = 2$ and in the limit of large separations for general k. Although far from the case of astrophysical interest suggested originally by Dennis, and in an approximation in which gravitational radiation is neglected, I feel that it is note-worthy that the original programme of Einstein and Rosen can be carried sufficiently far that equations of motion for solitons can can actually be worked out in considerable detail. One may even consider the non-relativistic quan-

tum mechanics of the solitons by studying a Schroedinger equation on the moduli space (Gibbons and Ruback 1986, Ferrell and Traschen 1992). The quantum mechanics of these solitons from the supergravity point of view has been treated in great detail in a series of papers by the Viennese group (Aichelburg and Embacher 1988).

3.4 Quantum Corrections

I want now to turn to the quantum properties of these supersymmetric solitons. The first and most obvious point to reiterate is that since they have zero surface gravity and hence zero Hawking temperature and because the graviton, photon and gravitini carry no charge, they are stable against thermal evaporation in. In fact a general non-extreme Reissner-Nordstrom black hole can only reduce its mass but not its charge by emitting Hawking radiation. If the ratio M^2/Z^2 exceeds 3/4 it will get hotter as it does so but as soon as this ratio falls below 3/4 the specific heat at constant charge becomes positive and thereafter the evolution slows down and the hole coasts towards the extreme state.

Another question concerns the quantum corrections to the geometry and whether or not the formula for the entropy for example, which may be calculated using a functional integral approach in which one evaluates the partition function for quantum gravity, is still given by one quarter of the area of the event horizon. Since the Papapetrou-Majumdar solutions belong to the general class of Israel-Wilson metrics which include the self-dual multi-centre metrics and all of these metrics admit Killing spinors, it always seemed to me quite likely that the quantum corrections should vanish. Indeed one motivation for studying this type of black hole, despite its lack of obvious astrophysical application, has always been that in supergravity theories the divergence difficulties which make it so difficult to make any kind of definitive prediction about the ultimate outcome of the thermal evaporation process in pure general relativity might to some extent be obviated. This optimism has been recently been reinforced by the work of Renata Kallosh (1992) who has shown, using superfield techniques, that indeed the quantum corrections do cancel in rather the same fashion that they do around self-dual backgrounds.

3.5 Black Hole Pair Creation

This still leaves open the question of why one should not simply ignore solitons of this sort. One might feel that since they appear to be absolutely stable, if the world contained none to begin with they would be irrelevant. The answer to this argument in conventional flat space theories is well known, and is that if it is possible to pair-produce soliton–anti–soliton pairs during high energy collisions or in extremely strong external fields, then it would be *inconsistent* to ignore their existence. In particular closed loops describing virtual soliton-anti-soliton creation and annihilation should in principle be included as a contribution to all processes. This is not necessarily true for neutral black holes which it seems do not correspond to stable states and should not therefore be included in a complete set of intermediate states. Moreover it appears that they cannot be created in pairs in pure general relativity with a vanishing cosmological term, although there are some indications that this is possible if there is a positive cosmological term (Gibbons 1992). Therefore one should ask whether one can pair–produce electrically or magnetically charged black holes using strong electromagnetic fields.

3.6 Real Tunneling Geometries

To answer this question and others about the possibility of topology change via quantum tunneling, I would like to recall the main features of an approach to such tunneling in the path integral approach to quantizing gravity. The basic idea is that one should be able to approximate the path integral by classical paths, i.e. solutions of the classical equations of motion, which have a Lorentzian portion in the future but which may in general be complex. A particularly simple class of complex paths which seem to include all the cases they have arisen "in practice" are those which Hartle and I (Gibbons and Hartle 1990) have called "Real Tunneling Geometries" .

A Real Tunneling Geometry may be regarded as a triple $\{M_R, \Sigma, M_L\}$ consisting of a complete Riemannian manifold M_R with a totally geodesic boundary $\Sigma = \partial M_R$ and a Lorentzian manifold M_L for which Σ is a Cauchy surface with vanishing second fundamental form. By glueing two copies of the Riemannian manifold M_R across the boundary Σ to obtain the double $2M_R$ one obtains a a particular kind of Gravitational Instanton, one admitting a reflection map θ stabilizing the surface Σ. The simplest and most frequently encountered case (at least in the pages of some physics journals if not in the real world) is when M_R is half of the 4-sphere with its standard round metric and M_L is half of de-Sitter spacetime to the future of a surface of time symmetry. This is the basis for many discussions for example of the "creation of the universe from nothing" considered as some sort of tunneling event, with Σ playing the role of the beginning of time. It seems reasonable to take the existence of a solution of the classical equations of motion which has the form of a Real Tunneling Geometry for which the classical action of the the Riemannian portion M_R is finite, as some sort of a priori evidence that a tunneling process can take place in which a spacetime or portion of spacetime with spatial topology given by Σ suddenly materializes. By the same token, the non-existence of a solution for a particular topology for Σ may be interpreted as quantum mechanical selection rule forbidding that particular topological transition at the lowest semi -classical level.

3.7 The Ernst-Melvin Instanton

The question that we must now ask is whether there exists a Real Tunneling Geometries involving pairs of black holes. My first thought (Gibbons 1986) was to turn to the C-metrics. These represent uniformly accelerating black holes, charged or neutral, and possess both black hole horizons and acceleration horizons. They have two hypersurface orthogonal Killing fields, one $\frac{\partial}{\partial \phi}$ with closed spacelike orbits corresponding to rotations about the direction of motion. The other $\frac{\partial}{\partial t}$ is timelike near infinity but spacelike behind the two types of horizons. Near infinity it behaves like a boost Killing vector rather than a timelike translation. In general the surface gravities of the two types of horizon are different. However in the special case that the mass M and central charge $Z = \sqrt{(Q^2 + P^2)}/\sqrt{(4\pi G)}$ are equal, the two surface gravities are also equal (and non-zero).

It is not possible to choose a range for the angular coordinate ϕ so as to obtain a metric which is everywhere non-singular outside the horizons. No matter what range one picks there always remain conical singularities which may be interpreted as a cosmic string or a cosmic strut depending upon whether the deficit angle is positive or negative. The strut runs from the acceleration horizon to the south pole of the black hole, where south means rear-most with respect to the direction of the acceleration in this context. The string runs from the north-pole of the black hole horizon to infinity. It is possible, by appropriate choice of the range of ϕ, to eliminate either the strut or the string but not both. This is not surprising physically, and may be seen immediately by noting that the solutions has vanishing ADM mass by virtue of the boost invariance at infinity. By the positive mass theorem, generalized to include horizons (Gibbons, Hawking, Horowitz and Perry 1983), there can be no solution which is non-singular everywhere outside the horizons. If there were, we might try to construct an Instanton or Real Tunneling Geometry by adjusting the surface gravities of the acceleration horizon and the black hole horizon to be equal. We would then have a solution which could be analytically continued to purely imaginary values of the time coordinate t. Its existence would have indicated that flat Minkowski spacetime is unstable to the creation of pairs of charged black holes.

Because of the conical singularities, pair creation of black holes from empty Minkowski spacetime is forbidden. It seems more reasonable to expect that they could be created from some strong external field, so we must turn to solutions generalizing the C-metrics representing charged black holes accelerated by an an external electromagnetic field. Such solutions have been constructed by Ernst. If no black hole is present they reduce to the everywhere non-singular static, cylindrically symmetric Melvin solution which is the metric due to a self-gravitating Maxwell flux-tube. Both the metric and the spacetime are also invariant under boosts in the direction of the flux-tube. In fact it may be considered as a rather degenerate limiting case of the solution representing a super-massive Nielsen-Olesen cosmic string in which the Higgs field has decoupled because its electric charge vanishes.

Ernst was able to show that by an appropriate choice of the strength of the appended electromagnetic field all conical singularities could be eliminated for any allowed ratio of charge to mass. This gives a Lorentzian solution of the Einstein-Maxwell equations, non-singular outside the black hole and acceleration horizons, representing a uniformly accelerating black hole. From the point of view of the original programme of Einstein and Rosen this is a remarkable achievement. More significantly for our present purposes it allows us, by choosing the case $M = \sqrt{(Q^2 + P^2)}/\sqrt{(4\pi G)}$ to obtain a Real Tunneling Geometry with a complete Riemannian portion M_R. The surface Σ has the topology of a Wheeler wormhole, i.e. $S^1 \times S^3$. The solution has two arbitrary constants, the charge or mass of the hole and the strength of the external electromagnetic field.

The amplitude for the creation of pairs depends on the classical action of the Riemannian portion of the complex path. In the limit when the black hole is very small compared with the scale set by the background electromagnetic field one obtains a tunneling pre-factor for the amplitude which agrees with what one would expect if one thought of black holes as a point particles and applied the well known formula due to Schwinger for the rate of their creation (Garfinkle and Strominger 1991).

I believe that the results I have described above clearly indicate that although topology appears not to play an important role in pure general relativity, its effects can be significant in theories like Einstein-Maxwell theory admitting central charges. Moreover the phenomenon of pair creation means that it is inconsistent to ignore it at the quantum level. I have discussed this process above using some of the ideas of what is often called Euclidean Quantum Gravity, but physical intuition strongly suggests that if the semi-classical picture makes any sense at all the this will continue to be true regardless of what formalism one quantizes with.

3.8 Dilatons and Strings

The picture may change radically however if one includes the effects of massless scalar fields such as the dilaton and the axion which arise in supergravity theories, Kaluza-Klein theories, and superstring theory. Currently there is a great deal of activity in this area. I will confine myself to a few remarks which will only hint at the richness and variety of the new considerations that come into play. Because of the coupling of the dilaton and axion to the various abelian gauge fields in these theories, the scalar no-hair theorems no longer hold in the literal sense that the dilaton and axion fields must be spacetime independent. There are still central charges, and if these are non-zero then a black hole must have an associated scalar or pseudo-scalar field but its strength is entirely determined by the central charges. There are no independent degrees of freedom connected with the scalars (Gibbons 1982b). It turns out that the thermodynamic properties of this more general class of black holes can be very different from those of Einstein-Maxwell black holes. Moreover they are rather sensitive to the coupling constant governing the interaction between the dilaton and the abelian gauge fields. The value that arises in superstring theory is a critical case (Gibbons and Maeda 1988). This is particularly intriguing because one expects in any case to have to turn to string theory in order to make sense of the quantum theory. This indicates at the very least that its effects on the issue of topology change may well prove to be decisive but that is a topic for some future celebration looking back over the even more distinguished and influential career of Dennis Sciama.

REFERENCES

Aichelburg, P.C. & Embacher, F. (1988) *Physical Review* **D37** 338–348, 911-917, 1436-1443, 2132-2141

Bichteler, K. (1968) *Journal of Mathematical Physics* **9** 813–815

Brill, D. (1992) Maryland preprint

Eardley, D.M. & Ferrell, R.C.(1987) *Physical Review Letters* **59** 1617–1620

Eardley, D.M. & Ferrell, R.C. (1989) in *Frontiers in Numerical Relativity* eds. C.R. Evans, L.S. Finn and D.W. Habill (Cambridge University Press)

Einstein, A. & Rosen, N. (1935) *Physical Review* **48** 73

Ferrell, R.C. & Traschen, J. (1992) *Physical Review* **D45** 2628–2638

Garfinkle, D. & Strominger, A. (1991) *Physics Letters* **B256** 146–149

Geroch, R.P. (1967) *Journal of Mathematical Physics* **8** 782–786

Geroch, R.P. (1968) *Journal of Mathematical Physics* **9** 1739–1744

Geroch, R.P. (1970) *Journal of Mathematical Physics* **11** 343–348

Gibbons, G.W. (1972) *Communications in Mathematical Physics* **27** 87

Gibbons, G.W. (1981) *Springer Lecture Notes in Physics* **160** 145– 151

Gibbons, G.W. & Hull, C.M. (1982) *Physics Letters* **B109** 337–349

Gibbons, G.W. (1982a) in *Quantum Structure of Space and Time* eds. M.J. Duff and C.J. Isham (Cambridge University Press)

Gibbons,G W. (1982b) *Nuclear Physics* **B207** 337–349

Gibbons G.W., Hawking, S.W., Horowitz, G. & Perry, M.J. (1983) *Communications in Mathematical Physics* **88** 295–308

Gibbons, G.W.(1984) in *Supersymmetry, Supergravity and Related Topics* eds.F. del Aguila, J.A. de Azcarraga and L.E. Ibanez (World Scientific, Singapore)

Gibbons, G.W. & Ruback, P.J. (1986) *Physical Review Letters* **57** 1492–1495

Gibbons, G.W. (1986) in *Fields and Geometry* ed. A Jadczyk (World Scientific, Singapore)

Gibbons, G,W. & Maeda, K. (1988) *Nuclear Physics* **B296** 741–775

Gibbons, G.W. & Hartle, J.B. (1990) *Physical Review* **D42** 2458–2468

Gibbons, G.W.(1992) in *Recent Developments in Field Theory* ed. Jihn E. Kim (Min Eum Sa, Seoul)

Gibbons, G.W. & Hawking, S.W. (1992a) *Communications in Mathematical Physics* **148** 1–8

Gibbons, G.W. & Hawking, S.W. (1992b) DAMTP preprint

Kallosh, R. (1992) *Physics Letters* **B282** 80-88

Wheeler J.A. (1961) *Geometrodynamics*, (Academic Press, New York)

Whitt, B. (1985) *Annals of Physics* **161** 244–253

Decoherence of the Cluttered Quantum Vacuum

DEREK J. RAINE

We review the properties of the cluttered Minkowski vacuum. In particular we discuss the example of a uniformly accelerated quantum oscillator in the Minkowski vacuum showing that it does not radiate. Equivalently, the presence of the oscillator does not lead to decoherence (i.e. the emergence of classical probabilities). Mach's Principle was related originally by Einstein to the non-existence of (classical) vacuum cosmological models. We speculate that Mach's Principle may acquire a quantum role as a condition for decoherence of the universe.

1 INTRODUCTION

Following Hawking's announcement (Hawking 1974,1975) of his result that black holes radiate a thermal flux, Davies (1975) applied an analogous technique to the spacetime of a uniformly accelerated observer in the Minkowski vacuum in the presence of a reflecting wall. He interpreted the result as a flux of radiation from the wall at a temperature $ha/4\pi^2 ck$, where a is the acceleration of the observer. Unruh (1976) independently showed that the Minkowski vacuum appears as a thermal state to any uniformly accelerated detector, the normal modes of which were defined with respect to its own proper time. There is no *flux* from the horizon but the detector is raised to an excited state with its levels populated according to a Boltzmann distribution at a temperature $ha/4\pi^2 ck$ as it would be in an inertial radiation bath at this temperature. In this section I shall outline these results which have been reviewed extensively and are uncontroversial. However, while all authors agree on the state of the uniformly accelerated detector there has been rather less agreement on the state of the field. In section 2, I outline the nature of this disagreement and its resolution. The conclusion can be summarised by saying that the presence of a uniformly accelerated detector does not bring about the decoherence of the Minkowski vacuum. The following 5 sections are devoted to some simple examples of the cluttered vacuum to explain and justify this remark. I then turn to the apparently unconnected subject of Mach's principle, to which I shall give an express guided tour. This will bring together the work on Mach's Principle, with which I started my research career under Dennis's guidance, and our most recent collaboration on the uniformly accelerated quantum

300

detector. The main point is that Einstein originally intended that Mach's Principle should be incorporated in general relativity as the non-existence of empty space solutions to the field equations. I shall speculate that Mach's principle may find a role as a decoherence condition in quantum gravity which would require non-vacuum solutions.

2 THERMAL PROPERTIES OF THE MINKOWSKI VACUUM

In this section I follow Takagi (1986). Amongst the many other reviews of this material are DeWitt (1975), Sciama (1979), Sciama, Candelas & Deutsch (1981), Birrell & Davies (1982). For the present purposes we restrict our discussion to one space dimension. Then the wave equation for a massless scalar field in Minkowski coordinates (x, t):

$$\frac{\partial^2 \phi}{\partial t^2} - \frac{\partial^2 \phi}{\partial x^2} = 0$$

can be written also in Rindler coordinates (ξ, η)

$$\frac{\partial^2 \phi}{\partial \eta^2} - \frac{\partial^2 \phi}{\partial (\ln \xi)^2} = 0$$

where $t = \xi \sinh \eta$ and $x = \xi \cosh \eta$ (Rindler, 1966; Fulling, 1973).The Rindler coordinates defined here cover the wedge R^+ of the Minkowski spacetime $x > 0, |t| < x$, but can be extended to cover the whole manifold. The worldline $\xi = 1$ is that of an observer with unit acceleration, for whom these are co-moving coordinates. We now have two alternative quantisations in R^+:

$$\phi = \int_{-\infty}^{\infty} (b_k e^{-i\omega t + ikx} + b_k^\dagger e^{i\omega t - ikx}) \frac{dk}{\sqrt{2\pi\omega}} \tag{1}$$

$$= \int_{-\infty}^{\infty} (A_K e^{-i\Omega\eta + iK \ln \xi} + A_K^\dagger e^{i\Omega\eta - iK \ln \xi}) \frac{dK}{\sqrt{2\pi\Omega}} \tag{2}$$

with $\omega = |k|$ and $\Omega = |K|$.

We can extract the physics from this by the following short cut: expressions (1) and (2) are certainly valid on $\xi = 1$. Putting $\xi = 1$ we can obtain the Bogoliubov transformation connecting b_k and A_K as a Fourier transform. One can then show by explicit calculation that

$$\langle 0_M | A_K^\dagger A_K | 0_M \rangle = 1/(e^{2\pi\Omega} - 1).$$

That is, there is a Planck spectrum at temperature $T = ha/2\pi kc$ of Rindler particles in the Minkowski vacuum. (Unruh (1976) gives a more elegant method.)
We can go further. For each mode it can be shown that

$$|0_M\rangle = (1 - e^{-2\pi\Omega})^{1/2} \exp\{e^{-\pi\Omega} A_K^\dagger A_{-K}^{(-)\dagger}\} |0_M\rangle$$

(Takagi 1986). This says that each Minkowski mode is a coherent superposition of a particle in R^+ with an anti-particle of opposite momentum in the wedge R^- ($x < 0, |t| > |x|$). So for an observable $\mathcal{O}$ restricted to R^+

$$\langle 0_M | \mathcal{O} | 0_M \rangle = tr\rho\mathcal{O}$$

where

$$\rho = \frac{e^{-2\pi\Omega A^\dagger A}}{tr(e^{-2\pi\Omega A^\dagger A})}.$$

That is, for measurements restricted to R^+, the system behaves as a thermal bath with Hamiltonian $H = \Sigma\Omega A^\dagger_K A_K$. In particular, a uniformly accelerated quantum system will get hot, but, once it reaches equilibrium, as seen by the inertial observer, does it radiate?

Unruh claimed that from the point of view of an inertial observer, even the *excitation* of the detector was accompanied by the *emission* of a photon, the energy for both the excitation and the emission coming from the external agent effecting the acceleration. This point of view was further elaborated by Unruh & Wald (1984), who related it to the apparent increase in energy in the field when a photon is detected in an ordinary inertial heat bath, and by Kolbenstvedt (1987) who calculated the angular distribution of the *emitted* radiation.

However, Grove (1986a) pointed out a number of problems with this interpretation. First, the apparent increase in energy associated with the detection of a photon in an inertial heat bath is the result of quantum mechanical 'reduction of the wave function' not a causal consequence of the detector excitation. (The expectation value of the energy in a state which is the superposition of an excited detector and an unexcited one is less than that - after a measurement - in the excited detector state.) By analogy we expect the same to be true of the uniformly accelerated detector and this is borne out by detailed calculation (Grove 1986a). In addition, the extra photon appears concentrated in the region of spacetime causally disconnected from the accelerated observer. Its presence cannot therefore be the result of an emission process. Grove also pointed out that, since there is no radiation reaction the energy cannot not come from the external agent, a point that is reinforced by the isotropy of the radiation field along the axis of acceleration (Grove & Ottewill 1985). (This leaves open the possibility that radiated energy might arise if the coupled system is unphysical and has no ground state (Ford et al. 1988).)

In order to resolve the issue it is necessary to calculate the dynamical evolution of the field as well as of the accelerated detector. One must then ask a definite question about the state of the field. The simplest such question is that of the behaviour of a second, *inertial* detector. If this can be excited then we should say that the uniformly accelerated system is radiating. This was investigated by Raine et al. (1991). They found no radiation. However, the issue is complicated by the fact that Raine et al. appear to imply that there is no effect at all on the inertial detector, hence that the accelerated system has no affect at all on the vacuum. In fact, the presence of even an inertial system will disturb the vacuum - this is the Casimir effect: the total system has a different ground state because it has a different number of modes (Janes 1988). But Casimir energies do not excite a (De-Witt-Unruh type) 'particle' detector. Thus, this effect does not appear in Raine et al. but will be included here.

3 PRELIMINARIES

Consider a simple detector, for example an oscillator linearly coupled to a field $\phi(x)$, $x = (\mathbf{x}, t)$. If this detector moves on a curve $\mathbf{x} = \mathbf{x}(t)$ the probability that it makes a transition between states with energy difference Ω is given by

$$\mathcal{F}(\Omega) = \int_{-\infty}^{\infty} dt \int_{-\infty}^{\infty} dt' e^{-i\Omega(t-t')} \langle \phi(x(t))\phi(x(t')) \rangle. \tag{3}$$

For a stationary system this gives a transition probability per unit time of

$$\int_{-\infty}^{\infty} dT e^{-i\Omega T} \langle \phi(t+T)\phi(t) \rangle. \tag{4}$$

Thus, this is a detector of the 'noise', $\langle \phi(t)\phi(t') \rangle$, in the field. All of our detectors will be of this type in what follows.

A simple example is the Casimir effect which gives us a direct manifestation of the zero point noise in a quantum field. As an illustration, consider a highly excited atom (which plays the role of our detector) placed in a small cavity. The effect of the exclusion of modes of the field with wavelengths larger than the cavity is to suppress the zero point noise reducing the spontaneous emission rate and extending the lifetime of the state. (See Meschede (1992) for a recent review of experimental work.)

4 THE STATIC MIRROR

To set the scene we begin with the simple problem of a mirror in two spacetime dimensions located at $x = 0$. The massless field is

$$\phi(x, t) = \int_0^{\infty} dk(u_k b_k + u_k^* b_k^\dagger)$$

where

$$u_k = \frac{1}{\sqrt{\pi\omega}} e^{-i\omega t} \sin kx \qquad \text{for } k > 0 \tag{5a}$$

$$= \frac{-i}{2\sqrt{\pi\omega}} (e^{-i\omega v} - e^{-i\omega u}) \tag{5b}$$

with $u = t - x$ and $v = t + x$. Let us now put an inertial detector a distance h from the mirror. The noise power $\langle \phi(x, t)\phi(x', t')\rangle$ is

$$\langle \phi(x)\phi(x')\rangle = \int_0^\infty dk(u_k(x)u_k^*(x')\langle b_k b_k^\dagger \rangle + u_k(x')u_k^*(x)\langle b_k^\dagger b_k \rangle). \tag{6}$$

So in the mirror vacuum the additional noise due to the presence of the mirror is

$$\Delta\langle \phi(t)\phi(t')\rangle = \int_0^\infty \frac{d\omega}{\pi\omega} e^{-i\omega(t-t')} \cos 2\omega h, \tag{7}$$

and the transition rate is

$$\Delta\mathcal{F} = \begin{cases} 0 & \text{if } \Omega > 0 \\ -\frac{2}{\Omega} \cos 2\Omega h & \text{if } \Omega < 0. \end{cases} \tag{8}$$

We have spelt out these details to emphasize the following points. First, the presence of the mirror gives rise to a 'density of states' factor [the $u(x)u^*(x')$ in equation (6) or the $\cos 2\omega h$ in equation (7)]. This changes the decay rate of a noise detector in equation (8) by contributing to the negative frequency noise, but cannot excite the detector. Or, to put it another way, a stationary mirror in the Minkowski vacuum does not radiate!

The energy-momentum tensor for a conformally coupled massless field is

$$\langle T_{\mu\nu}\rangle = \lim_{x \to x'} (\partial_\mu \partial_{\nu'} - \frac{1}{2} g_{\mu\nu} \partial_\rho \partial^{\rho'})\langle \phi(x)\phi(x')\rangle$$

which gives $\langle T_{\mu\nu}\rangle = 0$; so the change in the negative frequency structure of the vacuum is not reflected in a change in the vacuum energy. Finally, from equation (5) we see that, for an observer to the left of the mirror the Casimir effect arises from an interference between the right moving vacuum fluctuations and the left moving fluctuations reflected from the mirror.

If the mirror is placed in a thermal bath at temperature T we again get a modified vacuum noise in the negative frequency component to add to the Planck spectrum:

$$\Delta\mathcal{F} = \frac{2}{e^{|\Omega|/T} - 1} - \cos(2\Omega h)\Theta(-\Omega)\left(1 + \frac{2}{e^{|\Omega|/T} - 1}\right) \tag{9}$$

where Θ is the step function: $\Theta(x) = 0$ for $x < 0$, $\Theta(x) = 1$ for $x > 0$.

5 OSCILLATOR IN A HEAT BATH

I want to compare the above results for a mirror with the corresponding situation for an inertial oscillator. We choose the coupling of the oscillator to the massless scalar field ϕ to be that of scalar electrodynamics rather than a simple linear coupling. The resulting Hamiltonian is then positive definite and the coupled system has a ground state; in particular it can come into equilibrium. We shall only be interested in the equilibrium situation.

If the oscillator is at x_0 and has unit mass, natural frequency ω_0 and internal coordinates and momenta q, p, the equations of motion are

$$\frac{dq}{dt} = p + e\phi \tag{10}$$

$$\frac{dp}{dt} = -\omega_0^2 q \tag{11}$$

$$\frac{db_k}{dt} = -i\omega b_k - \frac{ie}{\sqrt{2\pi\omega}}(p + e\phi)e^{ikx_0} \tag{12}$$

where the field ϕ is given by (1). By eliminating b_k we obtain a Langevin equation for the oscillator:

$$\frac{d^2q}{dt^2} + e^2\frac{dq}{dt} + \omega_0^2 q = e\frac{d\phi_0}{dt} \tag{13}$$

where ϕ_0 is the free field. In equilibrium the oscillator is entirely driven by the field:

$$q(t) = \int_{-\infty}^{\infty} b_k e^{ikx_0 - i\nu t}\chi(-\nu)\frac{dk}{\sqrt{2\pi\nu}} + \int_{-\infty}^{\infty} b_k^\dagger e^{-ikx_0 + i\nu t}\chi(\nu)\frac{dk}{\sqrt{2\pi\nu}}$$

$$\phi(x,t) = \phi_0(x,t) - eq(t - |x - x_0|) \tag{14}$$

where $\nu = |k|$, $\gamma = e^2/2$ and

$$\chi(\omega) = ie\omega/(-\omega^2 + \omega_0^2 + i\gamma\omega). \tag{15}$$

The noise is again given by (6) with now

$$u_k = e^{-i\omega t + ikx} - e\chi(-\omega)e^{-i\omega(t-h) + ikx_0} \tag{16}$$

where we have put $x - x_0 = h > 0$. The 'density of states' factors $u(x)u^*(x')$ give rise to terms of the form $\phi_0\phi_0^*$, $\phi_0 q$, and qq. For $k > 0$ the first is just the unperturbed vacuum contribution, the second give a factor $e(\chi + \chi^*)$ and the last a factor $-e^2\chi\chi^*$. But

$$e(\chi + \chi^*) = e^2\chi\chi^*$$

so the $k > 0$ contributions cancel. This leaves the following negative frequency contribution to the additional noise due to the presence of the oscillator:

$$\Delta\langle\phi(t)\phi(t')\rangle = \frac{e^2}{\pi}\int_{-\infty}^{0} e^{i\omega(t-t')}[cos(2\omega h + \delta_\omega) - \omega\gamma]\frac{d\omega}{2\pi\omega}\langle b_k b_k^\dagger\rangle$$

where

$$tan\,\delta_\omega = \frac{\omega_0^2 - \omega^2}{\omega\gamma}.$$

The important point is the similarity between the oscillator and the mirror. If $x_0 = 0$ the perturbed mode functions (16) are of the same form as (5) except that the oscillator reflects the field only in the vicinity of its natural frequency and with a cross-section less than unity (and, of course, with a phase shift which differs from $\pi/2$). The phase δ_ω in the last equation has the effect of spreading out the Casimir effect of a mirror (at frequency ω) over a Lorentz type profile centred on ω_0 (neglecting the small Lamb shift in the oscillator frequency). This is a reflection of the fact that the vacuum has been modified by adding one extra mode. Such a modification will change the decay rate of a further detector but will not excite it. In other words, one cannot find an equilibrated oscillator in a thermal heat bath by looking for its radiation!

6 THE ACCELERATED MIRROR

Accelerated mirrors have been discussed widely in the literature since they provide tractable models of modified vacua (e.g Fulling and Davies 1976, Birrell & Davies 1982, Grove 1986b). All authors agree that a uniformly accelerated mirror in the Minkowski vacuum does not radiate. What does the co-moving observer see? We can assume the mirror is in the wedge R^+ and work in the Rindler basis (ξ, η). In these coordinates the mirror is stationary at $\ln \xi = 0$ so the calculation is formally the same as in the inertial case. Since $\langle A_k^\dagger A_k \rangle$ is equal to the Planck function we obtain precisely the thermal Casimir effect. So, the uniformly accelerated observer cannot use his mirror to distinguish locally an inertial heat bath in a gravitational field from uniform acceleration in the vacuum. (This is not to say the two are indistinguishable: the correlations of the field on separated worldlines in the vacuum are not the same as in the inertial bath.)

What about the inertial observer? To the right of the mirror the mode functions at the point with null coordinates $u = t - x, v = t + x$ are

$$u_k = i \left(\frac{1}{4\pi\omega} \right)^{1/2} (e^{-i\omega v} - e^{-i\omega v_0}) \tag{17}$$

where $k = \omega > 0$ and the point $(u, v_0) = (u, v_0(u))$ is the intersection of the ray $u =$constant with the mirror trajectory (see e.g. Birrell & Davies 1982). Explicitly, we have for a constantly accelerated mirror following a trajectory $z(t) = (1 + (1 + a^2 t^2)^{1/2})/a$,

$$v_0(u) \equiv t(u) + z(u) = 2t(u) - u = u/(1 + au). \tag{18}$$

For the vacuum noise at $x = x'$ we obtain

$$\langle \phi(t)\phi(t') \rangle = \int_0^\infty \frac{d\omega}{4\pi\omega} \left[e^{-i\omega(t-t')} - e^{i\omega(v_0'-v)} - e^{-i\omega(v_0-v')} + e^{i\omega(v_0'-v_0)} \right]. \tag{19}$$

From equations (18) and (3) the frequency structure of the first three terms of this expression is clear: there is no contribution to $\mathcal{F}(\Omega)$ if $\Omega > 0$ since the time dependence is of the form $e^{-i\omega't}$ with $\omega' > 0$. These terms are therefore purely negative frequency. The fourth term is also pure negative frequency. This can be seen by expanding $(v_0' - v_0)$ in terms of $(t' - t - i\epsilon)$ using (18) with the following transformation:

$$\epsilon' = \epsilon/(1 + at)(1 + at').$$

For $t, t' > -1/a$, ie in the causal future of the mirror trajectory, this term is analytic in the lower half $t' - t$ plane. Further details can be found in Grove (1986b). (The appendix of this paper gives the relation between his expressions and ours.)

It is well known that the renormalised energy flux from a constantly accelerated mirror vanishes (Fulling and Davies 1976). Only if the constant acceleration is limited to a finite period is there a flux associated with the switching on and off of the acceleration. (The Casimir energy here is lost in the renormalisation procedure.) The above calculation confirms that there is also no particle flux except from the switching (Grove, 1986b). The mirror does not radiate but does carry its appropriate Casimir energy with it.

7 THE UNIFORMLY ACCELERATED OSCILLATOR

It should now be clear that we must obtain a precisely similar result for the uniformly accelerated oscillator and by an analogous calculation. For the co-moving observer we use Rindler modes and reproduce the thermal Casimir effect of Section 5.

The question of what an inertial observer sees was considered by Grove (1986a) and by Raine et al. (1991). The latter computed explicitly the response of an inertial particle detector; they concluded that, since the positive frequency noise is unaffected by the presence in the vacuum of the uniformly accelerated oscillator, the inertial detector is unexcited. Note that this does not conflict with the now accepted result that a uniformly accelerated charge does radiate. The charge carries with it a long range field as a consequence of which radiation appears in the wedge of spacetime beyond the future horizon of the co-moving observer (Boulware 1980). This does however illustrate the need to consider the question of radiation from the viewpoint of the inertial observer for whom there is no apparent horizon. We shall use here a simpler method than in Raine et al. and include the complete expression for the negative frequency noise. Our point will be that the result is precisely analogous to the case of the mirror.

The equations of motion for the uniformly accelerated oscillator at $x = x_0(\tau), t =$

$t_0(\tau)$, where τ is the proper time along the oscillator trajectory, are

$$\frac{dq}{d\tau} = p + e\phi \tag{20}$$

$$\frac{dp}{d\tau} = -\omega_0^2 q \tag{21}$$

$$\frac{db_k}{d\tau} = -i\omega b_k \frac{dt}{d\tau} - \frac{ie}{\sqrt{2\pi\omega}}(p + e\phi)e^{ikx_0(\tau)}. \tag{22}$$

The oscillator frequency is defined with respect to proper time and this introduces the redshift factor $dt/d\tau$ in the proper time description of the field modes. The Langevin equation for the oscillator coordinate resembles (13) except that the coordinate time t is replaced by proper time τ. The oscillator therefore comes into equilibrium with the field with its internal noise in balance with that along its worldline in the external field (Sciama 1981) i.e. it is thermally excited. Solving for $q(\tau)$ we obtain:

$$q(\tau) = \int_{-\infty}^{\infty} \frac{dk}{\sqrt{2\pi\omega}} \left[b_k e^{ikx_0(\tau)-i\omega t(\tau)}\chi(-\omega) + b_k^\dagger e^{-ikx_0(\tau)+i\omega t(\tau)}\chi(\omega) \right]. \tag{23}$$

For the field, we get

$$\phi(x,t) = \phi_0(x,t) - eq(t - |x - x_0(\tau)|) \tag{24}$$

$$= \int_{-\infty}^{\infty} \frac{dk}{\sqrt{2\pi\omega}}(b_k u_k + b_k^\dagger u_k^*), \tag{24}$$

where

$$u_k = e^{-i\omega v} - e\chi(-\omega)e^{i\omega u} \qquad \text{for } k > 0$$
$$= e^{-i\omega u} - e\chi(-\omega)e^{i\omega v_0(u)} \qquad \text{for } k < 0 \tag{26}$$

We can now compute the noise power on an inertial worldline, which, for the sake of argument, we take to the right of the oscillator. Just as we found for the inertial heat bath the contributions from the $k > 0$ terms cancel; the $k < 0$ terms give:

$$\langle \phi(x)\phi(x') \rangle = \int_0^{\infty} \frac{d\omega}{2\pi\omega} \left[e^{-i\omega(t-t')} - e\chi^* e^{i\omega(v_0'-v)} \right.$$
$$\left. - e\chi e^{-i\omega(v_0-v')} + e^2|\chi|^2 e^{i\omega(v_0'-v_0)} \right]. \tag{27}$$

This expression is again analogous to that for the uniformly accelerated mirror (19) except for the appearance of the oscillator susceptibility which limits the effect of the oscillator on the vacuum to a restricted set of modes. The noise is negative frequency with respect to the Minkowski vacuum so does not excite a standard DeWitt-Unruh (particle) detector. Although the oscillator disturbs the vacuum, here by adding an extra Rindler mode, it does not radiate – just like the mirror.

In a slightly different language we can conclude that the addition of a finite number of uniformly accelerated oscillators or mirrors does not bring about the decoherence of the vacuum, even though the vacuum appears as a mixed state to the uniformly accelerated observer. This is a result of the fact that both the equilibrated uniformly accelerated system and the the inertial detector are driven by the vacuum noise (Sciama *et al* 1981; Senitzky (1959, 1960, 1961) was the first to discuss the inertial case.) They are therefore correlated with the vacuum and hence with each other just so as to bring about the cancellation of the $k > 0$ terms and to maintain a pure state. Usually one assumes the availability of a detector that is not correlated with the system one is trying to measure i.e. that the system can decohere. We turn to this next.

8 DECOHERENCE

We can think of decoherence as equivalent to the emergence of classical behaviour. What is the problem here? Suppose we start with a quantum system in a state $|\psi(0)\rangle$. This system evolves as a superposition of the possible paths $|\psi_n(t)\rangle$:

$$|\psi(t)\rangle = \sum c_n |\psi_n(t)\rangle. \tag{28}$$

Since these paths can be brought together to interfere we cannot assign to them a probability $|c_n|^2$; rather we have a density matrix

$$\rho = |\psi(t)\rangle\langle\psi(t)| = \sum_{n,m} c_n c_m^* |\psi_n(t)\rangle\langle\psi_m(t)|.$$

Classical behaviour requires

$$\rho_{class} = \sum |c_n|^2 |\psi_n(t)\rangle\langle\psi_n(t)|. \tag{29}$$

So decoherence is equivalent to the removal of the off-diagonal terms. There is now an extensive literature on this subject especially in the context of the emergence of classical behaviour in quantum cosmology (e.g. Hartle 1986, articles by Gell-Mann & Hartle, Halliwell, and Zeh in Zurek 1988; an excellent brief review in Paz & Sinha 1991). I would like to add a speculation.

9 MACH'S PRINCIPLE

Elsewhere in this volume, Tod gives a measured review of Mach's Principle and its possible embodiment in a quantum theory of gravity. For our express tour it is sufficient to look at it in the context of the question: Why is spacetime not the (quantum) vacuum? For the particular universe in which we find ourselves we can ask specifically: Why is $G\rho t^2 \sim 1$? (where ρ is the mean density of the universe at time t.) The possible answers can be arranged in terms of the three ways in which we can view this equation: as an equation for G, or for ρ or for t.

9.1 $G^{-1} = \rho t^2$

Sciama in his 1953 paper introduced the idea of inertial induction through the gravitational field (Sciama 1953, 1969). In this model, inertial forces are induced by the acceleration of (distant) matter. The model is consistent if gravity has the appropriate strength, for which we require $G^{-1} \sim \rho t^2$.

This toy model was shown to have an interpretation in general relativity by Sciama, Waylen & Gilman (1969) as an integral equation for the spacetime metric. In the formulation of Raine (1981) we write

$$g_{\mu\nu}(x) = \frac{G}{16\pi c^2} \int M_{\mu\nu}^{\alpha'\beta'} K_{\alpha'\beta'} \sqrt{|g'|} d^4 x' - \int M_{\mu\nu;\gamma'}^{\alpha'\beta'} g_{\alpha'\beta'} dS^{\gamma'} \tag{30}$$

where $M_{\mu\nu}^{\alpha'\beta'}$ is an appropriately defined Green function on the manifold with metric $g_{\mu\nu}$ and $K_{\mu\nu} = T_{\mu\nu} - 1/2 g_{\mu\nu} T$. A cosmological model satisfies Mach's Principle if the final term in (30) (the surface integral) vanishes in the limit that the volume integral includes the whole manifold. (See Tod's article in this volume.) Taking the trace in a Machian model appears to give an expression for G^{-1} in terms of the matter density:

$$G^{-1} = \frac{1}{16\pi c^2} \int (M_{\mu\nu}^{\alpha'\beta'} g^{\mu\nu}) K_{\alpha'\beta'} \sqrt{|g'|} d^4 x'.$$

But this equation is, in fact, an identity which places no restriction on the matter density. This result of the toy model does not therefore carry over to the full theory.

An alternative approach to this type of explanation was proposed by Dicke (1964) in which G^{-1} is introduced as an independent field with its own field equations. The theory is, however, consistent with empty space and in any case is effectively ruled out by observations.

9.2 $\rho \sim 1/Gt^2$

Reading this equation as the requirement that ρ be the critical density leads to inflationary models. For simple cosmological models the above critical densities are associated with spatial closure and hence to no-boundary proposals for the universe. The first of these appears to be Einstein (1917). This was developed by Wheeler and his students (Wheeler 1962). The closure property is, of course, a sufficient condition for the Sciama-Waylen-Gilman approach. The most recent no-boundary proposal is that of Hawking.

9.3 $t \sim 1/\sqrt{G\rho}$

The now widely known anthropic principle (Carter, 1974, 1983, Barrow & Tipler, 1986) stipulates that the universe in which we participate must be able to grow us; in particular there must be sufficient time for sufficiently complex systems to evolve. In fact, if the universe has a quantum origin, certain conditions must be satisfied if it is going to have a classical evolution of any sort (Raine, 1987). In particular, conditions for decoherence must be satisfied.

I shall take a simple model as the basis for an order of magnitude estimate. The model will be a quantum harmonic oscillator of mass m, natural frequency ω_0 in a radiation field at temperature T. The condition for decoherence of the field at points separated by distance d to occur in time t is (Caldiera & Leggett, 1983)

$$t > \frac{1}{\gamma}\left[\frac{h}{d\sqrt{2mkT}}\right]^2 \tag{31}$$

where γ is the oscillator decay rate. Assuming the oscillators interact gravitationally we have $\gamma = (Gh/c^5)\omega_0^3$, using the gravitational analogue of the fine structure constant, and ω_0^3/c^3 oscillators per unit volume. The mass density in the radiation field follows from the Rayleigh-Jeans law as $u_\nu\nu \sim kT(\omega_0/c)^3$. The distance we are interested in is the Compton wavelength of the oscillator $d \sim h/mc$ which we take to be the size of the Universe at time t. From (31) we derive the inequality $G\rho t^2 \gtrsim 1$.

Of course, our experience with the uniformly accelerated oscillator leads us to believe that this decoherence will have to result from some underlying physical mechanism. There the fact that (29) arises from a mathematical restriction to a certain class of observations was not in itself a sufficient condition to ensure decoherence. The natural speculation here, of course, is that quantum gravity will do the job (and would presumably also lead to an equality rather than a lower limit). Then we should be able to read our equation in a fourth way as an equation for $1 (= G\rho t^2)$: that is, as expressing the Unity of the Universe which arises from the quantum vacuum in accordance with Mach's Principle.

I am grateful to Dennis Sciama for his constant encouragement and enthusiasm over the years. I thank Peter Grove for sharing his insights and for his considerable help in preparing this review.

REFERENCES

Barrow, J. D. & Tipler, F. J., 1986, *The Anthropic Principle*, (Oxford University Press).

Birrell, N. D. & Davies, P. C. W., 1982, *Quantum Fields in Curved Space* (Cambridge University Press).

Boulware, D. G., 1980, *Ann. Phys.*, **124**, 169.

Caldiera, A. O. & Leggett, A. J., 1983, *Physica*, **121A**, 587.

Carter, B., 1974, in: *IAU Symposium 63: The Confrontation of Cosmological Theories with Observational Data*, ed. M. Longair, (Dortrecht).

—. 1983, *Phil. Trans. R. Soc. Lond.* **310**, 347.

Davies, P. C. W., 1975, *J. Phys. A*, **8**, 609.

DeWitt, B. S., 1975, *Phys. Rep.*, **19**, 295.

Dicke, R. H., 1964, *The Theoretical Significance of Experimental Relativity* (Blackie).

Einstein, A., 1917, *Sitzungsberichte der Preussischen Akademie der Wissenschaften*, 142; translated in Einstein *et al.*, *The Principle of Relativity* (Dover, 1952).

Ford, G. W., Lewis, J. T. & O'Connell, R. F., 1988, *Phys. Rev. A*, **37**, 4419.

Fulling, S. A., 1973, *Phys. Rev. D*, **7**, 2850.

Fulling, S. A. & Davies, P. C. W., 1976, *Proc. Roy. Soc. Lond. A*, **348**, 393.

Grove, P. G., 1986a, *Class. Quantum Grav.*, **3**, 801.

—. 1986b, *Class. Quantum Grav.*, **3**, 793.

Hartle, J. B., 1987, in: *Gravitation & Astrophysics, Cargèse 1986*, eds. B. Carter & J. B. Hartle, (Plenum Press).

Hawking, S. W.,1974, *Nature*, **284**, 30.

—. 1976, *Commun. Math. Phys.*, **43**, 199.

Janes, E. T., 1988, in: *Complexity, Entropy & the Physics of Information* ed. W. H. Zurek, (Addison-Wesley).

Kolbenstvedt, H., 1987, *Phys. Rev. D*, **38**, 1118.

Meschede, D., 1992, *Phys. Rep.*, **211**, 201.

Paz, J. P. & Sinha, S., 1991, *Phys. Rev. D*, **44**, 1038.

Raine, D. J., 1981, *Rep. Prog. in Phys.*, **44**, 1151.

—. 1987, in: *Gravitation & Astrophysics, Cargèse 1986*, eds. B. Carter & J. B. Hartle, (Plenum Press).

Raine, D. J., Sciama, D. W. & Grove, P. G., 1991, *Proc. Roy. Soc. Lond. A*, **435**, 205.

Rindler, W., 1966, *Am. J. Phys.*, **34**, 1174.

Sciama, D. W., 1953, *Mon. Not. R. astr. Soc.*, **113**, 34.

—. 1969, *The Physical Foundations of General Relativity* (Heinemann).

—. 1979, in: *Relativity, Quanta & Cosmology*, eds. M. Pantaleo & F. De Finis (Johnson Reprint Corporation) Vol. II.

—. 1981, in: *Quantum Gravity 2* eds. C. J. Isham, R. Penrose & D. W. Sciama, (Oxford University Press).

Sciama, D. W., Candelas, P. & Deutsch, D., 1981, *Adv. Phys.*, **30**, 327.

Sciama, D. W, Waylen, P. C. & Gilman, R. C., 1969, *Phys. Rev.*, **187**, 1762.

Senitzky, I. R., 1959, *Phys. Rev.*, **115**, 227.

—. 1960, *Phys. Rev.*, **119**, 670.

—. 1961, *Phys. Rev.*, **124**, 642.

Takagi, S., 1986, *Prog. Theoretical Phys.*, **88**, 1.

Unruh, W. G., 1976, *Phys. Rev. D*, **14**, 870.

Unruh, W. G. & Wald, R. M., 1984, *Phys. Rev. D*, **29**, 1047.

Wheeler, J. A., 1962, *Geometrodynamics* (Academic Press).

Zurek, W. H., 1988, (ed.) *Complexity, Entropy & the Physics of Information* ed. W. H. Zurek, (Addison-Wesley).

Quantum Non-Locality and Complex Reality

ROGER PENROSE

Although I am one of the very few people represented here who was never technically a student of Dennis Sciama's (or a student's student or a student's student's student), I was, on the other hand, very much a student of his in a less formalized sense. He was a close personal friend when I was at Cambridge as a research student, and then a little later as a Research Fellow. Although my Ph.D. topic was in pure mathematics, Dennis took me under his wing, and taught me physics. I recall attending superb lecture courses by Bondi and by Dirac, when I started at Cambridge, which in their different ways were inspirations to me, but it was Dennis Sciama who influenced my development as a physicist far more than any other single individual. Not only did he teach me a great deal of actual physics, but he kept me abreast with everything that was going on and, more importantly, provided the depth of insight and excitement - indeed, *passion* - that made physics and cosmology into such profoundly worthwhile and thrilling pursuits.

I first encountered Dennis at the Kingswood Restaurant, in Cambridge, somewhat before I went up there as a research student, where I was introduced to him by my brother Oliver. I tried to explain to him some geometrical thoughts that I had had in relation to the steady-state universe, as a result of hearing Fred Hoyle's BBC radio talks on the Nature of the Universe, and was impressed by Dennis's keen interest in what even a total unknown such as myself had to say. Already at that time Dennis's sense of excitement about the cosmos came through to me, and it never left me since. In our later discussions, when I was officially installed at Cambridge, Dennis and I had long discussions about relativity, unified field theory, cosmology, quantum theory, particle physics, and especially Mach's principle. I think that apart from gaining a great many necessary insights into existing physical theory from Dennis, especially into general relativity, what I took from those discussions were two things that were particularly important to my own individual (and peculiar?) approaches towards furthering physical understanding. These insights were (i) that some kind of non-locality must be a fundamental feature of the geometry of the actual physical world, and (ii) that there must be an underlying geometric structure to space-time in which the complex numbers of quantum theory play a role at least on a par with the real numbers of conventional relativity theory.

The importance of non-locality became clear to me as a result of the many discussions that we had concerning Mach's principle. "Suppose", Dennis would often say to me, "that the particles of the distant universe could be eliminated one-by one until only the earth survived. How would a Foucault pendulum then behave?" He would argue, powerfully, that since there would be no distant galaxies to determine the non-rotating frame with respect to which the earth is imagined to be still spinning, it is the earth itself that determines the only frame that is physically present, so no rotation with respect to it is possible and the Foucault pendulum would appear *not* to be deflected. I remember feeling uneasy about this, and we began wondering what would happen if things were taken to even greater extremes, so that there would be only one electron left. What would happen to its spin? No doubt it would have to retain its total spin, so that it would remain a spin 1/2 particle, but what about its spin *direction* It seemed reasonable to me that this concept would become divested of its meaning when there was nothing to compare the electron with, but suppose there were just two electrons. Then each could, in effect, measure its spin in relation to the other. There would be just two possibilities, the total spin of the pair could be 0 ("singlet state") or 1 ("triplet state"). I tried to develop this kind of idea to situations where several particles are involved, in the hope that well-defined directions of spin could arise when large numbers of particles are present. I was eventually able to make such a scheme work (the theory of "spin–networks" cf. Penrose 1971, 1972). One lesson from this study became particularly clear to me: the essential non-locality of physics when the effects of quantum entanglement are considered. It seemed to me that any deep theory of the way that the geometry of the physical world is structured must come seriously to terms with this non-locality.

Passing, now, to the second insight, such a geometric theory, if it were to represent a genuine union between quantum theory and space-time structure would also have to take into account the fact that quantum theory is based on *complex* numbers, rather than the real numbers that underlie the conventional space-time descriptions. Dennis himself, at that time, had been developing a generalization of standard general relativity in which complex numbers played an essential role (Sciama 1958), and we had many discussions together about this and related matters. Partly as a result of such discussions, my own thoughts drove me in the direction of basing space-time geometry on 2-component (complex) spinors - and eventually twistors, where ideas from non-locality also played an important part. (See Penrose and Rindler 1984, 1986; Penrose 1986.)

In the following sections I shall give brief descriptions of two fairly recent ideas that I have had, one concerning quantum non-locality and the other concerning the relationship between twistor theory and general relativity. These will give some idea of my latest thinking on those topics - where Dennis's inspiration has had a profound initial influence on the direction, for good or for bad, of my own work.

1 MAGIC DODECAHEDRA; A NEW EXAMPLE OF QUANTUM NON-LOCALITY

Let me describe a little story - and a puzzle. Imagine that I have recently received a beautifully made regular dodecahedron. It was sent to me by a company of superb credentials, known as "Quintessential Trinkets", who inhabit a planet orbiting the star Betelgeuse. They have also sent another identical dodecahedron to a colleague of mine who lives on a planet orbiting α-Centauri, and his dodecahedron arrived there at roughly the same time as mine did here. Each dodecahedron has a button that can be pressed, on each vertex. My colleague and I are to press individual buttons independently on our respective dodecahedra at some time and in some order that is completely up to our own choosing. Nothing may happen when one of the buttons is pressed, in which case all we do is to proceed to our next choice of button. On the other hand, a bell may ring, when one of the buttons is pressed, accompanied by a magnificent pyrotechnic display which destroys that particular dodecahedron!

Enclosed with each dodecahedron is a list of guaranteed properties that relate what can happen to my dodecahedron and to my colleague's. First, we must be careful to orient our respective dodecahedra in a very precise corresponding way. Detailed instructions are provided by Quintessential Trinkets as to how our dodecahedra are to be aligned, in relation to, say, the centres of the Andromeda galaxy and M-87, etc. The important thing is only that my dodecahedron and that of my colleague must be perfectly aligned with one another. The list of guaranteed properties is, perhaps, fairly long, but all we shall need from them is something quite simple. We must bear in mind that Quintessential Trinkets have been producing things of this nature for a very long time - of the order of a hundred million years, let us say, and they have never been found to be wrong in the properties that they guarantee. The very excellent reputation that they have built up over a million centuries depends upon this, so we can be quite sure that whatever they claim will turn out actually to be true. What is more, there is a stupendous CASH prize for anyone who *does* find them to be wrong!

The guaranteed properties that we shall need concern the following type of sequence of button-pressings. My colleague and I each independently select one one of the vertices of our respective dodecahedra. I shall call these the SELECTED vertices. We do *not* press those particular buttons; but we *do* press, in turn, and in some arbitrary order of our choosing, each of the three buttons that inhabit vertices *adjacent* to the SELECTED one. If the bell rings on one of them, then that stops the operation on that particular dodecahedron, but the bell need not ring at all. We shall require just two properties. These are:

(a) if my colleague and I happen to have chosen diametrically *opposite* vertices as our respective SELECTED ones, then the bell can ring on one of the ones I press (adjacent to my SELECTED one) if and only if the bell rings on the diametrically opposite one of his - irrespective of the particular orders in which either of us may choose to press our respective buttons;

(b) if my colleague and I happen to have chosen exactly *corresponding* vertices (i.e. in the *same* directions out from the centre) as our respective SELECTED ones, then it cannot be the case that the bell does not ring at all for either of us, on the total of six button-presses that we together make.

Now I want to try to deduce something about the rules that my *own* dodecahedron must satisfy independently of what happens on α-Centauri, merely from the fact that Quintessential Trinkets are able to make such strong guarantees without having any idea as to which buttons either I or my colleague are likely to press. The key assumption will be that there is no long-distance "influence" relating my dodecahedron to my colleague's. I shall suppose that our two dodecahedra behave as separate, completely independent objects after they have left the manufacturers.

My deductions are:

(c) each of my own dodecahedron's vertices must be pre-assigned as either a bell-ringer (colour it RED) or as silent (colour it BLUE), where its bell-ringing character is independent of whether it is the first, second, or third of the buttons pressed adjacent to the SELECTED one;

(d) no two next-to-adjacent vertices can be both bell-ringers (i.e. both RED);

(e) no set of six vertices adjacent to a pair of antipodal ones can be all silent (i.e. all BLUE).

We deduce (c) from the fact that my colleague *might* happen to choose, as his SELECTED one, the diametrically opposite one to my own SELECTED one; at least, Quintessential Trinkets will have no way of knowing that he will not. Thus, if one of my three button-presses happens to ring the bell, then we can be sure (by the absence of "influence" assumption) that Quintessential Trinkets must have pre-arranged that particular vertex to a bell-ringer in order to ensure no conflict with (a). Likewise, (d) follows from (a) also. For suppose two next-to-adjacent vertices on my dodecahedron are both bell-ringers. I might have chosen their common neighbour as my SELECT-ED one and my colleague might have chosen opposite to this as his SELECTED one; moreover my colleague might have elected to press the button first on his dodeca-hedron which is opposite to the particular bell-ringer on my dodecahedron which is next-to-adjacent to the one that I *actually* first press. Both our bells would have to ring, which contradicts (a). This establishes (d). Finally, (e) follows from (b), together with what we have now established. For suppose that my colleague happens to choose, as his SELECTED one, the vertex *corresponding* to my own SELECTED choice. If none of my three buttons adjacent to this choice is a bell-ringer, then, by (b), one of my colleagues three must be a bell-ringer. It follows from (d) that my own vertex opposite to my colleagues bell-ringer must also be a bell-ringer. This establishes (e).

Now comes the puzzle. Try to colour each of the vertices of a dodecahedron either RED or BLUE consistently according to the rules (d) and (e). You will find that no matter how hard you try you cannot succeed. A better puzzle, therefore, is to provide a *proof* that there is no such colouring. Can it be that, for the first time in a hundred centuries, Quintessential Trinkets, has made a mistake? Recalling the stupendous CASH prize, we eagerly wait the four years, or so, that is required for my colleague's message to arrive, describing what he did and when and whether his own bell rang; but when his message arrives, all hopes of CASH vanish, for Quintessential Trinkets has turned out to be right again!

How do Quintessential Trinkets - or "QT", as they are known for short - actually do it? Of course "QT" really stands for Quantum Theory, and what they have done is to arrange for an atom of spin 3/2 to be suspended at the centre of each of our dodecahedra. These two atoms were produced on Betelgeuse in an initial combined state of total spin 0, and then carefully separated and isolated, so that their combined total spin value remains at 0. Now when either my colleague or I press one of the buttons at a vertex of one of our dodecahedra, a (partial) spin measurement is made, in the direction out from the centre of that particular vertex. More specifically, what is performed is a measurement to determine whether the m-value of the spin, in that direction, is +1/2 rather than one of the other three possible values +3/2, −1/2, or −3/2. If the value is indeed found to be +1/2, then the bell rings and the pyrotechnic display follows shortly afterwards. If the value is found not to be +1/2, then the atom is not observed, and the three unobserved possibilities are carefully recombined without disturbing their relative phases. Then the process is repeated in another direction, as determined by the next choice of button.

I shall not go into the full details, here, of why (a) and (b) must hold. The essential points are, first, that the $m = +1/2$ states for two next-to-adjacent vertices of a dodecahedron are orthogonal (orthogonal in the Hilbert-space sense, that is, not perpendicular directions); and, second, that the state which is *orthogonal* to a triplet of mutually next-to-adjacent vertex states is itself orthogonal to the state correspondingly related to the antipodal triplet. For details, see Penrose (1992); and for further details, Zimba (1992).

The final conclusion, of course, is that the assumption of no long-distance influence is violated by Quantum Theory and indeed by experiment, according to the conclusions of Aspect and his colleagues (cf. Aspect and Grangier 1986). A virtue of the present scheme (as shared by a number of other, somewhat earlier, recent ideas) is that the puzzling features of this hypothetical experiment are of an entirely yes/no character, with no probabilities arising. For overviews of the earlier work of this nature, and some philosophical discussion, see Redhead (1987) and Brown (1992). Non-locality seems to be here to stay.

2 TWISTORS, MASSLESS FIELDS OF HELICITY 3/2, AND THE EINSTEIN EQUATIONS

It would take too long to provide an overview of twistor theory here, but I should point out some of its salient features. (For detailed accounts, see Huggett and Tod 1985, Penrose and Rindler 1986, and Ward and Wells 1989; for a historical account, see Penrose 1986.) Basically the original idea was to find a way of describing space-time geometry consistently with the above physical requirements: that it should be fundamentally non-local and fundamentally complex. As a first approximation to twistor theory, we may think of twistors as *light rays* (null geodesics) in space-time. If the space-time is flat (or at least conformally flat) then the space of such light rays can be described as a real–five–dimensional submanifold PN of a complex–three–dimensional projective space PT. In fact, PN divides PT into two open pieces PT$^+$ and PT$^-$. It is also possible to interpret the points of PT which do *not* lie in PN, by taking into account the fact that physical photons have spin (of fixed absolute value, giving a helicity parallel or antiparallel with the momentum) and *energy*, with the implication that they should be considered as not being exactly localized along any light ray, even though still being treated classically. The points of PT - PN ($=$PT$^+$ $\cup$ PT$^-$) then represent such "photons with energy and helicity". The photons of positive helicity are represented as points of PT$^+$, and those of negative helicity as points of PT$^-$. This gives us our second approximation to twistor theory. As third and fourth approximations, we can physically interpret the points of the non-projective space T, which is the complex–four–dimensional vector space from which the projective space PT arises as its space of one-dimensional linear subspaces. The points of T- or *twistors* - can be interpreted, up to a phase multiplier, as providing the energy-momentum structure of any massless particle (third approximation) or as helicity-raising operators (fourth approximation). (There are also *non*-projective versions N, T$^+$, and T$^-$ of the respective "projective" spaces PN, PT$^+$, and PT$^-$.)

The reason that I have used the terminology "approximation" to twistor theory, here, is that I believe that we have not yet found the fully appropriate physical form of the theory - despite the numerous and multifarious applications of twistor ideas that have already come to light. Thus, I am arguing, what has emerged so far constitutes, indeed, merely a set of conceptual approximations to a putative yet–unknown theory. The essential physical difficulty with the twistor theory that presently exists is that it does not fully apply to *general relativity*. Most of the twistor concepts that I have alluded to above depend upon the space-time being flat, or at least conformally flat. Only the first approximation (in terms of light rays) will apply in a general curved space-time, but the trouble here is that the essential *complex* structure disappears, in general, unless the space-time is indeed conformally flat. There are certain twistor concepts that do apply to general curved space-time (local twistors, asymptotic twistors, hypersurface twistors, 2–surface twistors) but these are too limited to provide a global alternative to the conventional space-time description.

There is one line of development that should be mentioned in this context, however. If we allow our "space-time" to become *complex* (with a complex–analytic metric), then its conformal (Weyl) curvature is seen to split into two halves: its *self-dual* and the *anti-self-dual* parts. The self-dual part may be thought of as describing the positive-helicity aspects of the "graviton field" and the anti-self-dual part, its negative-helicity aspects. It is also appropriate to consider "complex vacuum space-times" for which one or the other of these halves of the Weyl tensor is put equal to zero (the Ricci tensor being assumed zero). Such complex 4–manifolds are referred to as *right-flat* (if just the anti-self-dual part of the Weyl curvature survives) or *left-flat* (if just the self-dual part survives). It may be counted as one of the more successful achievements of existing twistor theory that a global complex twistor space T can be constructed for any left-flat complex space-time M and, moreover, the full metric structure of M, including its automatic satisfaction of the Einstein vacuum equations, is fixed in terms of the global structure of the complex manifold T - according to a procedure sometimes known as the "non-linear graviton construction". The (curved) twistor space T can itself be constructed from *free* complex-analytic data (see Penrose 1976). Thus, in a clear sense, twistor theory provides a complete description, in terms of free functions, of the general left-flat (vacuum) space-time.

There has, for a long time, been the tantalizing hope that something of a somewhat similar nature might perhaps be possible for the *general* vacuum equations, when the anti-self-dual condition is removed. Up to this point, attempts towards this goal had been concentrated in one of two directions: that afforded by "ambitwistors" (basically the complexified version of the "light ray" concept of a twistor) and the "googly" approach, where one attempts to describe the *right*-flat spaces using a twistor space rather than a dual twistor space. (The latter would be obtained, trivially, by merely forming the complex conjugate of the non-linear graviton construction.) The hope, with this googly approach, is that the general solution of the Einstein vacuum equations might be obtainable somehow by "adding together", in a suitably non-linear way, twistor spaces for the left-flat and right-flat "parts" of the vacuum space-time, using the original non-linear graviton construction for the former "part" and the googly construction for the latter "part". See Lebrun (1990) for an account of the current status of the ambitwistor approach, and Penrose (1990) for the status of the googly approach. Both approaches encounter fundamental difficulties. There is a basic awkwardness in the ambitwistor approach because, as yet, it has to be phrased in terms of "nth order extensions". Moreover, it is unclear how the construction could be reduced to one involving only free functions. The googly approach, though in principle closer to the original objectives of twistor theory, remains significantly further from a solution than does the ambitwistor approach. The direct googly constructions are as yet somewhat unconvincing as to their "naturalness"; more serious is a lack of any understanding of the non-linear "adding" that would be necessary.

It is for reasons such as these that we may feel encouraged by the existence of a completely fresh approach towards a twistorial description of the Einstein equations. This approach springs from two observations concerning massless fields of helicity $+3/2$. These observations are:

(A) the appropriate field equations governing massless fields of helicity $3/2$ are consistent if and only if the space-time is Ricci-flat;

(B) in Minkowski space-time, the space of conserved charges for massless fields of helicity $3/2$ is twistor space T.

Thus, it would seem that all we need do, in order to construct the elusive twistor space T for a general Ricci-flat space-time M, is to find the space of charges for massless fields of helicity $3/2$ on M. The hope would be that, as was the case for the non-linear graviton construction, the complex space T would be completely equivalent to the space-time M, although related to it only in a non-local way. One might also hope that the general such T could be constructed by the use of free functions, and that twistor theory would thereby establish itself as providing a general procedure for analysing the Einstein vacuum equations. If these aims could be realized, then there would be a powerful case for considering that the physically "correct" twistor concept had at last been found.

Some words of explanation are needed to clarify the meanings of (A) and (B) more fully. In the first place, the "appropriate" field equations referred to in (A) are equations that relate to the (first) *potential* for the field, subject to a gauge freedom. These equations can be taken either in the Dirac (1936) form or (essentially) the Rarita-Schwinger (1941) form. In the Dirac form (using the 2-component spinor notation, cf. Penrose and Rindler 1984), we have a potential $\gamma^C_{A'B'}$, symmetric in A' and B', subject to the field equation

$$\nabla^{AA'}\gamma^C_{A'B'} = 0,$$

where there is the gauge freedom

$$\gamma^C_{A'B'} \rightarrow \gamma^C_{A'B'} + \nabla^C_{A'}\nu_{B'}$$

for which $\nu_{A'}$ satisfies the neutrino equation

$$\nabla^{AA'}\nu_{A'} = 0.$$

In the Rarita–Schwinger form (in 2–component spinor notation), no symmetry condition is imposed on $\gamma^C_{A'B'}$, and we have the pair of field equations

$$\nabla^{A'(A}\gamma^{C)}_{A'B'} = 0, \quad \nabla_{C(C'}\gamma^{C'C}_{A')} = 0,$$

with the same gauge freedom as above, except that $\nu_{A'}$ is not now restricted to satisfy any equation. (Round brackets around indices always denote symmetrization.) The Dirac description is equivalent to that of Rarita–Schwinger, in a special choice of gauge.

The fact that the Ricci-flat condition is necessary and sufficient for the consistent propagation of these equations follows from the work of Buchdahl (1958), Deser and Zumino (1976), Julia (1982), and others.

In *flat* space-time, we can define a field $\varphi_{A'B'C'}$, totally symmetric in its indices, by

$$\varphi_{A'B'C'} = \nabla_{CC'}\gamma^C_{A'B'}$$

in the Dirac case, and by

$$\varphi_{A'B'C'} = \nabla_{C(C'}\gamma^{C}_{A')B'}$$

in the Rarita–Schwinger case, which is *gauge invariant*, and satisfies the massless free field equation

$$\nabla^{AA'}\varphi_{A'B'C'} = 0,$$

but these properties fail in a general Ricci-flat space-time. (I am using the term "Ricci-flat" here rather than just "vacuum" to emphasize that the cosmological constant must vanish in this discussion.)

With regard to (B), we must bear in mind that the "charges" being referred to are allowed to be "origin-dependent" quantities, like angular momentum - which would provide examples of the charges for massless fields in the case of spin 2. Since we are now considering Minkowski space-time, we can use the gauge-invariant description of the helicity $+3/2$ field. To see how twistors arise as the charges for such fields, we introduce a spinor $\mu^{A'}$ subject to the "dual twistor equation"

$$\nabla^{A(A'}\mu^{B')} = 0,$$

which has exactly four independent solutions. The solution space may be identified as the *dual* T* of the twistor space T(cf. Penrose and Rindler 1986). Now this equation implies the important property that if we form the quantity

$$\chi_{A'B'} = \varphi_{A'B'C'}\mu^{C'},$$

we find that it satisfies the massless free field equation

$$\nabla^{AA'}\chi_{A'B'} = 0,$$

but now for helicity 1 (so $\mu^{A'}$ acts as a helicity–lowering operator). Thus, $\chi_{A'B'}$ is the spinor form of a free self–dual *Maxwell* field. If we consider a world–tube in Minkowski space-time which is surrounded by a $\chi_{A'B'}$ satisfying this equation, then we can perform the appropriate Gauss integral

$$\int \chi_{A'B'}dx^{A'}_{A} \wedge dx^{AB'}$$

over any 2–surface surrounding the tube, in order to obtain its charge. But this charge is a (complex–)linear expression in $\mu^{C'}$, so the "charge" for the original field must be an element of the *dual* space to the space of the quantities $\mu^{C'}$, i.e. it must belong to *twistor* space T.

This establishes (B), but if we try to generalize this procedure to a curved Ricci-flat space-time, we run into all sorts of problems. (See Penrose 1992, for a taste of what is involved.) The root of the difficulty seems to be the fact that most concepts of "conserved charge" lead to objects that constitute a vector space, which is not what we expect, in general here. Even in the anti-self-dual case, the twistor space T is a particular kind of complex manifold, not a complex vector space. The way around this difficulty seems to be to generalize the linear object described by $\gamma^C_{A'B'}$ to something non-linear, most likely to a connection of some kind. The most promising idea in this direction seems to be to take this to be a connection on an SL(3,C)–bundle over the space-time. This is still "work in progress" and it remains to be seen where it will all lead.

REFERENCES

Aspect, A. and Grangier, P. (1986) "Experiments on Einstein–Podolsky–Rosen–type correlations with pairs of visible photons", in *Quantum Concepts in Space and Time* eds. R.Penrose and C.J.Isham (Oxford University Press, Oxford).

Brown, H.R. (1992) "Bell's other theorem and its connection with nonlocality. Part I.", preprint; lecture at Cesena conference.

Buchdahl, H.A. (1958) "On the compatibility of relativistic wave equations for particles of higher spin in the presence of a gravitational field", *Nuovo Cim.*, **10**, 96-103.

Deser, S. and Zumino, B. (1976) "Consistent supergravity", *Phys. Lett.*, **62B**, 335-7.

Dirac, P.A.M. (1936) "Relativistic wave equations", *Proc. Roy. Soc. (Lond.)*, **A155**, 447-59.

Huggett, S.A. and Tod, K.P. (1985) "An Introduction to Twistor Theory", London Math. Soc. student texts (L.M.S. publ.)

Julia, B. (1982) "Système linéaire associé aux équations d'Einstein", *Comptes. Rendus. Acad. Sci. Paris, Sér. II*, **295**, 113-6.

Lebrun, C.R. (1990) "Twistors, ambitwistors and conformal gravity", in *Twistors in Mathematics and Physics*, eds T.N.Bailey and R.J.Baston (Cambridge Univ. Press, Cambridge), 71-86.

Penrose, R. (1971) "Angular momentum: an approach to combinatorial space-time", in *Quantum theory and Beyond*, ed. Ted Bastin (Cambridge University Press, Cambridge).

Penrose, R. (1972) "On the nature of quantum gravity", in *Magic without Magic*, ed. J.R.Klauder (Freeman, San Francisco).

Penrose, R. (1976) "Non-Linear gravitons and curved twistor theory", *Gen. Rel. Grav.*, **7**, 31-52.

Penrose, R. (1986) "On the origins of twistor theory", in *Gravitation and Geometry*, (I. Robinson festschrift volume), eds. W. Rindler and A. Trautman (Bibliopolis, Naples).

Penrose, R. (1990) "Twistor theory after 25 years - its physical status and prospects", in *Twistors in Mathematical Physics*, eds T.N.Bailey and R.J. Baston (Cambridge Univ. Press, Cambridge).

Penrose, R. (1992a) "Twistors as spin 3/2 charges", in *Gravitation and Modern Cosmology* (P.G.Bergmann's 75th birthday vol.) eds. A. Zichichi, N. de Sabbata, and N. Sanchez (Plenum Press, New York).

Penrose, R. (1992b) "On Bell non-locality without probabilities: some curious geometry", in CERN volume in honour of J.S Bell, ed. J. Ellis.

Penrose, R. and Rindler, W. (1984) *Spinors and Space-Time, Vol. 1: Two-Spinor Calculus and Relativistic Fields* (Cambridge University Press, Cambridge).

Penrose, R. and Rindler, W. (1986) *Spinors and Space-Time, Vol. 2: Spinor and Twistor Methods in Space-Time Geometry*, (Cambridge University Press, Cambridge.

Rarita, W and Schwinger, J. (1941) "On the theory of particles with half-integer spin", *Phys. Rev.*, **60**, 61- .

Readhead, M.L.G. (1987) *Incompleteness, Nonlocality, and Realism* (Clarendon Press, Oxford).

Sciama, D.W. (1958), *Nuovo Cim.*, **8**, 417.

Ward, R.S. and Wells, R.O. Jr (1989) *Twistor Geometry and Field Theory* (Cambridge University Press, Cambridge).

Zimba, J. (1992) Oxford preprint.

The Different Levels of Connections Between Science and Objective Reality

NICOLÒ DALLAPORTA

I had the privilege of collaborating with Dennis Sciama for a few years here in Trieste in building up the Astrophysical Sector of SISSA; and I am glad to tell him today that it has been for me an enjoyable and wonderful experience.

Now, first of all, I feel in some sense obliged to justify the subject of my contribution by saying that, at an age over eighty, it becomes much easier making some philosophical reflections about science than bringing some significant scientific consideration; that is why, in order to take part actively to this conference, intended to convey to Dennis all our wishes for further important scientific achievements, I have found myself confined to presenting only some epistemological puzzles. I was told by the organizers that this could be considered as tolerable; so that I have now only to ask for kindly forgiving me such a deviation from the main line of this meeting.

The second thing to do is to clarify what I mean in the title by "reality". If scientists and philosophers are quite aware of the almost endless meanings that can be given to this word, at different levels of philosophical depth, this is not so for plain people, who generally stick to our immediate feeling that reality is what we perceive through our senses in our surroundings. And there is little doubt, at a first stage, that this type of reality constitutes the spontaneous background on which science has been constructed at its beginnings, let us say at Galileo's times; and even much later, models of the microworld such as the kinetic theory of matter or the earliest atomic schemes were still conceived with exactly the same features we directly observe at the macroscopic scale. This of course is the reason why positivistic philosophy, developing quite naturally from such a conception of science, has become nowadays the widespread background on which modern western civilization has been almost entirely constructed, and almost all the mass-media are sticking to such a view. Instead nowadays, almost all scientists know the situation to be entirely different, and of course much more complicated, but this is rarely mentioned nor diffused at a popular level; so that ordinary people remain widely attached to the positivistic picture, assumed to be extensible at all scales, from atomic to cosmological dimensions.

A first and well-known limitation to such an extension has been reached from the atomic side in the second decade of this century by understanding quantum mechanics as ruling the microworld. This branch of science met with such well-known successes concerning the properties of matter at the atomic scale and below, that its acceptance as being "truly real" requires only the recognition of matter as being undescribable at this scale by our imagination possibilities; and the fact that human intuition should be adequate to catch realities only at the human scale could in some sense, philosophically speaking, be expected. With such an acknowledgement, the ensemble of galilean physics and quantum mechanics, each at its own scale, forms a compact body of standard physics; and the main philosophical problem left consists in testing up to what point this main body can be extended to the other extreme of the dimensional scale, that is to cosmology.

With the aim of exploring the possible meanings taken by science in such an extension, let us first choose to call "real physics" only the ensemble galilean physics, including its well–known models, its nineteenth century conceptual and theoretical extensions, plus quantum mechanics; and in contrast, according to etymology, consider as "metaphysics" everything outside its boundaries. The conclusion of such a procedure will be that, taken in such a positivistic sense, the domain of physics is bounded by limits, beyond which a positivistic knowledge of the world is no longer possible; so that, in such a sense, the domain of science appears as confined to more or less sharp boundaries.

In order to focus this line of thought, let us begin with two remarks: we presume by induction that whatever physical theories have been able to explain up to now in a given physical surrounding will be equally valid in the future in the same surrounding. The never belied evidence of such a fact has rooted in our mind the idea of the unchangingness and universality of physical laws. However, let us observe that such an induction has been tested only in the case when applied to surroundings of similar characteristics as the one from which these physical laws have been deduced, while we have no proof that their validity is conserved without modifications for a physical medium whose conditions are completely outside the reach of our experimental apparatuses.

I would like also to remark that the boundaries of the physical domain testable by our experimental tools do not coincide at all with those of the whole of what we may call the world of matter. The confusion between these two kinds of concepts is current as long as we refer to bodies to be found on earth, in direct contact with us, while the difference appears immediately in cosmology: we have just to quote the fact that we have no doubts concerning the existence of bodies outside our light cone, outside therefore from the limits inside which we may now observe them acting. This is enough to show that the domain of matter is far more extended than the physical domain, as it was defined above; and therefore, that a science which wants to remains inside the limits of experimentation must be prepared to forsake the claim to knowledge of the whole domain of matter.

If now "metaphysics" is everything lying outside the domain of physics, we have still to distinguish in it, from one side, what, according to the normal use of the word, I would like to call "true metaphysics", forming the main content of religions and ancient philosophies, related to the non–bodily layers of the universe: and, from the other side, what I would now call "physico–metaphysics", related to the bodily layer of cosmos, but located outside our reach of observation and experimentation. From such a distinction depends of course what has to be meant under the general name of cosmology: in its wider sense, cosmology includes everything belonging to the cosmos, that is both the bodily and the non–bodily layers of it; and such was in fact the domain of the ancient cosmologies; nowadays, in its restricted sense corresponding to the purely scientific viewpoint, ignoring all non-bodily considerations, cosmology however still includes, beyond its purely physical part, a metaphysical portion inherent to the "physico–metaphysics". And anybody knows that, if a vast part of cosmologists work on the construction of physical cosmology, a no less important portion of scientists has turned its main interest towards such a metaphysical domain of this science.

At this point, it is perhaps convenient to observe that all the ancient or "true" meta-physics are, on one side, very likely right in their general metaphysical outlooks, as is in some sense tested by the fact that all of them – either Hindu, Chinese, Abraham-ic, Christian or Islamic – do lie on the same fundamental background of universal principles, their relative differences being secondary, as due to a peculiar stressing or focusing, by each of them, on one or another of the fundamental principles; while, on the other side, their deductions from these same principles concerning the physical world are notoriously erroneous, and often contradicted by experimental evidence. This is a strong indication that there are, not a single one, but two different ways, in a sense orthogonal to each other, for gathering the total knowledge necessary for a universal understanding of the cosmos: the scientific method, adequate for con-necting realities on the same "horizontal" cosmic level, as the physical level; and the metaphysical outlook, for connecting "vertically" realities located on different cosmic layers, as are the physical, the psychical, and the spiritual ones. Now, if such is the case, it appears very likely, due to the orthogonality of these two ways of learning, that the scientific method, rooted in observation and experimentation, is inadequate to gather knowledge on the metaphysical domain, as much as the metaphysical view-point is inadequate to learn physics.

If now we adopt such a general view as to what concerns modern cosmology, we have to take care, not only that it has of course no connection with what I have termed "true metaphysics", but also that, while its physical part is strongly rooted in observation and experimentation, - as are in fact numerical simulations-, the "physico-metaphysical" one, not having a direct access to the objects of its speculations, can only reach its main results by extrapolating known physical laws towards domains in which their validity cannot be tested, as such domains differ completely from anything that can be observed in our surroundings, so that the objective value of their results is on a quite different footing with respect to those of the observable cosmology. Such a different gradation, which is quite obvious to any scientist working in these fields, is normally overlooked in almost all presentations to the vast public through mass media, either newspapers, popular books and even schoolbooks; and this confusion, to my mind, is really unwholesome, not so much for cosmology itself, to which it might at most subtract for watchful spirits some credibility for some of its extrapolations, but rather through its impact on other domains of learning, mainly humanistic, mistakenly assuming for experimentally proven what might be just speculations. It is mostly this consideration that has led me to tackle this subject; which, after having been presented up to this point under its general aspect, I would like now to clarify with some examples concerning the definite boundaries or more dilute transitions between true physics and "physico–metaphysics", in some of the most discussed extrapolations of cosmology.

I think there is no better conceptual frame in the whole domain of physics than the Big Bang in order to test the gradual transition from phases tightly grounded on experimental physics towards extrapolations of a more and more metaphysical character when moving from the present epoch to the past, starting from well observed data concerning faraway galaxies, passing then to regions where the state of matter is deduced from high energy elementary particle data, and finally ending in zones where there is presently no hope of ever being able to get any kind of observable information. It is of course in the initial fractions of seconds of the whole process that the strongest doubts, concerning the correspondence between our views and reality, are gathering.

A) As a first question mark, is there any physical sense in dealing with times of 10^{-43}sec, temperatures of 10^{32}K, or densities of 10^{93}gr/cm^3? Of course, mathematics has no limits towards infinities or infinitesimals but up to what point is it allowed to proceed in real physics? even into domains where not even indirect measurements will ever likely be possible? It is generally admitted that, whenever we can correlate two or more present day observable data through some theory such as cosmic inflation going back to the very primordial times, this can be considered as an indirect proof of the validity of such a theoretical scheme. But validity for being a tool adequate to formulate further predictions is obviously not equivalent to reality. And this is confirmed by the fact that often several models proposed are able to explain the same data, or when a given model explaining n data appears unable to explain the $n + 1st$ one. Therefore, to my mind such theories belong to the domain of "physico-metaphysics"; and in fact, while some of them, as Dirac's theory of the variation of the gravitational constant as $G \approx t^{-1}$ do satisfy the Popper criterion of falsifiability, many of them do not; which should mean, according to him, that they do not belong to physics.

B) A further consideration reinforces the preceding conclusion: theories of the kind of inflation are generally worked out, as nothing else can be done, just by extrapolating the known physical laws, as has been said before, into domains entirely different from those from which they have been derived. Now this is exactly the reverse of what practical experience has been teaching us. Newtonian space–time and mechanics did once appear as the most solid stronghold of the whole of physics; however, they had to be replaced: for high velocities by special relativity, in the neighbourhood of strong gravitational fields by general relativity, and for the microworld by quantum mechanics. Such past experience strongly suggests, to my mind, when moving towards the early Big Bang, that we could likely meet new modifications of the laws ruling such domains, and not only quantum gravity beyond Planck's time, as can be foreseen with our present knowledge, but also other unpredictable ones, as taught by past history.

C) My strongest doubts concerning the validity of these extrapolations are related to quantum mechanics. It is frequently assessed that quantum-mechanical laws have to be taken as being universally valid, and from such an assumption some very paradoxical consequences, as the birth of the universe from a quantum fluctuation, are then derived. However, it appears to me that there would not be much sense in applying quantum-mechanical considerations to masses larger than the Planck mass M_p. We know in fact that for $m < M_p$, the Compton wave–length λ_c is larger than the Schwarzschild radius R_S so that quantum phenomena prevail over gravitational ones, and general relativity corrections can be neglected; viceversa, for $m > M_p$, the Compton wave–length becomes smaller than the Schwarzschild radius; and as λ_c is indicative of the distance between interference minima, the "dimensions" of the particles are confined inside the Schwarzschild radius, so they cannot but be mini black–holes: - and those with mass $m \lesssim 10^{15}$gr would have had time to explode since the origin of the universe. Even if likely, such a consequence has not to be taken literally, it probably shows that for elementary particle masses larger than M_p the normal concepts of quantum mechanics lose their usual significance.

Finally I could not easily refrain from quoting as the climax of "metaphysics invading physics" any consideration related to ensembles of universes, either finite or infinite, generally conceived as being without communications among each other. Apart from the fact that, if the physical universe, be it by itself finite or infinite, has to be the totality of the physical world, it would be self-contradictory speaking of more than a single universe, and with the exception of those portions of the universe presently outside our horizon, but included in it in some future time, I would think that we could not easily consider as belonging to the physical domain anything that by definition could not be able to reach us by any kind of signal.

I think these few examples are enough in order to clarify the meanings I have attributed to the concept of "physico- metaphysics", so we can now directly turn to draw the general conclusion of the present contribution. If one insists that "true science" is rooted only in experiment - and this in some sense constitutes the basis of the positivistic philosophical attitude - one must then admit that, in respect to the totality of knowledge required for the understanding of the whole cosmos, its capacity of grasping "reality" is clearly limited to portions of the universe not too dissimilar from our own surroundings, and that any attempt of speculation overpassing these limits is not truly "scientific", but only "metaphysical".

If however one is reluctant to limit to such an extent the concept of "science", one must then admit that a wider range attributed to "science" cannot be associated to positivistic philosophy; so that, if one wishes to include the domain of "physico-metaphysics" as belonging to "true science", then "science" is only partly based on observation or experiment; as much, let us say, as are poems, sagas, or epics, while it includes vast portions of imagination, fantasy and intelligence, which make it akin with art and poetry in the largest sense. This is by no means a criticism, as art, as well known, includes sometimes much more truth than anything else. Reality, then, becomes unfathomable, and science acquires, besides its objective aspects which of course it maintains, a subjective character, which makes it more akin with human nature in its more general aspects than pure rationality.

Such a conclusion has nothing in it of course which might surprise scientists working in such abstract fields. What is basically important to my mind, in such a kind of consideration, are not its scientific, but its philosophical implications; as, from whichever side you wish to take it one sees now that real knowledge cannot be reduced to the frame of positivism, whose outlook appears now as quite insufficient to grasp the deeper layers of reality. As generally occurs in the domain of art, the understandable parts of any work are only relatively small portions of the whole, so that conceptual science today has a quite different meaning in respect to what it was in the *XIXth* century. And this, to my mind, should be made clear at the level of mass media, as it could have a not negligible influence in orienting the average mentality of our times.